M.I.T · BE 38.

Rocket and Spacecraft Propulsion
Principles, Practice and New Developments (Second Edition)

Martin J. L. Turner

Rocket and Spacecraft Propulsion

**Principles, Practice and New Developments
(Second Edition)**

Published in association with
Praxis Publishing
Chichester, UK

Professor M. J. L. Turner, CBE
Department of Physics and Astronomy
University of Leicester
Leicester, UK

SPRINGER–PRAXIS BOOKS IN ASTRONAUTICAL ENGINEERING
SUBJECT *ADVISORY EDITOR*: John Mason B.Sc., M.Sc., Ph.D.

ISBN 3-540-22190-5 Springer-Verlag Berlin Heidelberg New York

Springer is part of Springer-Science + Business Media (springeronline.com)

Bibliographic information published by Die Deutsche Bibliothek

Die Deutsche Bibliothek lists this publication in the Deutsche Nationalbibliografie; detailed bibliographic data are available from the Internet at http://dnb.ddb.de

Library of Congress Control Number: 2004111933

Apart from any fair dealing for the purposes of research or private study, or criticism or review, as permitted under the Copyright, Designs and Patents Act 1988, this publication may only be reproduced, stored or transmitted, in any form or by any means, with the prior permission in writing of the publishers, or in the case of reprographic reproduction in accordance with the terms of licences issued by the Copyright Licensing Agency. Enquiries concerning reproduction outside those terms should be sent to the publishers.

© Praxis Publishing Ltd, Chichester, UK
Second edition published 2005
First edition published 2001
Printed in Germany

The use of general descriptive names, registered names, trademarks, etc. in this publication does not imply, even in the absence of a specific statement, that such names are exempt from the relevant protective laws and regulations and therefore free for general use.

Cover design: Jim Wilkie
Typesetting: Originator, Great Yarmouth, Norfolk, UK

Printed on acid-free paper

Contents

Preface to the second edition xiii

Preface to the first edition xv

Acknowledgements xvii

List of figures xix

List of tables xxiii

List of colour plates xxv

1 History and principles of rocket propulsion 1
   1.1 The development of the rocket 1
          1.1.1 The Russian space programme 6
          1.1.2 Other national programmes 6
          1.1.3 The United States space programme 8
          1.1.4 Commentary 13
   1.2 Newton's third law and the rocket equation 14
          1.2.1 Tsiolkovsky's rocket equation 14
   1.3 Orbits and spaceflight 17
          1.3.1 Orbits 18
   1.4 Multistage rockets 25
          1.4.1 Optimising a multistage rocket 28
          1.4.2 Optimising the rocket engines 30
          1.4.3 Strap-on boosters 31
   1.5 Access to space 34

2 The thermal rocket engine 35
2.1 The basic configuration 35
2.2 The development of thrust and the effect of the atmosphere 37
    2.2.1 Optimising the exhaust nozzle 41
2.3 The thermodynamics of the rocket engine 42
    2.3.1 Exhaust velocity 43
    2.3.2 Mass flow rate 46
2.4 The thermodynamic thrust equation 51
    2.4.1 The thrust coefficient and the characteristic velocity 52
2.5 Computing rocket engine performance 56
    2.5.1 Specific impulse 57
    2.5.2 Example calculations 58
2.6 Summary 60

3 Liquid propellant rocket engines 61
3.1 The basic configuration of the liquid propellant engine 61
3.2 The combustion chamber and nozzle 62
    3.2.1 Injection 63
    3.2.2 Ignition 64
    3.2.3 Thrust vector control 67
3.3 Liquid propellant distribution systems 69
    3.3.1 Cavitation 71
    3.3.2 Pogo 71
3.4 Cooling of liquid-fuelled rocket engines 72
3.5 Examples of rocket engine propellant flow 74
    3.5.1 The Aestus engine on Ariane 5 74
    3.5.2 The Ariane Viking engines 75
    3.5.3 The Ariane HM7 B engine 77
    3.5.4 The Vinci cryogenic upper-stage engine for Ariane 5 78
    3.5.5 The Ariane 5 Vulcain cryogenic engine 78
    3.5.6 The Space Shuttle main engine 81
    3.5.7 The RS 68 engine 83
    3.5.8 The RL 10 engine 85
3.6 Combustion and the choice of propellants 86
    3.6.1 Combustion temperature 86
    3.6.2 Molecular weight 87
    3.6.3 Propellant physical properties 88
3.7 The performance of liquid-fuelled rocket engines 90
    3.7.1 Liquid oxygen–liquid hydrogen engines 91
    3.7.2 Liquid hydrocarbon–liquid oxygen engines 93
    3.7.3 Storable propellant engines 93

4 Solid propellant rocket motors 97
4.1 Basic configuration 97
4.2 The properties and the design of solid motors 99

4.3	Propellant composition	100
	4.3.1 Additives	102
	4.3.2 Toxic exhaust	103
	4.3.3 Thrust stability	103
	4.3.4 Thrust profile and grain shape	104
4.4	Integrity of the combustion chamber	106
	4.4.1 Thermal protection	107
	4.4.2 Inter-section joints	109
	4.4.3 Nozzle thermal portection	110
4.5	Ignition	110
4.6	Thrust vector control	111
4.7	Two modern solid boosters	111
	4.7.1 The Space Shuttle SRB	111
	4.7.2 The Ariane MPS	113

5 Launch vehicle dynamics ... 115

5.1	More on the rocket equation	115
	5.1.1 Range in the absence of gravity	117
5.2	Vertical motion in the Earth's gravitational field	120
	5.2.1 Vehicle velocity	120
	5.2.2 Range	122
5.3	Inclined motion in a gravitational field	123
	5.3.1 Constant pitch angle	124
	5.3.2 The flight path at constant pitch angle	127
5.4	Motion in the atmosphere	128
	5.4.1 Aerodynamic forces	129
	5.4.2 Dynamic pressure	130
5.5	The gravity turn	131
5.6	Basic launch dynamics	133
	5.6.1 Airless bodies	133
5.7	Typical Earth-launch trajectories	135
	5.7.1 The vertical segment of the trajectory	136
	5.7.2 The gravity turn or transition trajectory	136
	5.7.3 Constant pitch or the vacuum trajectory	137
	5.7.4 Orbital injection	137
5.8	Actual launch vehicle trajectories	139
	5.8.1 The Mu-3-S-II launcher	139
	5.8.2 Ariane 4	141
	5.8.3 Pegasus	143

6 Electric propulsion ... 145

6.1	The importance of exhaust velocity	145
6.2	Revived interest in electric propulsion	146
6.3	Principles of electric propulsion	147
	6.3.1 Electric vehicle performance	148

x   Contents

        6.3.2   Vehicle velocity as a function of exhaust velocity 149
        6.3.3   Vehicle velocity and structural/propellant mass 150
  6.4   Electric thrusters 152
        6.4.1   Electrothermal thrusters 152
        6.4.2   Arc-jet thrusters 155
  6.5   Electromagnetic thrusters. 157
        6.5.1   Ion propulsion 158
        6.5.2   The space charge limit 160
        6.5.3   Electric field and potential 162
        6.5.4   Ion thrust 163
        6.5.5   Propellant choice 164
        6.5.6   Deceleration grid 166
        6.5.7   Electrical efficiency 166
  6.6   Plasma thrusters 168
        6.6.1   Hall effect thrusters 170
        6.6.2   Radiofrequency thrusters 177
  6.7   Low-power electric thrusters 179
  6.8   Electrical power generation 180
        6.8.1   Solar cells 180
        6.8.2   Solar generators 181
        6.8.3   Radioactive thermal generators 181
        6.8.4   Nuclear fission power generators 184
  6.9   Applications of electric propulsion 187
        6.9.1   Station-keeping 189
        6.9.2   Low Earth orbit to geostationary orbit 190
        6.9.3   Nine-month one-way mission to Mars 191
        6.9.4   Gravity loss and thrust 191
  6.10 Deep Space 1 and the NSTAR ion engine 193
  6.11 SMART-1 and the PPS-1350 194

7 Nuclear propulsion. 197
  7.1   Power, thrust, and energy 197
  7.2   Nuclear fission basics 199
  7.3   A sustainable chain reaction 202
  7.4   Calculating the criticality 203
  7.5   The reactor dimensions and neutron leakage 205
  7.6   Control 209
  7.7   Reflection 210
  7.8   Prompt and delayed neutrons 211
  7.9   Thermal stability 212
  7.10 The principle of nuclear thermal propulsion 213
  7.11 The fuel elements 214
  7.12 Exhaust velocity of a nuclear thermal rocket 216
  7.13 Increasing the operating temperature 218

7.14	The nuclear thermal rocket engine.	221
	7.14.1 Radiation and its management	221
	7.14.2 Propellant flow and cooling	224
	7.14.3 The control drums	226
	7.14.4 Start-up and shut-down	227
	7.14.5 The nozzle and thrust generation	228
7.15	Potential applications of nuclear engines.	229
7.16	Operational issues with the nuclear engine	230
7.17	Interplanetary transfer manoeuvres	231
7.18	Faster interplanetary journeys	232
7.19	Hydrogen storage	234
7.20	Development status of nuclear thermal engines	235
7.21	Alternative reactor types	241
7.22	Safety issues	243
7.23	Nuclear propelled missions	246

8 Advanced thermal rockets 249

8.1	Fundamental physical limitations	249
	8.1.1 Dynamical factors	249
8.2	Improving efficiency	252
	8.2.1 Exhaust velocity	252
8.3	Thermal rockets in atmosphere, and the single stage to orbit	255
	8.3.1 Velocity increment for single stage to orbit	256
	8.3.2 Optimising the exhaust velocity in atmosphere	258
	8.3.3 The rocket equation for variable exhaust velocity	259
8.4	Practical approaches to SSTO	261
	8.4.1 High mass ratio	261
8.5	Practical approaches and developments	263
	8.5.1 Engines	264
8.6	Vehicle design and mission concept	272
	8.6.1 Optimising the ascent	272
	8.6.2 Optimising the descent	272
8.7	SSTO concepts	273
	8.7.1 The use of aerodynamic lift for ascent	274

Appendix 1   Orbital motion 277
A1.1 Recapitulation of circular motion 277
A1.2 General (non-circular) motion of a spacecraft in a gravitational field   278

Appendix 2   Launcher survey 283
A2.1 Launch site. 283
A2.2 Launcher capability 284

A2.3 Heavy launchers 284
A2.4 Medium Launchers 285
A2.5 Small launchers. 285

Appendix 3   Ariane 5 299
A3.1 The basic vehicle components. 300
A3.2 Evolved Ariane 5. 300

Appendix 4   Glossary of symbols 303

Further reading. 307

Index 309

Preface to the second edition

In the period since the publication of the first edition, rocket propulsion and launcher systems have experienced a number of major changes. The destruction of the Space Shuttle *Columbia*, on re-entry, and the tragic loss of seven astronauts, focused attention on NASA, its management systems, and on the shuttle programme itself. This led to a major re-direction of the NASA programme and to the plan to retire the Space Shuttle by 2010. At the same time, President Bush announced what was effectively an instruction to NASA to re-direct its programme towards a return of human explorers to the Moon, and to develop plans for a human Mars expedition. This has significant implications for propulsion, and, in particular, nuclear electric and nuclear thermal propulsion seem very likely to play a part in these deep space missions. The first example is likely to be the Jupiter Icy Moons Orbiter, to be powered by a nuclear electric thruster system.

I have thought it wise therefore to include a new chapter on nuclear thermal propulsion. This is based on the work done in the 1960s by both NASA and the Russian space agencies to develop and test nuclear rocket engines, with updates based on the latest thinking on this subject. There are also major revisions to the chapters on electric propulsion and chemical rocket engines. The rest of the book has been revised and updated throughout, and a new appendix on Ariane 5 has been provided. The planned update to the Space Shuttle sections has been abandoned, given its uncertain future.

Since its publication, this book has modestly fulfilled the hope I had for it, that it would prove useful to those requiring the basics of space propulsion, either as students or as space professionals. As a replacement for the now out of print first edition, I venture to hope that this second edition will prove equally useful.

Martin J. L. Turner
Leicester University, June 2004

Preface to the first edition

Rockets and launch vehicles are the keys to space exploration, space science and space commerce. Normally, the user of a launcher is several steps removed from the launcher itself; he may not even be present during spacecraft–launcher integration, and is usually far away at the moment of launch. Yet the few minutes of the launch can either fulfil the dreams and aspirations that have driven the mission for many years, or it can destroy them. As a space scientist I have worked on some half dozen missions in different space agencies; but it was not until I was present for the launch of Ginga, on a Japanese Mu-3-S rocket, that I actually came close to the vehicle and met the designers and engineers responsible for it. The Ginga launch was perfect, and I had agreeable discussions with the designer of the Mu rocket. I realised that I knew little about this most important component of a space mission; I had little idea of the engineering of rocket engines, and little knowledge of launch vehicle dynamics. In seeking to rectify this lamentable ignorance I found very few books on rockets which were accessible to non-specialists and yet were not trivial. Most of the work on rocket design was undertaken in the 1950s and 1960s, and many of the engineering books were published during that period. Moreover, since engineers care about numerical accuracy and precise detail (they have to) many of the books are extremely difficult for the non-specialist. It seemed, therefore, that there might be a place for a book dealing with the subject in a non-trivial way, but simplifying the mass of detail found in books intended for professional rocket engineers. I have never met a 'rocket scientist'.

This book, then, is the result. I have tried to examine rockets and rocket engines from the point of view of a non-specialist. As a physicist I am inclined to look for the physical principles and for accessible explanations of how the rocket works. This necessarily requires some mathematics, but I have included as many graphs of functions as possible, to enable those who would prefer it, to eschew the formulae, and yet gain some feeling for the dependence of a rocket's performance on its design. Whether or not I have succeeded, the reader will judge. To illustrate the principles I have used examples of real engines and launch vehicles, although the inclusion or exclusion of a particular engine or vehicle has been governed by convenience for explanation, rather than the excellence or currency of the item itself. Appendix 2 includes a table of present-day launch vehicles, although this is not exhaustive, and new vehicles are constantly appearing.

My early research for this book indicated that the development of modern rockets took place mostly during the middle years of the last century, and that we were in the mature phase. The Space Shuttle had been around for 20 years, and was itself the epitome of rocket design; this is still true, but the closing years of the twentieth century have seen a renaissance in rocketry. While engines designed in the 1960s are still in use, new engines are now becoming available, and new vehicles are appearing in significant numbers. This seems to be driven by the rapidly growing commercial demand for launches, but is also the result of the opening up of Russian space technology to the world. I have tried to reflect this new spirit in the last two chapters, dealing with electric propulsion – now a reality – and the single stage to orbit, which is sure to be realised very soon. However, it is difficult to predict beyond the next few years where rocket design will lead us. The SSTO should reduce space access costs, and make space tourism possible, at least to Earth orbit. Commercial use of space will continue to grow, to support mobile communication and the Internet. These demands should result in further rocket development and cheaper access to space. Progress in my own field of space science is limited, not by ideas, but by the cost of scientific space missions. As a space scientist I hope that cheaper launches will mean that launches of spacecraft for scientific purposes will become less rare. As a human being I hope that new developments in rocket engines and vehicles will result in further human exploration of space: return to the Moon, and a manned mission to Mars.

This preface was originally written during the commissioning of the XMM–Newton X-ray observatory, which successfully launched on Ariane 504 in December 1999. The Ariane 5 is the latest generation of heavy launcher, and the perfection of its launch, which I watched, is a tribute to the rocket engineers who built it. But launching is still a risky business, however carefully the rocket is designed and assembled. There is always that thousand to one chance that something will go wrong; and as space users we have to accept that chance.

Martin J. L. Turner
Leicester University, March 2000

Acknowledgements

I have received help in the preparation of this book from many people, including my colleagues in the Department of Physics and Astronomy at Leicester University and at the Space Research Centre, Leicester, and members of the XMM team. I am particularly grateful to the rocket engineers of ISAS, Lavotchkin Institute, Estec, and Arianespace, who were patient with my questions; the undergraduates who attended and recalled (more or less satisfactorily) lectures on rocket engines and launcher dynamics; and, of course, my editor, Bob Marriott. While the contents of this book owe much to these people, any errors are my own.

I am grateful to the following for permission to reproduce copyright material and technical information: Société National d'Etude et Construction de Moteurs d'Aviation (SNECMA), for permission to reproduce the propellant flow diagrams of Ariane engines (Figures 3.5, 3.6, 3.7 and 3.9 in the colour section); Boeing–Rocketdyne and the University of Florida, for permission to reproduce the SSME flow diagram (Figure 3.8, colour section) and the aerospike engine (Figure 7.11); NASA/JPL/California Institute of Technology, for permission to reproduce the picture of the Deep Space 1 ion engine (Figure 6.16 and cover); the European Space Agency, for the picture of the XMM–Newton launch on Ariane 504 (cover); and Mark Wade and *Encyclopaedia Astronautica*, for permission to use tabular material which appears in Chapters 2 and 3 and Appendix 2. Figure 6.15 is based on work by P.E. Sandorf in *Orbital and Ballistic Flight* (MIT Department of Aeronautics and Astronautics, 1960), cited in Hill and Peterson (see Further Reading). Other copyright material is acknowledged in the text.

List of figures

1.1	Konstantin Tsiolkovsky	3
1.2	Herman Oberth	4
1.3	Robert Goddard	5
1.4	The J-2 engine used for the upper stages of Saturn V	10
1.5	The launch of the Space Shuttle *Atlantis*, 3 October 1985	11
1.6	Tsiolkovsky's rocket equation	15
1.7	Spacecraft movement	19
1.8	Orbit shapes	20
1.9	Injection velocity and altitude	23
1.10	Multistaging	26
1.11	Launch vehicle with boosters	32
2.1	A liquid-fuelled rocket engine	36
2.2	A solid-fuelled rocket motor	36
2.3	Forces in the combustion chamber and exhaust nozzle	38
2.4	Gas flow through the nozzle	39
2.5	Static force due to atmospheric pressure	40
2.6	P–V diagram for a heat engine	43
2.7	Gas velocity as a function of the pressure ratio	45
2.8	Mass flow in the nozzle	46
2.9	Variation of flow density through the nozzle	48
2.10	Area, velocity and flow density relative to the throat valves as a function of the pressure ratio	50
2.11	Expansion ratio as a function of the pressure ratio for changing γ	53
2.12	Thrust coefficient plotted against expansion ratio for different atmospheric pressures	55
2.13	Characteristic velocity as a function of the combustion temperature and molecular weight	56
3.1	Schematic of a liquid-propellant engine	62

List of figures

3.2	Injection and combustion	63
3.3	Types of injector	65
3.4	The impinging jet injector	66
3.5	The Aestus engine on Ariane 5	75
3.6	The pump-fed variant Aestus engine firing. In this test the long nozzle extension has been removed	76
3.7	The Vinci cryogenic upper-stage engine	79
3.8	The Vulcain 2 under test	80
3.9	The SSME on a test stand	82
3.10	The RS 68 engine firing	84
3.11	The RL 10 engine	85
3.12	The variation of exhaust velocity, temperature and molecular weight for different propellant combinations	88
4.1	Schematic of a solid-fuelled rocket motor	98
4.2	Cross-sections of grains	105
4.3	Thermal protection	108
4.4	The Ariane MPS solid booster	114
5.1	Velocity function as a function of mass ratio	117
5.2	Range as a function of mass ratio	119
5.3	Gravity loss: velocity gain and thrust-to-weight ration	122
5.4	Thrust and pitch angle	124
5.5	Gravity loss: velocity gain and pitch angle	126
5.6	Flight path angle as a function of time and pitch angle	128
5.7	The aerodynamic forces acting on a rocket	129
5.8	Dynamic pressure, velocity and altitude as functions of mass ratio	130
5.9	Flight path angles and velocity as functions of time for a gravity turn	133
5.10	Velocity, acceleration and altitude as functions of time	140
5.11	Dynamic pressure and pitch angle as functions of time	140
5.12	Ariane 4 dynamic parameters	142
5.13	Pegasus dynamic parameters	143
6.1	Vehicle velocity and payload fraction as a function of exhaust velocity	146
6.2	Vehicle velocity as a function of exhaust velocity and burn time	150
6.3	Vehicle velocity as a function of payload/propellant mass and exhaust velocity	151
6.4	Vehicle velocity as a function of power supply efficiency and exhaust velocity	151
6.5	Schematic of an electrothermal thruster	153
6.6	Schematic of an arc-jet thruster	156
6.7	A schematic diagram of the NSTAR ion thruster	159
6.8	The NSTAR engine mounted on Deep Space 1 for testing	160
6.9	Electric field and potential in space charge limit	163
6.10	Thrust per unit area as a function of quiescent field for an ion thruster	164
6.11	Exhaust velocity and ion species for an ion thruster	165
6.12	Thrust-to-power ratio for various ions as a function of exhaust velocity	168
6.13	Two ion engines that were used on the ESA Artemis spacecraft to raise the perigee	168
6.14	Principle of the plasma thruster	169
6.15	Principle of the Hall effect thruster	172

List of figures   xxi

6.16	Schematic of the Hall thruster	172
6.17	The Russian SP-100 Hall effect thruster	174
6.18	A Russian D-100 TAL Hall thruster with a metallic anode layer	175
6.19	The concept of the VASIMIR radiofrequency plasma thruster	178
6.20	A complete RTG, cutaway	182
6.21	A single section of a RTG heat generator	183
6.22	A Stirling cycle mechanical electricity generator	184
6.23	An early United States designed nuclear fission power generator	185
6.24	An early design for a spacecraft with nuclear electric generation	186
6.25	The JIMO mission concept, powered by a fission reactor electrical system	188
6.26	The propellant efficiency as a function of the ratio of the vehicle velocity to the exhaust velocity	189
6.27	Velocity increment loss factor as a function of thrust-to-weight ratio for electric propulsion	192
6.28	The NSTAR ion thruster operating on the Deep Space 1 spacecraft	194
7.1	Test firing of a nuclear rocket engine at Jackass Flats in Nevada	199
7.2	Schematic graph of the cross section for neutron interactions in natural uranium	201
7.3	The fission chain	205
7.4	The NRX-NERVA nuclear rocket engine at the test stand	210
7.5	The principle of nuclear thermal propulsion	213
7.6	Fuel element assembly from the KIWI reactor core	216
7.7	The KIWA A-Prime reactor on its test stand	217
7.8	Cutaway drawing of a NERVA nuclear rocket engine	224
7.9	Close-up of the propellant delivery part of the NERVA engine	225
7.10	The hot bleed cycle	226
7.11	The Earth–Mars minimum energy transfer orbit	233
7.12	A short flight to Mars	233
7.13	The transit time to Mars as a function of initial delta-V and orbit eccentricity	234
7.14	The KIWI reactor	236
7.15	The NERVA family of engines	241
7.16	The scheme for approval of the use of Radioactive Thermal Generators on spacecraft for launch in the United States	245
8.1	Separation of two masses	250
8.2	Propulsion efficiency as a function of mass ratio	251
8.3	Thrust coefficient *in vacuo* as a function of pressure ratio	254
8.4	Velocity increment and mass ratio necessary to reach orbit, as a function of burn time	257
8.5	Instantaneous thrust coefficient as a function of pressure through the atmosphere for fixed and variable ratios	259
8.6	Normalised vehicle velocity as a function of mass ratio for fixed and variable expansion	260
8.7	Flow separation in a nozzle	265
8.8	Principle of the plug nozzle	266
8.9	Plug nozzle exhaust streams for varying atmospheric pressure	267

8.10	Principle of the aerospike nozzle	268
8.11	The linear aerospike engine	269
A1.1	Circular motion	277
A1.2	Non-circular motion	278

List of tables

1.1	The Saturn V rocket	31
3.1	Combustion temperature and exhaust velocity for different propellants	87
3.2	Liquid oxygen engines	92
3.3	Storable propellant engines	94
4.1	Two modern solid boosters	112
6.1	Development status and heritage of some Hall effect thrusters	176
7.1	Melting/sublimation points of some common constituents of nuclear rocket cores	214
7.2	Complete nuclear thermal rocket engine schemes based on the NERVA programme	220
7.3	The tests carried out for the NERVA programme up to 1972	238
A.2.1	Launchers	286

List of colour plates (between pages 150 and 151)

1. The Ariane 5 Aestus engine.
2. The Ariane Viking engine.
3. The Ariane HM7 B engine.
4. The Space Shuttle main engine.
5. The Ariane 5 Vulcain cryogenic engine.
6. The launch of Apollo 16 on the Saturn V rocket.
7. Test firing of RL 10 engine.
8. Firing test of the Space Shuttle main engine.
9. Testing the thrust vector control system on a Space Shuttle main engine while firing.
10. Titan IV launcher.
11. The NSTAR ion engine mounted on Deep Space 1 prepared for testing in a vacuum.
12. The NSTAR ion engine firing in a vacuum tank.
13. Artists impression of Deep Space 1.
14. The PPS 1350 Hall effect engine used for SMART-1, under test.
15. Artists impression of the SMART-1 spacecraft on its way to the Moon.
16. Experimental ion propulsion system under test.
17. The NERVA Nuclear Thermal Rocket Engine.
18. A NERVA programme engine on the the test stand.
19. A possible Mars expedition vehicle powered by three nuclear thermal rocket engines (artists impression).
20. An exploded view of the Ariane 5 launcher—see Appendix 3.

1

History and principles of rocket propulsion

Human development has always been closely linked with transportation. The domestication of the horse and the invention of the wheel had a dramatic effect on early civilisation—not always beneficial. Most of the past millennium has been strongly influenced by sailing ship technology, both for war and commerce; in the twentieth century, motor vehicles and aircraft have revolutionised transport. At the beginning of the twenty-first century the rocket may be seen as the emerging revolution in transport. So far, only a few humans have actually travelled in rocket-propelled vehicles, but a surprising amount of commercial and domestic communication is now reliant on satellites. From telephone calls, through news images, to the Internet, most of our information travels from one part of the world to another, through space. The proposed return to the Moon, and new plans to send humans to Mars indicate a resurgence of interest in space exploration for the new millenium.

Rocket propulsion is the essential transportation technology for this rapid growth in human communication and exploration. From its beginnings in ancient China through its rapid development during the Cold War, rocket propulsion has become the essential technology of the late twentieth century. It influences the lives and work of a growing number of people, who may wish to understand at least the principles behind it and its technical limitations. In most cases, users of space transportation are separated from the rocket technology which enables it; this is partly because of the mass of engineering detail and calculation which is essential to make such a complex system work. In what follows, we shall attempt to present the basic principles, and describe some of the engineering detail, in a way that exposes the essential physics and the real limitations to the performance of rocket vehicles.

1.1 THE DEVELOPMENT OF THE ROCKET

Hero of Alexandria (*c.* 67 AD) is credited with inventing the rocket principle. He was a mathematician and inventor and devised many machines using water, air pressure, and steam, including a fire engine and a fountain. His *aeolipile* consisted of a metal boiler in which steam was produced, connected by a pipe through a rotating joint, to a pivoted jet system with two opposing jets. The steam issuing from the jets caused the system to rotate. It is not clear if Hero understood the cause of the rotation; but this was the earliest machine to use the reaction principle—the theoretical basis of the rocket. The real inventor of the rocket was certainly Chinese, and is sometimes said to be one Feng Jishen, who lived around 970 AD. Most dictionaries insist on an Italian derivation of the word meaning 'a small distaff'; and this may be correct, although a derivation from the Chinese, meaning 'Fire Arrow', is also plausible. The invention was the practical result of experiments with gunpowder and bamboo tubes; it became, as it still is, a source of beauty and excitement as a firework. There seem to have been two kinds. One was a bamboo tube filled with gunpowder with a small hole in one end; when the gunpowder was ignited, the tube ran along the ground in an erratic fashion, and made the girls scream. The second, like our modern fireworks, had a bamboo stick attached for stability, and took to the sky to make a beautiful display of light and colour.

The rocket was also used as a weapon of oriental war. Kublai Kahn used it during the Japanese invasion of 1275; by the 1300s rockets were used as bombardment weapons as far west as Spain, brought west by the Mongol hordes, and the Arabs. They were also used against the British army in India, by Tipoo Sultan, in the 1770s. Shortly afterwards, Sir William Congreve, an artillery officer, realised the rocket's potential, and developed a military rocket which was used into the twentieth century. Congreve rockets were used at sea during the Napoleonic Wars, with some success. They appeared famously at the siege of Fort McHenry in the American War of Independence, and feature in song as 'the rocket's red glare'. At about the same time, rockets came into standard use as signals, to carry lines from ship to ship, and in the rescue of shipwrecked mariners—a role in which they are still used.

The improvements in guns, which came about during the late 1800s, meant that the rocket, with its small payload, was no longer a significant weapon compared with large-calibre shells. The carnage of the First World War was almost exclusively due to high-explosive shells propelled by guns. Military interest in the rocket was limited through the inter-war years, and it was in this period that the amateur adopted a device which had hardly improved over six centuries, and created the modern rocket.

The names of the pioneers are well known: Goddard, Oberth, von Braun, Tsiolkovsky, Korolev. Some were practical engineers, some were mathematicians and others were dreamers. There are two strands. The first is the imaginative concept of the rocket as a vehicle for gaining access to space, and the second is the practical development of the rocket. It is not clear to which strand the (possibly apocryphal) seventeenth century pioneer Wan Hu belongs. He is said to have attached 47 rockets to a bamboo chair, with the purpose of ascending into heaven.

Konstantin Tsiolkovsky (1857–1935) (Figure 1.1), a mathematics teacher, wrote about space travel, including weightlessness and escape velocity, in 1883, and he wrote about artificial satellites in 1895. In a paper published in 1903 he derived the rocket equation, and dealt in detail with the use of rocket propulsion for space travel; and in 1924 he described multi-stage rockets. His writings on space travel were

Figure 1.1. Konstantin Eduardovich Tsiolkovsky.
Courtesy Kaluga State Museum.

soundly based on mathematics—unlike, for example, those of Jules Verne—and he laid the mathematical foundations of space flight.

Tsiolkovsky never experimented with rockets; his work was almost purely theoretical. He identified exhaust velocity as the important performance parameter; he realised that the higher temperature and lower molecular weight produced by liquid fuels would be important for achieving high exhaust velocity; and he identified liquid oxygen and hydrogen as suitable propellants for space rockets. He also invented the multi-stage rocket.

Tsiolkovsky's counterpart in the German-speaking world was the Rumanian, Herman Oberth (1894–1992) (Figure 1.2). He published his (rejected) doctoral thesis in 1923, as a book in which he examined the use of rockets for space travel, including the design of liquid-fuelled engines using alcohol and liquid oxygen. His analysis was again mathematical, and he himself had not carried out any rocket experiments at the time. His book—which was a best-seller—was very important in that it generated huge amateur interest in rockets in Germany, and was instrumental in the foundation of many amateur rocket societies. The most important of these was the Verein für Raumschiffarht, to which Oberth contributed the prize money he won for a later book, in order to buy rocket engines. A later member of the VfR was Werner von Braun.

The people mentioned so far were writers and mathematicians who laid the theoretical foundations for the use of rockets as space vehicles. There were many engineers—both amateur and professional—who tried to make rockets, and who had the usual mixture of success and failure. Most of these remain anonymous, but in the United States, Robert Goddard (1882–1945) (Figure 1.3), a professor of

Figure 1.2. Herman Oberth.
Courtesy NASA.

physics at Clark University in Massachusetts, was, as early as 1914, granted patents for the design of liquid-fuelled rocket combustion chambers and nozzles. In 1919 he published a treatise on rocket vehicles called, prosaically, *A Method of Reaching Extreme Altitudes*, which contained not only the theory of rocket vehicles, but also detailed designs and test results from his own experiments. He was eventually granted 214 patents on rocket apparatus.

Goddard's inventions included the use of gyroscopes for guidance, the use of vanes in the jet stream to steer the rocket, the use of valves in the propellant lines to stop and start the engine, the use of turbo-pumps to deliver the propellant to the combustion chamber, and the use of liquid oxygen to cool the exhaust nozzle, all of which were crucial to the development of the modern rocket. He launched his first liquid-fuelled rocket from Auburn, Massachusetts, on 16 March 1926. It weighed 5 kg, was powered by liquid oxygen and petrol, and it reached a height of 12.5 metres. At the end of his 1919 paper Goddard had mentioned the possibility of sending an unmanned rocket to the Moon, and for this he was ridiculed by the Press. Because of his rocket experiments he was later thrown out of Massachusetts by the fire officer, but he continued his work until 1940, launching his rockets in New Mexico. In 1960 the US government bought his patents for two million dollars.

The way Goddard was regarded in the United States—and, indeed, the way in which his contemporaries were treated in most other countries—is in marked contrast to the attitude of the German and Russian public and government to rocket amateurs. Tsiolkovsky was honoured by both the Tsarist and the Communist

Sec. 1.1] The development of the rocket 5

Figure 1.3. Robert Goddard.
Courtesy NASA.

governments, and in Germany the serious public interest in rockets was mirrored by the government. Thus by the time of the Second World War, Russia and Germany were well ahead of other nations in rocketry—Germany most of all. The amateur German rocket societies were noticed by the military, and they soon came under pressure to turn their talents to the military sphere. Von Braun was one of the enthusiastic engineers who took this step, joining the military research station at Peenemunde. Here was developed the A4 liquid-fuelled rocket which became the notorious V2 weapon, and which, from its launch site in Germany, carried a 1,000-pound bomb into the centre of London.

The A4 embodied many concepts similar to those patented by Goddard, and it was the first practical and reproducible liquid-fuelled rocket. It brought together the ideas of the amateurs and dreamers, and these were developed within the discipline and urgency of a military programme. When the war ended, there was a race to reach Peenemunde, both by the Russians (informed by Churchill) and the Americans. In the end, as is well known, the Russians took Peenemunde and its contents, and the US took von Braun and key members of his team.

The military weapon and space races, which followed, have been well documented. To some extent, the old attitudes to rocketry were reflected in the relative progress of the United States and Russia. The Apollo programme will be seen as the most significant achievement of the twentieth century, and nothing can take that away from the United States. It did, however, represent a concerted effort to catch up with, and overtake the earlier, dramatic Russian successes.

1.1.1 The Russian space programme

In general terms, the Russian space programme has been the most active and focused in history: the first artificial satellite, the first man in space, the first spacecraft on the Moon, the first docking of two spacecraft, and the first space station. All of these are the achievements of Russia (or, rather, the Soviet Union). In the period from 1957 to 1959, three satellites and two successful lunar probes had been launched by the USSR, ironically fulfilling Goddard's prophesy. In 1961, Yuri Gagarin became the first man in space, and at the same time, several fly-bys of Mars and Venus were accomplished. In all there were 12 successful Russian lunar probes launched before the first Saturn V. Apart from the drive and vision of the Soviet engineers—particularly Sergei Korolev—the reason for this success lay in the fact that the Russian rockets were more powerful, and were better designed. The pre-war Russian attitude to rocketry had found a stimulus in the captured German parts, leading to the development of an indigenous culture which was to produce the best engines. It is significant that the Saturn V was the brainchild of Werner von Braun, a German, and the Vostok, Soyuz, and Molniya rockets were the brainchildren of Korolev and Glushko, who were Russian.

This Russian inventiveness has continued, and it is interesting to note that, following the end of the Cold War, Russian rocket engines for new launchers are being made under licence in the United States; and that Hall effect electric thrusters, developed in Russia, are one of the key technologies for future exploration of space. The collapse of the Iron Curtain is, in this specific sense, the analogue of the collapse of the walls of Constantinople, in generating a renaissance in space propulsion.

As the epitome of the practical engineer, Sergei Korolev (1906–1966) and his colleague Valentin Glushko (1908–1989) should be credited with much of the Soviet success. Glushko was the engine designer, and Korolev was the rocket designer. Glushko's engines, the RD 100, 200 and 300 series, were and still are used in Russian launchers. It is significant that the 100 series, using liquid oxygen and alcohol, was a Russian *replacement* for the A4 engine. The desire to use a purely Russian engine was already strong. In fact, in the 1930s Glushko had developed liquid-fuelled engines which used regenerative cooling, turbo-pumps, and throttles. Korolev, as chairman of the design bureau, led the space programme through its golden age.

1.1.2 Other national programmes

Before turning to the United States' achievements in rocketry, we should remember that a number of other nations have contributed to the development of the present-day portfolio of launchers and space vehicles. There are active space and launcher programmes in the Far East, where China, Japan, India, and Pakistan all have space programmes. China and Japan both have major launcher portfolios.

In the Middle Kingdom, the invention of the rocket was followed by a long sleep of nearly 1,000 years. It was not until Tsien Hsue-Shen (1911–) was deported from the United States in 1955 that China began the serious development of modern rockets. He had won a scholarship to MIT in 1935, and later became Robert Goddard Professor of Jet Propulsion at CalTech. It is ironic that this supposed communist had been assigned the rank of temporary colonel in the US Air force in 1945, so that he could tour the German rocket sites, and meet Werner von Braun. He became, in effect, the Korolev or Werner von Braun of the Chinese space programme. Work on modern rockets began in China in 1956, and by the end of 1957, through an agreement with the USSR, R-1 and R-2 rocket technology had been transferred to the Chinese. Understandably these were old Russian rockets, and bore more resemblance to the German A4 than to the then current Russian launchers. Following the breach with the USSR in 1960, the Chinese programme continued, with an indigenous version of the R-2 called Dong Feng, or East Wind. Engulfed by the Cultural Revolution, the programme struggled through, with the support of Zhou Enlai, to the design of a new rocket—the Chang Zheng, or Long March. This was ultimately used to launch China's first satellite in 1970, a year after Apollo 11. China has continued to launch satellites for communications and reconnaissance, using versions of the Long March. Since 1990, this vehicle has been available as a commercial launcher. Tsien continued to play a major part in the programme, but fell into disfavour in 1975. Nevertheless he is still considered as the father of the modern Chinese space programme, and was honoured by the government in 1991. The Long March used a variety of engines, all developed in China, including those using liquid hydrogen and liquid oxygen. Despite the setbacks caused by political upheaval, China has succeeded in establishing and maintaining an indigenous modern rocket technology. Recent developments in China have placed the country as the third in the world to have launched a man into space. China is likely to develop a strong manned space programme.

Japan is a modern democracy, and rockets were developed there in an exclusively non-military environment. In fact, Japan's first satellite, Osumi, was launched by a rocket designed and built by what was essentially a group of university professors. The heritage of this remarkable success is that Japan had two space agencies: the Institute of Space and Astronautical Science, depending from the Ministry of Culture, or Monbusho; and the National Space Development Agency, depending from the Ministry of Industry. ISAS was founded in the mid-1950s, and has developed a series of indigenous, solid-fuelled launchers used exclusively for scientific missions. These have ranged from small Earth satellites, to missions to the Moon, Mars and to comets. ISAS launched Japan's first satellite in February 1970, after the US and France, and before China and the United Kingdom. Although small, ISAS has continued to develop advanced rocket technology including liquid hydrogen and liquid oxygen engines; and experiments on electric propulsion and single stage to orbit technology are in progress. NASDA is more closely modelled on NASA and ESA, and is concerned with the development of heavy launchers and the launching of communication and Earth resources satellites. It has an ambitious space programme, and is a partner in the International Space Station. After its foundation in 1964, NASDA began work, using US prototype technology, to produce the N series of heavy launchers. It has now developed entirely Japanese rocket technology for the H series of launchers. Japan was the fourth nation to launch a satellite, and with its two space agencies ranks as a major space-faring nation.[1]

India began space activities in 1972, when its first satellite was launched by the USSR; but development of a native launcher—the SLV rocket, which launched the satellite Rohini in 1980—took longer. There is now a substantial launcher capability with the ASLV and PSLV rockets.

Following the devastation of the Second World War, European nations entered the space age belatedly, with satellite launches by France, and later Britain. The National Centre for Space Studies (CNES) was founded in France in 1962, and retains responsibility for an active and wide-ranging national space programme. Using the Diamant rocket, it launched the first French satellite, Asterix, in 1965. Britain also developed a launcher to launch the Prospero satellite in 1971. Given the size of the US and USSR space programmes, individual nations in Europe could not hope to make a significant impact on space exploration. This was recognised by the creation of the European Space Agency in 1975.

ESA enabled the focusing of the technology programmes of the individual nations into a single space programme, and has been remarkably successful. It has succeeded in the creation of a coherent space programme, in which is combined the co-operative efforts of 14 member-states. This is evident in the many satellites which have been launched, and major participation in the International Space Station; but more so in the development of the Ariane European heavy launcher. Beginning with Ariane 1 in 1979, some 84 launches had been completed by 1996, the versions advancing to Ariane 4. Ariane has continued to develop, and in 1998 the first successful launch of the Ariane 5 vehicle took place. This is all-new technology, with a main-stage engine fuelled by liquid hydrogen and liquid oxygen, solid boosters, and the most modern control and guidance systems. The scale of the Ariane effort can be appreciated from the fact that engine production numbers exceed 1,000 (for the Viking, used on Ariane 4). Thus Europe has the most up-to-date rocket technology, and is in serious contention with the United States for the lucrative commercial satellite launcher market. Amongst the European nations, France and Germany take the lead in the Ariane programme, as in much else in Europe. The Ariane V has recently increased its capacity to 10 tonnes in geostationary transfer orbit.

1.1.3 The United States space programme

The achievement of the United States in realising humanity's dream of walking on the Moon cannot be overrated. Its origin in the works of Tsiolkovsky and Oberth, its national expression in the dream of Robert Goddard, and its final achievement through the will of an American president and people, is unique in human history. From what has gone before it is clear that the ambition to walk on the Moon was universal amongst those who could see the way, and did not belong to any one

---

[1] The two agencies have now been merged to form a new agency called JAXA.

nation or hemisphere. Nor was the technology exclusive. In fact, the Soviet Union came within an ace of achieving it. But it rested with one nation to achieve that unity of purpose without which no great endeavour can be achieved. That nation was the United States of America.

After the Second World War, the United States conducted rocket development based on indigenous technology, and the new ideas coming from the German programme, involving von Braun. Inter-service rivalry contributed to the difficulties in achieving the first US satellite launch, which was mirrored by the divisions between different design bureaux in the Soviet Union, although there it was less costly. The Army developed the Redstone rocket, basically improving upon the A4, in the same way that the R series developed in the USSR. The Navy had its own programme based on the indigenous Viking.

Finally there developed two competing projects to launch a satellite—one involving von Braun and the Redstone, and the other the Naval Research Laboratory with the Viking. The Redstone version, approved in 1955, was empowered to use existing technology to launch a small satellite, but only three months later, the more sophisticated Vanguard project from the NRL, was put in its place. In the event, this rocket—with a Viking first stage, and Aerobee second stage and a third stage, not yet developed—lost the race into space.

After two successful sub-orbital tests, the first satellite launch was set for December 1957—just after the successful Sputnik 1 flight. It exploded 2 seconds after launch. A second attempt in February 1958 also failed, and it was not until March 1958 that a 1-kg satellite was placed in orbit. This was too late, compared with both the Soviet programme and the rival Redstone programme. The latter had been restarted after the first Vanguard loss, and on 31 January 1958 it launched the United States' first satellite, which weighed 14 kg. The first stage was a Redstone, burning liquid oxygen and alcohol, and the upper stage was a Jupiter C solid motor. The satellite was both a programmatic and a scientific success: it discovered the Van Allen radiation belts.

After this, the von Braun concept held sway in the US space programme. Although the folly of competing programmes was not completely abandoned, NASA was set up on 1 October 1958 and began looking at plans for a Moon landing. Immediately after the first sub-orbital flight of an American in May 1961, and one month after Yuri Gagarin's orbital flight, John F. Kennedy made his famous announcement to Congress: 'I believe that this nation should ... .'

But there were still competing concepts from the different organisations. Von Braun conceived the Saturn under the aegis of the Army. The debate on whether to refuel in orbit, or to complete the mission through a direct launch from Earth, continued for some time; and a similar debate took place in Russia a few years later. The latter concept required a huge, yet-to-be-designed rocket called Nova; but finally the concept of Apollo, using the lunar orbiter, emerged as the most practical solution. This was based on an original Russian idea published in the 1930s, and was elaborated by John Houbolt, from NASA Langley. Von Braun's support was crucial in the final acceptance of this idea. The launcher needed to be huge, but not as big as the Nova and it eventually emerged as the Saturn V. The lunar lander

J-2 ENGINE FACT SHEET

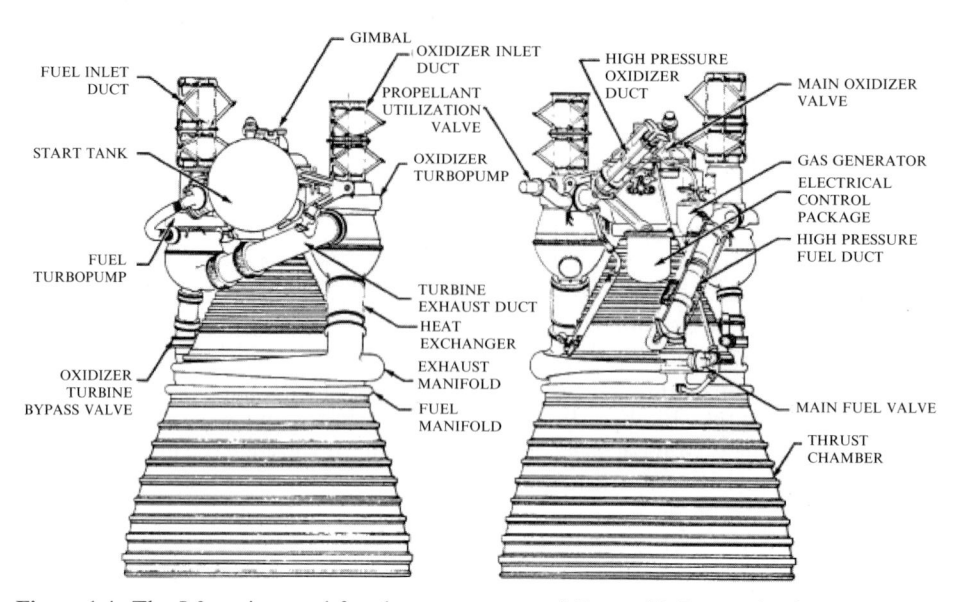

Figure 1.4. The J-2 engine used for the upper stages of Saturn V. It was the first production engine to use liquid hydrogen and liquid oxygen. It was also re-startable, a remarkable development for the 1960s. Note the lower part of the nozzle is not shown here.

would be a separate spacecraft, which would need only to journey between the Moon and lunar orbit. The lunar orbiter would be designed to journey between Earth and lunar orbit. This separation of roles is the key to simple and reliable design, and it contributed to the success of the Apollo programme. The concept was fixed in July 1962, after which work began on the Saturn.

In its final form, as the Apollo 11 launcher, the Saturn V (Plate 6) was the largest rocket ever built. It needed to be, in order to send its heavy payload to the Moon, in direct flight from the surface of the Earth. It needed powerful high-thrust engines to lift it off the ground, and high exhaust speed to achieve the lunar transfer trajectory. The lower stage was based on the liquid oxygen–kerosine engines, which had emerged, via the Redstone rocket, from the original German A4 engine that used liquid oxygen and alcohol. To achieve sufficient thrust to lift the 3 million-kg rocket off the pad, five F-1 engines—the largest ever built—provided a thrust of 34 MN, using liquid oxygen and kerosene. The exhaust velocity of these engines was $2,650 \, \text{m s}^{-1}$, but they had a very high total thrust. The important innovation for the second and third stages was the use of liquid hydrogen. It was the first operational use of this fuel and was vital in achieving the necessary velocity to reach the Moon. The second stage had five J-2 engines (Figure 1.4), burning liquid oxygen and liquid hydrogen, and providing a total thrust of 5.3 MN, with an exhaust velocity of $4,210 \, \text{m s}^{-1}$. The third stage had a single J-2 engine, providing a thrust of 1.05 MN.

The first manned operational launch took place in December 1968, and the first

Figure 1.5. The launch of the Space Shuttle *Atlantis*, 3 October, 1985.
Courtesy NASA via Astro Info Service.

lunar landing mission was launched on 16 July 1969. The mission took eight days, and the astronauts returned safely, having spent 22 hours on the Moon.

As a milestone in technology, the Saturn V was unique in the twentieth century; and as a human achievement, Apollo 11 was unique in the history of the planet. There was a strong hope that Apollo 11 would be the first step in a concerted effort towards human exploration of space—in particular, the planets. However, the shock of achievement left a sense of anticlimax, and the incentive to continue the programme, in the United States, began to diminish almost at once. The NASA budget fell from around $20 billion in 1966 to $5 billion in 1975, and since then there has been a slow rise to around $13 billion. The planetary programme continued with unmanned probes, which have been very successful, and which have provided us with close-up views of all the planets, and some comets. These probes were launched on Atlas and Titan rockets—considerably smaller than the Saturn V.

The main technical advance since the Saturn V has been the development of the Space Shuttle. The idea of a 'space plane' originates from at least as early as the 1920s, when it was proposed by Friedrich Tsander, and elaborated ten years later by Eugene Sanger. The latter devised A4 propelled rocket planes at Peenemunde. The US Air Force had a design called the Dynosoar at the time Saturn was selected as the lunar vehicle, and, interestingly, this concept had been worked on by Tsien.

It became clear, fairly soon after Apollo 11, that the budget for space could not be sustained at a high level, and that a more cost-effective way of continuing the manned programme was needed. Manned missions need to be very reliable, and this means that the components and construction have to be of the very best quality; multiple subsystems have to be provided, so that there is always a backup if one fails. The Saturn V had all of this, but it was used only once, and then discarded. All the expensive component manufacture and the huge effort to make a reliable vehicle had to be repeated for the next launch. The Shuttle concept was to bring the main vehicle back to Earth so that it could be re-used.

The use of a space plane had obvious advantages, and designs were emanating from the drawing board as early as 1971, two years after Apollo 11. The final selection—a delta-winged vehicle—was made partly on the basis of the Air Force's need to launch military payloads into polar orbits. The large wing meant that the Shuttle could return from a polar launch attempt to Vandenburg AFB in the event of an abort. As embodied in the first orbiter, *Columbia*, the Shuttle concept enabled re-use of the engines and control systems; propellant was carried in a disposable drop-tank, and solid boosters were used, which could be re-charged with propellant for re-use. After a ten-year development programme, the first Shuttle flew in April 1981. The total mass was two thousand tonnes, and the thrust at lift-off was 26 MN—both around 60% of the Saturn V. The height was 56 m, about half that of the Saturn V. The orbiter was fitted with three liquid hydrogen–liquid oxygen engines (SSME) which provided the high exhaust velocity needed. The propellant for these was contained in the external drop tank. At launch, the main engines were ignited, and after a few seconds, when their thrust was stable, the twin solid boosters were fired. The boosters dropped off when exhausted, and the orbiter continued into space under the power of the main engines. When the propellant was exhausted, the external tank was discarded. The orbiter was fitted with smaller storable propellant engines for orbital manoeuvring, and the all-important de-orbiting. Most of the kinetic energy from the orbit was dumped as heat during re-entry to the atmosphere. Once the velocity had dropped to a low enough value for the orbiter to behave like an aircraft, it was flown as a supersonic glider to its landing strip.

There were 24 successful flights before the loss of the *Challenger* due to a failed gas seal on one of the boosters. This underlined the need for continuous vigilance and attention to detail when dealing with such powerful forces. Shuttle flights were resumed in 1988, and have continued to the present day. The Shuttle is the primary means of launching the components of the International Space Station. The payload capability of 24 tonnes to low Earth orbit is not matched by any of the current expendable launchers, although the Ariane 5 has a capability similar to the Shuttle for geosynchronous orbit.

The recent accident with *Columbia*, and the loss of seven astronauts, has underlined the difficulties and dangers of human spaceflight. The Space Shuttle programme, and NASA itself have been subject to major review, and very significant changes to the NASA programme and organisational structure have been implemented. The Space Shuttle will be retired, and all non-space station flights have been cancelled. New safety requirements make this essential. The Space Shuttle will be replaced with a *Crewed Exploration Vehicle*, probably launched on an expendable rocket, and using Apollo style re-entry rather than the complex, and—as has been proven so tragically—dangerous, Shuttle system of tiles. The programme of NASA seems likely to be re-directed towards a return to the Moon and ultimately a human expedition to Mars.

1.1.4 Commentary

This brief summary of the history of rockets brings us to the present day when, after a period of relative stagnation, new rocket concepts are again under active consideration; these will be discussed later. A number of noteworthy points emerge from this survey. The invention of the rocket preceded the theory by 1,000 years, but it was not until the theory had been elaborated that serious interest in the rocket as a space vehicle developed. The theory preceded the first successful vehicle, the A4, by about 50 years. This seems to be because there were serious engineering problems, which required solution before ideas could be put into practice. One of these problems was guidance. A rocket is inherently unstable, and it was not until gyroscopes were used that vehicles could be relied upon to remain on course. Rocket engines are very high-power devices, and this pushes many materials and components to their limits of stress and temperature. Thus, rocket vehicles could not be realised until these problems were solved. And this required the materials and engineering advances of the early twentieth century.

It is also noteworthy that the basic ideas were universal. Looking at the developments in different countries, we see parallel activity. Goddard's patents were 'infringed', but not through theft; the basic ideas simply led to the same solutions in different places. It is, however, remarkable that the A4 programme should have played such a seminal role. It seems that the solutions arrived at on Peenemunde provided just that necessary step forward needed to inspire engineers around the world to apply their own knowledge to the problems of rocket vehicles. The use to which the A4 rocket was put, and its means of manufacture, were dreadful, but cannot be denied a position in the history of rocket engineering.

1.2 NEWTON'S THIRD LAW AND THE ROCKET EQUATION

As we have seen, the rocket had been a practical device for more than 1,000 years before Tsiolkovsky determined the dynamics that explained its motion. In doing so, he opened the way to the use of the rocket as something other than an artillery weapon of dubious accuracy. In fact, he identified the rocket as the means by which humanity could explore space. This was revolutionary: earlier, fictious journeys to the Moon had made use of birds or guns as the motive force, and rockets had been discounted. By solving the equation of motion of the rocket, Tsiolkovsky was able to show that space travel *was* possible, and that it could be achieved using a device which was readily to hand, and only needed to be scaled up. He even identified the limitations and design issues which would have to be faced in realising a practical space vehicle. The dynamics are so simple that it is surprising that it had not been solved before—but this was probably due to a lack of interest: perusal of dynamics books of the period reveals consistent interest in the flight of unpowered projectiles, immediately applicable to gunnery.

In its basic form, a rocket is a device which propels itself by emitting a jet of matter. The momentum carried away by the jet results in a force, acting so as to accelerate the rocket in the direction opposite to that of the jet. This is familiar to us all—from games with deflating toy balloons, if nothing else. The essential facts are that the rocket accelerates, and its mass decreases; the latter is not so obvious with a toy balloon, but is nevertheless true.

In gunnery, propulsion is very different. All the energy of a cannon ball is given to it in the barrel of the gun by the expansion of the hot gases produced by the explosion of the gunpowder. Once it leaves the barrel, its energy and its velocity begin to decrease, because of air friction or drag. The rocket, on the other hand, experiences a continuous propulsive force, so its flight will be different from that of a cannon ball. In fact, while the cannon ball is a *projectile*, the rocket is really a *vehicle*. The Boston Gun Club cannon, in Jules Verne's novel, was in fact the wrong method. To get to the Moon, or indeed into Earth orbit, requires changes in speed and direction, and such changes cannot be realised with a projectile. H. G. Wells' *cavorite*-propelled vehicle was closer to the mark.

1.2.1 Tsiolkovsky's rocket equation

Tsiolkovsky was faced with the dynamics of a vehicle, the mass of which is decreasing as a jet of matter is projected rearwards. As we shall see later, the force that projects the exhaust is the same force that propels the rocket. It partakes in Newton's third law—'action and reaction are equal and opposite', where 'action' means force. The accelerating force is represented, using Newton's law, as

$$F = mv_e$$

In this equation, the thrust of the rocket is expressed in terms of the *mass flow rate*, m, and the *effective exhaust velocity*, v_e.

So the energy released by the burning propellant appears as a fast-moving jet of

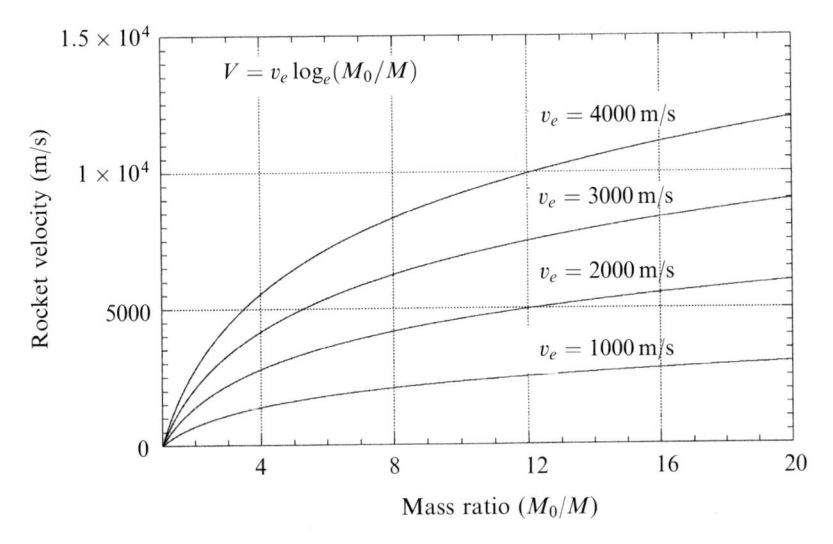

Figure 1.6. Tsiolkovsky's rocket equation.

matter, *and* a rocket accelerating in the opposite direction. Newton's law can be applied to this dynamical system, and the decreasing mass can be taken into account, using some simple differential calculus (the derivation is given at the beginning of Chapter 5). The resultant formula which Tsiolkovsky obtained for the vehicle velocity V is simple and revealing:

$$V = v_e \log_e \frac{M_0}{M}$$

Here M_0 is the mass of the rocket at ignition, and M is the current mass of the rocket. The only other parameter to enter into the formula is v_e, the effective exhaust velocity. This simple formula is the basis of all rocket propulsion. The velocity increases with time as the propellant is burned. It depends on the natural logarithm of the ratio of initial to current mass; that is, on how much of the propellant has been burned. For a fixed amount of propellant burned, it also depends on the exhaust velocity—how fast the mass is being expelled.

This is shown in Figure 1.6, where the rocket velocity is plotted as a function of the *mass ratio*. The mass ratio, often written as R, or Λ, is just the ratio of the initial to the current mass:

$$R = \frac{M_0}{M}$$

In most cases, the final velocity of the rocket needs to be known, and here the appropriate value is the mass ratio when all the fuel is exhausted. Unless otherwise stated, the final mass ratio should be assumed.

The rocket equation shows that the final speed depends upon only two numbers: the final mass ratio, and the exhaust velocity. It does not depend on the thrust, rather surprisingly, or the size of the rocket engine, or the time the rocket burns, or any other parameter. Clearly, a higher exhaust velocity produces a higher rocket velocity, and much of the effort in rocket design goes into increasing the exhaust velocity. As we shall see in Chapter 2, it happens that the exhaust velocity, within a narrow range of variability related to engine design, depends just on the chemical nature of the propellant. Gunpowder, and the range of propellants used for nineteenth century rockets, produced an exhaust velocity around 2,000 m s^{-1}, or a little more. The most advanced liquid-fuelled chemical rockets today produce an exhaust velocity of, at best, 4,500 m s^{-1}. There is nowhere else to go: this is close to the theoretical limit of chemical energy extraction.

To achieve a high rocket velocity, the mass ratio has to be large. The mass ratio is defined as the ratio of vehicle-plus-propellant mass, to vehicle mass. In these terms, a mass ratio of, say, 5 indicates that 80% of the initial mass of the rocket is fuel. This is very different from a car, for instance, which has a typical empty mass of 1.5 tonnes, and a fuel mass of 40 kg; a mass ratio of 1.003. So a rocket vehicle is nothing like any other kind of vehicle, because of the requirement to have a mass ratio considerably greater than 1. The most obvious feature about a rocket like the Saturn V, or the Space Shuttle, is its sheer size compared with its payload. The Saturn V carried three men on an eight-day journey, and weighed 3,000 tonnes. Most of this weight was fuel.

It can be seen in Figure 1.6 that the rocket can travel faster than the speed of its exhaust. This seems counter-intuitive when thinking in terms of the exhaust pushing against something. In fact, the exhaust is not pushing against anything at all, and once it has left the nozzle of the rocket engine it has no further effect on the rocket. All the action takes place inside the rocket, where a constant accelerating force is being exerted on the inner walls of the combustion chamber and the inside of the nozzle. So, while the speed of the rocket depends on the magnitude of the exhaust velocity, as shown in Figure 1.6, it can itself be much greater. A stationary observer sees the rocket and its exhaust passing by, both moving in the same direction, although the rocket is moving faster than the exhaust. The point at which the rocket speed exceeds the exhaust speed is when the mass ratio becomes equal to e, or 2.718, the base of natural logarithms. It should also be kept in mind that the accelerating force is independent of the speed of the rocket; however fast it goes, the thrust is still the same. So with a very large mass ratio, a very high speed can be attained. A big enough rocket could, in principle, reach α Centauri within a few centuries.

It is as well to mention here that a rocket carries both its fuel and its oxidiser, and needs no intake of air to operate, like, for example, a jet engine. It can therefore function in a vacuum—and in fact works better, because air pressure retards the exhaust and reduces the thrust. It also works, rather inefficiently, under water, provided that the combustion chamber pressure exceeds the hydrostatic pressure; those who have cast a weighted firework into water can vouch for this. There is a story, from the early days of rocketry, in which a rocket, launched from the beach, crashed into the sea. After fizzing around under the water for a while, it emerged and headed back to the beach, and to the terrified launch team who had gone to the water's edge to watch.

Tsiolkovsky also calculated how fast a rocket needs to travel to reach space, and he realised, from the rocket equation, that there was a limit. It is obvious from Figure 1.6 that after a certain point, increasing the mass of fuel has a diminishing effect on the velocity gain—notwithstanding what we have said about α Centauri. If we take the curve for an exhaust velocity of $1,000 \, \text{m s}^{-1}$—already about the speed of sound—we can see that a speed of $3,000 \, \text{m s}^{-1}$ is about the limit that can be reasonably achieved. A higher mass ratio would produce a higher velocity, but with a diminishing return. Figure 1.6 has a wildly optimistic ordinate: a mass ratio of 10 is almost impossible to achieve, particularly with a sophisticated high exhaust velocity engine. Those working, at the moment, on single stage to orbit rockets, would be happy to achieve a mass ratio of around 8. So while Tsiolkovsky was able to calculate the velocity achievable by a particular rocket, we would no doubt have been disappointed with the numbers that derived from his calculations. He knew that a velocity of $11 \, \text{km s}^{-1}$ was needed to escape the Earth's gravitational field. Faced with a gunpowder rocket, having at most about $2 \, \text{km s}^{-1}$ of exhaust velocity, the necessary mass ratio would have been wholly impossible to achieve.

Naturally, the first thing to do was to consider increasing the exhaust velocity. Tsiolkovsky knew that this was a matter of combustion temperature and molecular weight, which could be handled by nineteenth century chemistry. He quickly realised that liquid-fuelled rockets, using pure hydrogen and oxygen, could produce a considerable increase in exhaust velocity—in excess of $4,000 \, \text{m s}^{-1}$. Referring to the graph, escape velocity begins to appear possible. A mass ratio of about 14 is a less daunting task, but was still extremely difficult to achieve.

1.3 ORBITS AND SPACEFLIGHT

Leaving the problems of exhaust velocity and mass ratio for a moment, we shall turn, with Tsiolkovsky, to the question of how to get into space. This involves gravity, and the motion of vehicles in the Earth's gravitational field. Common experience, with a cricket ball for example, tells us that the faster a body is projected upwards, the further it goes. The science of ballistics tells us that a shell, with a certain velocity, will travel furthest in a horizontal direction, if projected at an initial angle of 45°. The equations of motion of a cricket ball, or a shell, can be solved using a constant and uniform gravitational field, with very little error. This is a matter for school physics. When we consider space travel, the true shape of the gravitational field becomes important: it is a radial field, with its origin in the centre of the Earth. Note that the gravitational field of a spherical object is accurately represented by assuming that it acts from the centre, with the full mass of the object. The flat Earth approximation is good enough for distances travelled which are small compared with the curvature of the Earth, but cannot be applied to space travel, where the distances are much greater.

The path of a ball may appear to be a parabola which begins and ends on the surface of the Earth, but in reality it is a segment of an ellipse with one focus at the centre of the Earth. If one imagines the Earth to be transparent to matter, then the ball would continue through the impact point, down past the centre of the Earth, and return upwards to pass through the point from which it was thrown. Without the drag caused by the atmosphere and by the solid Earth, it would continue to move in an elliptical orbit forever. Of course, the Earth would continue to rotate, so the points where the orbit passes through the surface would change for each cycle. This latter point is important to remember: a body moving in an orbit does not 'see' the rotation of the Earth; we do not notice this for normal projectiles, but it is important for rocket launches.

1.3.1 Orbits

Having introduced the topic of orbits we can now look into the motion of spacecraft, which always move in orbits and not in straight lines. Gravity cannot be turned off. A spacecraft does not 'leave' the Earth's gravity; it would be more correct to say that it 'gives in' to the Earth's gravity. So the motion of a spacecraft is that of a body, with a certain momentum, in a central gravitational field. The mathematics of this is (or should be) taught in school, and is given in Appendix 2. The path of the spacecraft can be calculated in terms of the total angular momentum it has in its orbit, which is a constant. The other defining parameter is the minimum distance of the orbit from its focus—the centre of the Earth.

The way to think of this is to imagine a spacecraft stationary at a certain altitude. The rocket fires, and gives it a certain velocity V, tangential to the gravitational field (parallel to the surface of the Earth). The kinetic energy of the spacecraft will be $\frac{1}{2}MV^2$; and its momentum and angular momentum about the centre of the Earth will be MV and MrV respectively. Here r is the distance from the centre of the Earth, not just the height above the surface.

The spacecraft moves under the combined effects of its momentum, given by the rocket, and the attraction of gravity towards the centre of the Earth. It will move in a curved path that can be represented by an equation of motion, and its solution. The solution to the equation of motion gives the radius r, of the orbit—the current distance of the spacecraft from the centre of the Earth, as a function of the angle made by the current radius vector to that of closest approach. This is the angle between r and r_0 in Figure 1.7. As time elapses the spacecraft will travel along the curve shown, initially becoming further from the Earth, while the angle increases. The expression for the path (derived in Appendix 1) is:

$$\frac{1}{r} = \frac{GM_\oplus M^2}{h^2}(1 + \varepsilon \cos\theta)$$

In this expression, G is the gravitational constant, which takes the value $6.670 \times 10^{-11}\,\text{Nm}^2\,\text{kg}^2$. The mass of the Earth is represented by $M_\oplus$, h is the (constant) angular momentum, and ε is the eccentricity of the orbit.

The eccentricity defines the shape of the orbit. For an ellipse, ε is the ratio of the distance between the foci, to the length of the major axis. For a circle, in which the foci coincide, ε becomes equal to zero. In order to understand how the orbit varies with the initial velocity of the spacecraft, the angular momentum and the

Sec. 1.3] Orbits and spaceflight 19

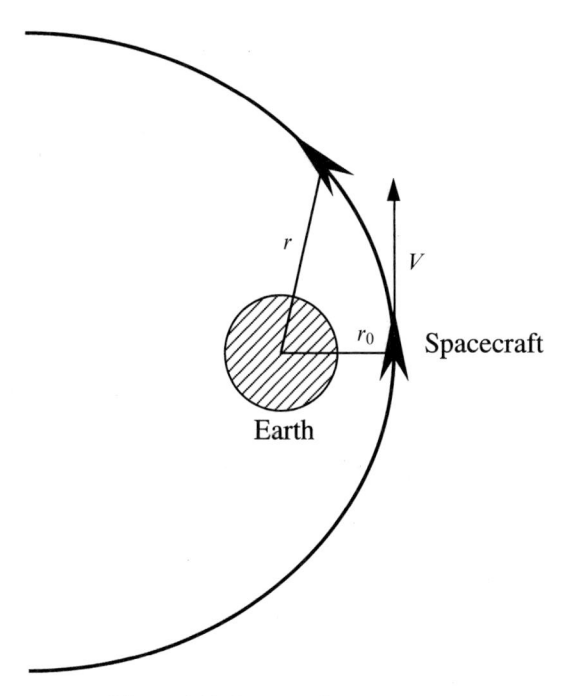

Figure 1.7. Spacecraft movement.

eccentricity have to be expressed in terms of useful parameters. They are given by the following formulae:

$$h = MrV$$

$$\varepsilon = \frac{h^2}{GM_\oplus M^2 r_0} - 1$$

Since h is constant throughout the orbit, it can be evaluated at the most convenient point (where the radius is at a minimum), and we know the velocity. This is just the initial velocity given to the spacecraft by the rocket. So $h = Mr_0V_0$, where V_0 is the initial velocity. Having fixed values for the initial radius and velocity, we can see that both the angular momentum and the eccentricity are fixed. Thus, the shape of the orbit depends only on the initial velocity and the distance from the centre of the Earth.

Figure 1.8 shows some orbit shapes of differing eccentricity: a circular orbit with eccentricity of zero, an elliptical orbit of eccentricity 0.65, and a parabolic orbit of eccentricity 1.0. For larger eccentricities the shape of the orbit becomes a hyperbola, which is more open than the parabola shown.

But what do these orbits mean, in practical terms, and how can they be predicted from the known parameters of the spacecraft? To understand this we need to express the eccentricity in terms of the initial velocity and height of the

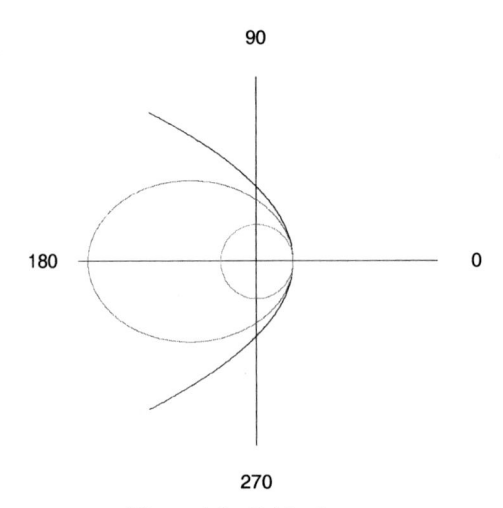

Figure 1.8. Orbit shapes.

spacecraft. Substituting the angular momentum expression in that for the eccentricity we obtain

$$\varepsilon = \frac{r_0 V_0^2}{GM_\oplus} - 1$$

We can see that if $r_0 V_0^2 = GM_\oplus$ then the eccentricity becomes zero; the orbit for this case is shown in Figure 1.4, as a circle. Or we can substitute zero in the orbit equation: the $\cos\theta$ term goes to zero and the radius is independent of the angle; that is, constant. Thus the orbit is circular. Since the condition for a circular orbit is that $r_0 V_0^2 = GM_\oplus$, it is easy to calculate the initial velocity, given the distance from the centre of the Earth.

$$V_0 = \sqrt{\frac{GM_\oplus}{r_0}}$$

The mass of the Earth is 5.975×10^{24} kg, and the mean radius is 6,371 km. Therefore, for an initial radius of 500 km above the Earth's surface, the initial velocity is 7.6 km s^{-1}.

This is not exactly the velocity needed to get into space, but it is the velocity necessary to stay there. The means of getting from the surface of the Earth to the injection point are discussed in Chapter 5, and, briefly, later in this chapter. For the present, let us continue with orbit shapes.

If the velocity given to the spacecraft is somewhat greater than the minimum value, so that the eccentricity is a little greater than zero, then the $\cos\theta$ term is finite, and the radius and the velocity depend on the location of the spacecraft in its orbit. In such an elliptical orbit, the product of velocity and radius has to remain constant, so that angular momentum is conserved. This means that when the radius is smallest, at the closest approach to Earth, the velocity is greatest; while, as the radius increases, the velocity drops, being lowest at the most distant point from the Earth. This agrees with intuition, and with energy arguments. Some of the kinetic energy given to the spacecraft is exchanged for potential energy, as it rises in the Earth's gravitational field. Elliptical orbits can be very eccentric, and spacecraft in such orbits can travel outwards many Earth radii before returning to pass through the injection point. Elliptical orbits are important because they are used to transfer a spacecraft from one circular orbit to another: for example, from low Earth orbit to geostationary orbit. From Figure 1.8, it can be seen that the apogee—the most distant point of the orbit—is opposite the point of injection, which becomes the perigee.

As the velocity given to the spacecraft increases, the eccentricity of the elliptical orbit becomes greater, and the apogee moves farther out. The ellipse in Figure 1.8 has an eccentricity of 0.65. If the eccentricity becomes equal to unity, the major axis of the ellipse and the separation of the foci are, by definition, equal. This can happen only if the far focus is at infinity; the near focus is at the centre of the Earth. In this case, the orbit ceases to be closed and the spacecraft never returns. In geometric terms, the ellipse becomes a parabola, and has the property that the trajectory becomes parallel to the axis at infinity. Substituting $\varepsilon = 1$ in the orbit equation, we see that the bracket containing $\cos\theta$ takes the value zero when $\theta = 180°$. The radius r is then infinite.

From the expression for eccentricity we see that for $\varepsilon = 1$,

$$\frac{r_0 V_0^2}{GM_\oplus} = 2$$

and

$$V_0 = \sqrt{\frac{2GM_\oplus}{r_0}}$$

This is the *escape velocity*—the minimum velocity that must be given to a spacecraft for it to escape from the Earth's gravitational field. Note that there is no boundary beyond which the field does not act. It continues to act to infinity, the spacecraft having zero velocity with respect to the Earth at that point.

There are two notable points about this escape trajectory. The first is that the velocity necessary to escape from the Earth is just $\sqrt{2}$ or 1.414 times the velocity necessary to remain in low Earth orbit. Once a rocket has achieved Earth orbit, it is comparatively easy to escape. The second point is that to escape from Earth, the initial direction of travel should be parallel to the Earth's surface, not perpendicular. Thus to enter Earth orbit, or to escape, a spacecraft must be given a large *horizontal* velocity. Vertical velocity helps only in bringing the spacecraft to a sufficient height in the atmosphere for drag to be minimal.

Elliptical transfer orbits

This horizontal acceleration is also used to move from one orbit to another, using an elliptical transfer orbit. When the Space Shuttle has to rendezvous from a low orbit, with, for instance, the Space Station, it accelerates *along* its orbit. Instead of travelling faster around the orbit, the orbit rises; it has in fact become a slightly elliptical orbit, the perigee being where the thrust was turned on. This elliptical orbit intersects with the Space Station orbit at the point where the Station is located when the Shuttle arrives there—if the burn was actuated at the correct time. Once at the Station, the Shuttle has to accelerate, again horizontally, in order to match its speed to that of the Station. This acceleration effectively puts the Shuttle into a circular orbit at the new altitude.

It should be clarified that the velocity of a spacecraft in a circular orbit decreases with the square root of the radius. So the Space Station is moving more slowly in its orbit than is the Shuttle in its lower orbit. However, for an elliptical orbit the velocity at apogee is lower than the intersecting circular orbit, because of the exchange of kinetic energy for potential energy as the spacecraft rises. At the same time, the velocity of the elliptical orbit is greater than that of the intersecting circular orbit at perigee. So to transfer from a low orbit to a higher orbit, two velocity increments are necessary: one at the perigee of the elliptical transfer orbit, and another at the apogee.

For convenient reference, some of the equations for the velocity of tangential circular and elliptic orbits are presented here.

$$V_0 = \sqrt{\frac{GM_\oplus}{r_0}} \quad \text{for a circular orbit}$$

$$V_{Escape} = \sqrt{\frac{2GM_\oplus}{r_0}} \quad \text{for a parabolic escape orbit}$$

$$\frac{V_1}{V_0} = \sqrt{1 + \frac{r_2 - r_0}{r_2 + r_0}}, \frac{V_2}{V_1} = \frac{r_0}{r_2} \quad \text{for elliptic orbits}$$

where r_0 is the perigee radius, r_2 is the apogee radius, V_1 is the elliptic orbit velocity at perigee, and V_2 is the elliptic orbit velocity at apogee. These are shown in Figure 1.9.

It will perhaps be obvious from the foregoing that the manoeuvre to 'catch up' with a spacecraft in the *same* orbit is quite complicated. Simply accelerating will not cause the Shuttle to move faster round the orbit; it will put it into an elliptical orbit, which will pass above the target. The Shuttle needs to decelerate and drop to a lower and faster orbit; and then, at the correct point in the lower orbit it should accelerate again to bring it up to the target. Both of these manoeuvres are elliptical orbit transfers. A further acceleration is then needed to match the target speed; that is, to circularise the orbit of the shuttle. These orbits need only be separated from one another by a few tens of kilometres, and the velocity changes are small.

A decrease in velocity when a spacecraft is in a circular orbit causes it to enter an elliptical orbit. The apogee is at the same altitude as the circular orbit, and the perigee is determined by the velocity decrease. At perigee, the spacecraft velocity is greater than the corresponding new circular orbit, so a further decrease is needed. This is the initial manoeuvre when a spacecraft returns to Earth: atmospheric drag

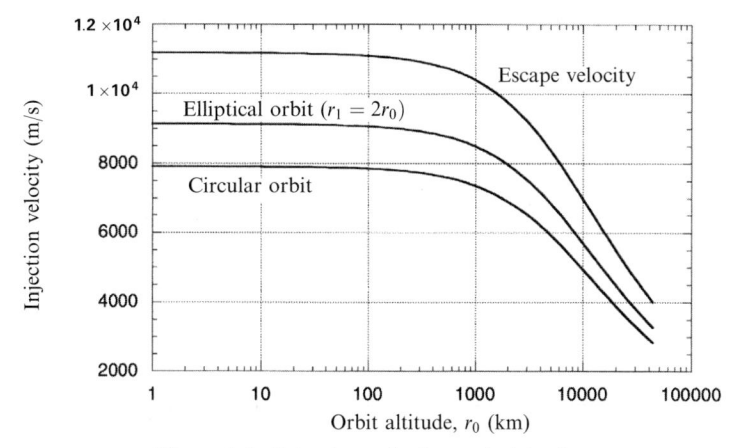

Figure 1.9. Injection velocity and altitude.

takes over before perigee is reached, and the resultant deceleration causes the trajectory to steepen continuously until it intersects with the Earth's surface.

On an airless body like the Moon, the elliptical orbit taken up should intersect with the surface at the desired landing point, and the vehicle has to be brought to rest using thrust from the motors. The Apollo 11 descent ellipse did not pass through the correct landing point, because the non-spherical components of the Moon's gravitational field were not known accurately. Armstrong had to take over control, and by using lateral thrust from the motors he guided it to the correct point.

Launch trajectories

Having discussed how a spacecraft can move from one orbit to a higher orbit, it is possible to see how a spacecraft can leave the surface of a planet and enter into a low orbit around that planet. For the Moon it is a reversal of the descent just described. For minimum energy expenditure, it would, in theory, be better to launch in a horizontal direction, regarding the launch point as being on a circular orbit at zero altitude. The velocity of the spacecraft, at rest on the surface, is wrong for the circular orbit, otherwise it would be hovering above the ground; so, during the burn, a significant vertical component of velocity is needed, while horizontal velocity is gained. There is also the question of mountains. The Apollo lunar module took off with a short vertical segment, before moving into a horizontal trajectory. Once in an elliptical orbit with an apogee at the required altitude, the spacecraft coasts towards it. At apogee, a further burn circularises the orbit.

For launch from Earth, the atmosphere is a significant problem. Although the density of air drops rapidly with height, the velocity of a spacecraft is large, and drag is proportional to the square of the velocity. Below 200 km there is sufficient drag to make an orbit unstable. In fact, to have a lifetime measured in years, an orbit needs to be above 500 km. This means that a significant proportion of the stored chemical energy in a rocket has to be used to raise the spacecraft above the atmosphere.

Atmospheric drag also slows the rocket in the lower atmosphere. For this reason, spacecraft are launched vertically; height is gained, and the drag in the dense lower atmosphere is minimised, because the rocket is not yet moving very fast. Once the densest part of the atmosphere is passed—at about 30 km—a more inclined trajectory can be followed. It cannot be horizontal because the velocity is not yet high enough. Ultimately, sufficient velocity is reached for an elliptical trajectory to the desired altitude to be followed, and the spacecraft can then coast. At apogee a further horizontal acceleration is needed to enter the circular orbit.

During a launch, the velocity is never high enough to attain a circular orbit at an intermediate altitude; this occurs only after the final injection has taken place. Intermediate orbits would require a still larger horizontal velocity, because of the inverse dependence of velocity on radius. Thus the motion of the rocket before final injection is along a trajectory which is always steeper than the free space ellipse, and it intersects the Earth's surface. When a launch fails, this becomes obvious. The non-optimum nature of the launch trajectory, compared with a transfer ellipse, means that a good deal of the energy of the rocket is lost to the Earth's gravitational field. This *gravity loss* is dealt with in Chapter 5.

The rotation of the Earth

There is also a small additional component of velocity, which can be gained from the rotation of the Earth. A spacecraft in orbit responds to the gravitational field as if all the mass of the Earth were concentrated at its centre. There is no means for the rotation of the Earth to affect the motion of an orbiting satellite. Conversely, seen from the frame of reference of the satellite in orbit, the rotation of the Earth's surface is seen as a real velocity. This means that the velocity of rotation of the Earth's surface at the launch site adds algebraically to the velocity of the satellite. The effect of this depends on the inclination of the orbit and the direction of motion of the satellite in its orbit. If the plane of the orbit is parallel to the equator (the planes of all orbits pass through the centre of the Earth) and the satellite travels in a west–east direction, then the speed of the Earth's rotation is added to the velocity given by the rocket. If the satellite travels in an east–west direction, then the speed is subtracted. If the orbit is at right angles to the equator, and the satellite travels over the poles, then the rotation speed of the Earth has no effect. The magnitude of the effect is simple to calculate. The Earth rotates once in 24 hours, and has a radius of 6,400 km, so the surface velocity is 40,212 km per day, or 465 m s^{-1}. This additional velocity can be used to increase the payload, and reduce the required mass ratio. For this reason, the majority of launch sites are located as close to the equator as possible. Strategic arguments may well mitigate against an actual equatorial launch site, but the major launch site in the United States, for example, is at Cape Canaveral—almost as far south as is possible on the US mainland. Russian launches take place from Baikanour in Kazakhstan, Japan uses the southern tip of Kyushu island, and Arianespace has its launch site on the equatorial coast of French Guyana.

For satellites to be launched into high inclination, or polar orbits, this effect is not very useful, and so the launch sites can be at any convenient latitude.

The velocity increment needed for launch

It is possible to calculate the total velocity increment required, without gravity loss, as follows, using the earlier formulae. Assume that the launch from the Earth's surface is the equivalent of a transfer from a circular orbit with a radius which is that of the Earth, via a transfer ellipse, to a 500 km circular orbit. The imaginary Earth-radius circular orbit would have a horizontal velocity of 7,909 m s^{-1} (see Figure 1.9). The transfer ellipse, with perigee at the Earth's surface and apogee at 500 km altitude, has a perigee velocity of 8,057 m s^{-1}. The apogee velocity is 7,471 km s^{-1}, and the necessary circular velocity is 7,616 m s^{-1}. Thus the total velocity increment is $(8057 + 7616 - 7471) = 8,203$ m s^{-1}. So the velocity cost of the launch, over and above that needed for a circular orbit injection at 500 km altitude, is 587 m s^{-1}. This would be true if all the velocity could be given to the rocket all at once, and there were no atmosphere, but because of the gravity loss we need to include an extra allowance of velocity. This depends on the trajectory; an approximate value is 500 m s^{-1}, and the total velocity increment required is approximately 8,700 m s^{-1}.

There is a distinction between velocity increment and the actual velocity of the vehicle. The velocity increment is the velocity calculated from the rocket equation, and is a measure of the energy expended by the rocket. The vehicle velocity is less than this, because of gravity loss, and the energy needed to reach orbital altitude. So the actual velocity of the vehicle in its 500-km circular orbit is 7.6 km s^{-1}, while the velocity increment is 8.7 km s^{-1}. The difference represents the energy expended against gravity loss and potential energy.

The mass ratio for such a velocity increment—especially with primitive rocket fuels, giving low exhaust velocity—is too high to achieve, even with modern construction methods. Tsiolkovsky realised this, and in 1924 he published a paper called *Cosmic Rocket Trains*, in which he proposed to solve the difficulty by using multistage rockets. This was the essential breakthrough which has enabled humanity, 1,000 years after the invention of the rocket, to travel in space.

1.4 MULTISTAGE ROCKETS

Once the concept has been suggested, it is easy to see intuitively that discarding, at least, the empty fuel tanks during the flight, is bound to improve the performance of a rocket. The thrust remains the same, but after the tanks have been dropped off, the mass of the rocket is smaller, so the acceleration will be greater. To calculate the effect, one can consider a rocket of given mass ratio, and then divide it into two smaller rockets whose combined mass is the same.

The payload—or that part which is to be put into orbit (the satellite or spacecraft) is the same in both cases. This means that for a fair test we need to separate the payload mass from the structural mass. The mass ratio of the single rocket can be defined as

$$R_0 = \frac{M_S + M_F + M_P}{M_S + M_P}$$

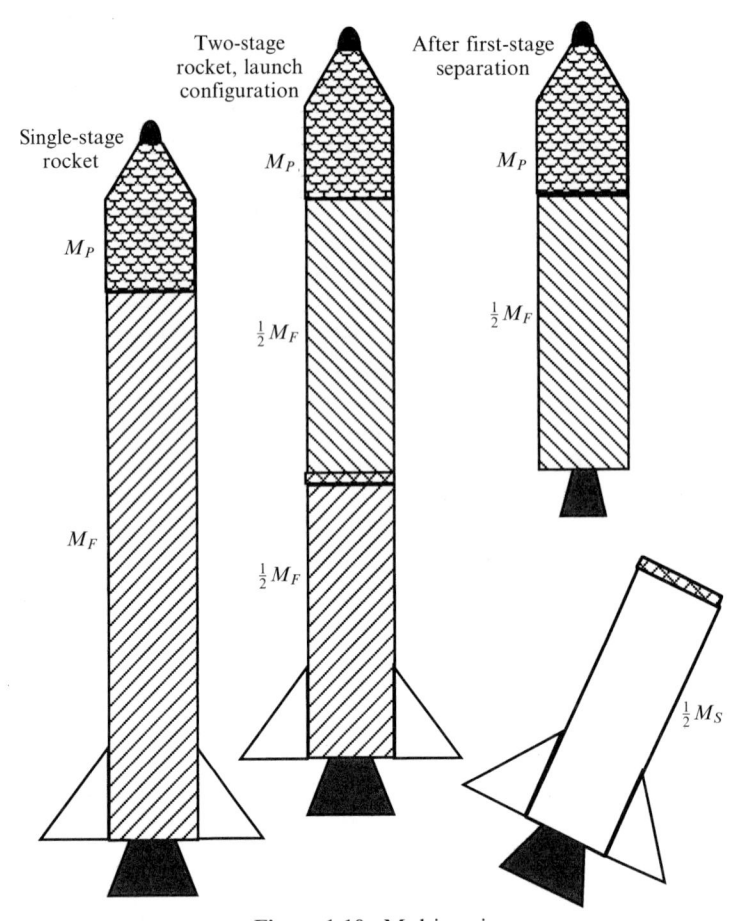

Figure 1.10. Multistaging.

This is the same definition as before, but with the structural mass separated out. Note that here it is convenient to use M_F as the propellant or fuel mass to distinguish it from M_P, the payload mass. The structural mass will include the mass of the engines, turbo-pumps, and fuel tanks, as well as the guidance and control electronics, and so on. Since, in general, we may expect that the structural mass will be kept to a minimum, we may reasonably assume that it is a constant fraction of the fuel mass, for stages using the same fuel.

The rocket is then divided into two rockets, each having half the fuel and stacked one on top of the other, as shown in Figure 1.10. The first rocket is ignited, and burns until all its fuel is exhausted. This gives the whole stack a velocity, defined by the rocket equation, with the mass ratio given by:

$$R_1 = \frac{M_S + M_F + M_P}{M_S + \frac{1}{2}M_F + M_P}$$

Sec. 1.4] Multistage rockets 27

Here the mass of fuel burned is half the single rocket fuel load. The lower rocket then drops off, and the upper rocket is ignited. It then gains additional velocity, defined by the rocket equation, with mass ratio defined by:

$$R_2 = \frac{\frac{1}{2}M_S + \frac{1}{2}M_F + M_P}{\frac{1}{2}M_S + M_P}$$

Here the upper rocket begins its burn with half the structural mass and half the fuel mass, and ends with half the structural mass and the payload. The final velocity will be the sum of the velocity increments produced by the two rockets. So to compare the performance of a single and a two-stage rocket, we have to compare the following:

$$V_0 = v_e \log_e R_0$$

$$V = v_e \log_e R_1 + v_e \log_e R_2$$

Any advantage of the two-stage approach will be revealed as $V > V_0$.

For an example calculation, we assume a rocket of total mass 100 tonnes, carrying a spacecraft of 1 tonne. The engines develop a constant exhaust velocity of $2{,}700 \, \text{m s}^{-1}$. The structural mass is assumed to be 10% of the fuel mass. Substituting in the first equation, we obtain for the velocity of the single-stage rocket:

$$V_0 = 2{,}700 \log_e \frac{10 + 89 + 1}{10 + 1} = 5{,}959 \, \text{m s}^{-1}$$

This corresponds to a mass ratio of 9.09, which is fairly optimistic. If the rocket is divided into two smaller ones, each with half the fuel, and the structural mass also shared equally, the payload being the same, then the velocity of the first stage is

$$V_1 = 2{,}700 \log_e \frac{10 + 89 + 1}{10 + 44.5 + 1} = 1{,}590 \, \text{m s}^{-1}$$

This corresponds to a low mass ratio of 1.8. However, the first-stage rocket can now be discarded, so that the remaining quantity of fuel has less dead mass to propel. The velocity increment of the second stage is

$$V_2 = 2{,}700 \log_e \frac{5 + 44.5 + 1}{5 + 1} = 5{,}752 \, \text{m s}^{-1}$$

Here the mass ratio is 8.42, and the velocity increment is close to that of the single-stage rocket. The final velocity is the sum of V_1 and V_2—$7{,}342 \, \text{m s}^{-1}$. Therefore, using the same quantity of fuel and dividing the structural mass between two smaller rockets, an extra $1{,}383 \, \text{m s}^{-1}$ is realised.

The next logical step is to divide the rocket into three stages. The velocity of the first stage is

$$V_1 = 2{,}700 \log_e \frac{10 + 89 + 1}{10 + 59.3 + 1} = 952 \, \text{m s}^{-1}$$

The mass ratio is of course even smaller, 1.42, but this time there are three rockets to share the load. The steps are the same as for two rockets, with the fuel and structural mass being shared equally amongst the three rocket stages. The total velocity achieved by the three-stage rocket is 8,092 m s^{-1}, so the extra stage has improved the velocity by another 749 m s^{-1}. The total gained by dividing a single rocket into three equal stages is 2,133 m s^{-1}.

1.4.1 Optimising a multistage rocket

The example given above assumes an equal division of fuel amongst the stages. It is valid to ask if this is the optimum ratio. It can be shown, by means of the calculus of variations, that for equal exhaust velocities and the same propellants, the optimum velocity increment occurs when the *payload ratios* of the stages are equal. If the structural efficiency of the stages is the same, then this is equivalent to declaring that the mass ratios should be equal.

The payload ratio of a single stage is given by

$$L = \frac{M_P}{M_S + M_F}$$

It is also convenient to express the structural efficiency by

$$\sigma = \frac{M_S}{M_F + M_S}$$

Neither of these ratios is a simple function of the mass ratio, R; they are related to R by

$$R = \frac{1 + L}{\sigma + L}$$

The payload ratio, L, is the ratio of the mass of the payload to the mass of the rest of the rocket; the structural coefficient, σ, is the ratio of the structural mass to the combined mass of the propellant and the structure (that is, the mass of the rocket, excluding the payload). The payload ratio is of course a measure of the usefulness of the rocket, and the structural coefficient is a measure of the degree of optimisation of the engineering design. Heavy engines, or propellant tanks which have thicker walls than necessary, will increase the structural coefficient, and, from the above, decrease the mass ratio. It can be seen that if L and σ are defined to be the same for each stage, as is required for optimum performance, then the mass ratio will also be the same for each stage.

Since the payload for the first stage is the combined mass of the second and third stages, and the payload for the second stage is the third stage, we can see that the lower stages must have a bigger share of the propellant than do the upper stages, in order for the mass ratios to be the same.

It is convenient to represent the fractional mass of each of the three stages as A, B, and C, respectively, beginning with the lower stage. The mass ratio of the first two stages, R, is then defined by:

$$\frac{A + B + C}{B + C} = \frac{B + C}{C} = R$$

and the mass ratio of the third stage is, of course, also R. Simple manipulations lead to

$$A = \frac{R-1}{R}; \quad B = \frac{R-1}{R^2}; \quad C = \frac{1}{R^2}$$

The value of R is still to be determined. For a fair test of the optimisation the value should be determined from the parameters of the original single-stage rocket. Using these parameters for the third stage, we can determine that

$$R = \frac{C}{M_P + \gamma M_S}$$

where γ is the fraction of propellant and structure to be assigned to the third stage. Further manipulation leads to

$$\gamma M_S = \frac{C}{R} - M_P$$

$$R \leq \sqrt[3]{\frac{1}{M_P}}$$

For the assumed parameters of our single-stage rocket, this limit is 4.64. Taking this as an approximate value of R, in order to determine the values of A, B, and C, we find

$$A = 0.7845$$
$$B = 0.1691$$
$$C = 0.0465$$

showing that the first stage should have nearly 80% of the total mass. From these values the multipliers for the mass of structure and propellant can be derived, and the mass ratio of the stages calculated more accurately:

$$A = \alpha(M_S + M_F)$$
$$B = \beta(M_S + M_F)$$

This leads to a more accurate value of 3.39 for the mass ratio of the stages. Since each stage now has the same mass ratio, the velocity increment is the same, and so the total velocity increment is

$$V = 3 \times 2{,}700 \log_e 3.39 = 9{,}889 \,\mathrm{m\,s^{-1}}$$

Comparing this with the velocity increment of the three-stage rocket with stages of equal mass, we see that the optimum mass distribution produces an increase of $1{,}797 \,\mathrm{m\,s^{-1}}$. The velocity increment is sufficient to launch a spacecraft into orbit. The margin of over $1 \,\mathrm{km\,s^{-1}}$ is even sufficient to allow a more gentle acceleration. Gravity loss worsens as the acceleration of the rocket is reduced.

1.4.2 Optimising the rocket engines

It will be apparent from the above calculations that increasing the number of stages beyond three is not efficient. Four-stage rockets have been built, but the gain in the velocity increment is not so great as the gain when moving from two to three stages. The counter-argument is the cost and complexity of additional stages. Each stage has to have its own engines, fuel pumps, and guidance systems. A multistage rocket is therefore more expensive than a single-stage rocket, and there is more that can go wrong. In one respect however, the multistage rocket offers a positive advantage over a single stage in the freedom that it provides in matching the rocket engines to the job they have to do.

A rocket in the lower atmosphere operates in conditions very different from those in the vacuum of space. In the initial stages of flight, the atmospheric pressure is high, and this affects the performance of the rocket. Atmospheric drag and other aerodynamic effects are also strong in the lower atmosphere. In space, the engines are operating in a vacuum, and there is no drag or lift. A rocket exhaust is strongly affected by the pressure at the exit of the exhaust nozzle, as the air pressure retards the exhaust stream. It will be shown in Chapter 2 that for optimum exhaust velocity, low in the atmosphere, the exhaust nozzle should be short, so that the exhaust does not expand too much. For a vacuum, the nozzle should be long, and the exhaust should be expanded as much as possible. This means that a rocket cannot be optimally designed for the whole journey into space.

The multistage rocket offers an ideal solution to the dilemma: the first stage can be designed for best performance in the lower atmosphere, while the upper stages can be designed to perform best in vacuum. This applies to the nozzle length, and it can also apply to the type of fuel used. The first stage has the task of lifting the rocket more or less vertically through the lower atmosphere. It needs to have high thrust, because it has to lift the entire mass of the multistage rocket. On the other hand, it cannot achieve a very high final velocity, because of its adverse mass ratio, and because of atmospheric drag, which depends on the square of the velocity. These requirements suggest the use of large engines producing high thrust; it should be obvious that the thrust needs to be greater than the total weight of the rocket, if it is to leave the launch pad. Because the mass ratio and efficiency of the nozzle are so poor, the use of a propellant combination giving high exhaust velocity is less important. This means that less demanding propellants can be used, which simplifies the design and operation of the large first-stage engines. The upper stages are lighter and need less thrust. The rocket is not working against gravity to the same extent, because its path is now inclined, and so smaller engines can be used. Because the mass ratio and nozzle efficiency are increased, the use of propellants such as liquid oxygen and liquid hydrogen is beneficial, leading to the high final velocity needed for injection into orbit. The added complexity of, for example, all-cryogenic propellants, is offset, to some extent, by the smaller size of the engines. The multistage rocket thus lends itself to optimum engine design.

The archetypal three-stage rocket is the Saturn V, described earlier in the chapter. The first stage had five liquid oxygen and kerosene engines; the second stage, five

Sec. 1.4]   Multistage rockets   31

Table 1.1. The Saturn V rocket.

	Stage 1	Stage 2	Stage 3	Payload (LEO)	Payload (LTO)
Launch mass (kg)	2,286,217	490,778	119,900	118,000	47,000
Dry mass	135,218	39,048	13,300	–	–
Propellant	LO_2/kerosene	LO_2/LH_2	LO_2/LH_2	–	–
Engines	5 F-1	5 J-2	1 J-2	–	–
Exhaust velocity	2,650 m s^{-1}	4,210 m s^{-1}	4,210 m s^{-1}	–	–
Mass ratio	3.49	2.63	–	1.81	3.95
Velocity increment	3,312 m s^{-1}	4,071 m s^{-1}	–	2,498 m s^{-1}	5,783 m s^{-1}

liquid oxygen and liquid hydrogen engines; and the third stage, a single liquid oxygen and liquid hydrogen engine. The mass ratio for the first stage burn was 3.49, and the exhaust velocity was about 2,650 m s^{-1}. The short nozzle, while optimum for low altitudes, still produces a relatively low exhaust velocity. This results in a velocity increment of 3,312 m s^{-1}. The actual velocity at first-stage burn out would have been considerably less because of the gravity and drag losses. The second-stage mass ratio is 2.63, although with the hydrogen fuel the exhaust velocity of 4,210 m s^{-1} produces a velocity increment, for the stage, of 4,071 m s^{-1}. Since by this point in the trajectory the atmospheric pressure is negligible, and the rocket motion closer to horizontal, drag and gravity losses are much smaller. More of this velocity appears as real vehicle velocity. The same is true for the third stage.

Table 1.1 shows the mass ratios and payloads for the third stage, both for LEO injection, and for injection into a lunar transfer orbit. The mass ratios of these three stages are not far from the optimum values derived above. The upper and middle stages both use liquid oxygen and liquid hydrogen. The exhaust velocity with this propellant combination is much higher than the oxygen–kerosene used for the first stage. The tankage requirement for the low-density hydrogen fuel will also alter the structural coefficient, and bigger tanks are required for the same mass of propellant. We might therefore expect a departure from the constant mass ratio derived for identical propellants. The total velocity increment of 9,881 m s^{-1} compares well with the velocity increment needed for LEO injection. The gravity loss for this mission will be large, because of the limitation on acceleration imposed by the human payload. Gravity loss is lower for a high thrust-to-weight ratio (see Chapter 5).

1.4.3 Strap-on boosters

The three-stage rocket, exemplified by the Saturn V, rather quickly evolved into the two-stage rocket, with strap-on boosters. This technique—which was used in Russia very early in the programme—has the advantage that the thrust of the first stage can be altered to account for an increased payload without changing the fundamental design of the main rocket. Up to six boosters were used with the R-7 rocket. Among modern launchers, the strap-on booster is a key feature, the largest being those used on the Space Shuttle and Ariane 5. Boosters are usually solid-fuelled, but liquid-fuelled boosters are also used, notably on the Ariane 4 launchers and several Russian rockets.

The approach is a variant on multistaging, and the calculation of velocity increment is carried out in the same way. Boosters can be used to improve the performance of a three-stage rocket, effectively making it a four-stage vehicle, or they can be used with a two-stage rocket (Figure 1.11). In either case, the boosters are ignited at lift-off and burn for part of the first-stage operation (Plate 10). In modern launchers such as the Space Shuttle and Ariane 5, the first stage is optimised for high altitude and high mass-ratio. This would produce insufficient thrust at low altitudes to lift the rocket off the launch pad and the nozzle, being optimised for high altitude, is also inefficient near sea level. The boosters provide the necessary high thrust for the early stages of flight.

When the propellant is exhausted, the boosters are separated from the rocket by the firing of explosive bolts, which drive the nose of the booster away from the rocket

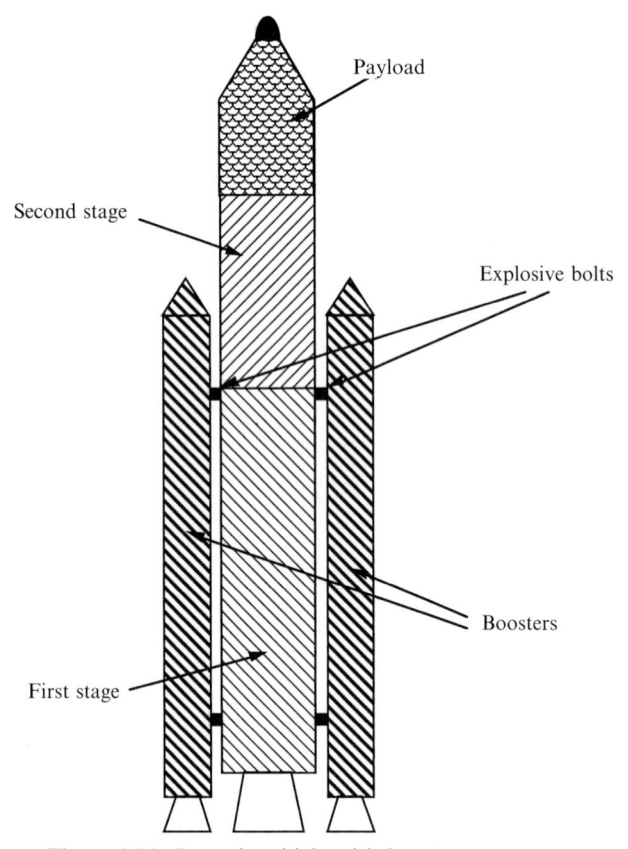

Figure 1.11. Launch vehicle with boosters.

axis. The lower attachment may also be released by means of explosive bolts, or maybe a simple latch which releases once the booster axis has rotated through a certain angle. It is important that the boosters do not collide with any part of the rocket during separation, and similar requirement applies during stage separation. Again, explosive charges, springs, or even small rockets are used to guide the empty stages away from the main rocket. The need for this becomes clear when it is realised that in the period between the shut-down of the first stage and the ignition of the second stage, the two parts of the rocket are essentially weightless. Small relative velocities, if not controlled, can cause a collision.

Boosters generally have a very high thrust, and therefore a high mass-flow rate, so they burn for a shorter time than the first stage. It is helpful to calculate the mass ratio and exhaust velocity in two parts. The first calculation is carried out for the total mass change of the two boosters, together with that part of the first stage propellant exhausted, up to the point of booster burn-out. This requires an estimate of the amount of propellant consumed by the first stage, and therefore knowledge of the burn time of the boosters and the mass flow rate of the first stage. It is usually acceptable to assume a constant mass-flow rate for a rocket engine, as the vast majority are not throttled—the SSME is a notable exception.

$$R = \frac{M_B + M_1 + M_2 + M_P}{M_{SB} + (M_1 - mt_B) + M_2 + M_P}$$

$$m = \text{mass flow rate} = \frac{M_1 - M_{S1}}{t_1}$$

Here the numerical subscripts refer to the stages, subscript B refers to the boosters, and subscripts $S1$ and SB refer to the dry mass of the first stage and boosters respectively. The burn time of the boosters is t_B, and the burn time of the first stage is t_1. This is adequate for an approximate calculation. The second part of the first-stage burn is treated in the same way, with the final mass of the first stage in the numerator of the mass ratio, and the mass of the first stage at booster burn-out in the denominator. The second- and third-stage mass ratios are calculated in the usual way.

There is a final element of the staging philosophy which has not so far been mentioned. The payload shroud, or nose-cone, is needed during the early part of the flight to protect the payload from atmospheric forces. At low altitudes these are the common forces of lift and drag which are experienced by aircraft, and result in the requirement for streamlining. At higher altitudes, the rocket has reached a high velocity—much more than the speed of sound—while the air is now very thin. The rapid motion of the air past the nose of the rocket now leads to heating. All of these effects will damage the payload, or require it to be made more strongly than is consistent with its use in space. Thus, all launchers have a shroud around the payload to protect it. The satellite or spacecraft is in fact attached securely to the top of the last-stage motor, and the shroud is mounted round it.

Once the rocket is sufficiently high in the atmosphere for aerodynamic and thermal effects to be negligible, the shroud is discarded, and the launch continues with the payload exposed. To avoid damage to the payload during this operation, the shroud splits in half, and springs or explosive charges are used to safely jettison the two parts. The effect of this decrease in dead-mass can be accounted for by splitting the calculation of the appropriate stage mass-ratio in two parts, in a similar way to the above. The design of the shroud requires that it should have a low mass (for obvious reasons), and that it should be able to withstand the heating and aerodynamic effects of high velocity. For heavy launchers it also has to provide a large enclosed volume in order to allow large spacecraft to be carried. These requirements generally lead to a composite construction for the shroud.

1.5 ACCESS TO SPACE

In this chapter we have reviewed the history and development of the rocket vehicle as the means by which human beings can explore space. We have also considered the physics of space exploration, showing how the rocket equation may be applied to launch vehicles and to the transfer of a spacecraft from one orbit to another. Human beings walked on the Moon within ten years of the first man orbiting the Earth. Since then the rate of progress has slowed, but much has been achieved. The immediate environment of the Earth and of many of the planets is now well understood; close-up images of all of the planets have been obtained; and comets have been visited by robot spacecraft. The International Space Station is under construction, and, as the first world space mission, points the way for future large co-operative ventures. Space commerce, while still in its infancy, has made great strides. Space is now central to electronic communication, and Earth observation from space is now a major tool with many applications.

Over the next decade there is certain to be a growth in the commercial use of space, and the demand for launches will continue to increase. The major technical challenge in space, for the next two decades, will be the human exploration of Mars. Once this has been achieved, the space age can truly be said to have arrived.

All these activities and challenges depend on rocket propulsion. The rocket is the oldest self-propelled vehicle, and, at the same time, the most modern. The golden age of chemical rocket development in the 1960s was succeeded by a retrenchment, during which attempts—mostly unsuccessful—were made to reduce the cost of space access, and there was little new development. Now we see a renewed interest in space propulsion. The race to develop a single-stage to orbit vehicle is under way, and electric propulsion—a dream of the early pioneers—is now a reality. It is even possible that nuclear propulsion will be developed in the next decade.

2

The thermal rocket engine

The rocket principle is the basis of all propulsion in space, and all launch vehicles. The twin properties of needing no external medium for the propulsion system to act upon, and no external oxidant for the fuel, enable rockets to work in any ambient conditions, including the vacuum of space. The thermal rocket is the basis of all launchers, and almost all space propulsion (although some electric propulsion uses a different principle). In this chapter we shall treat the rocket motor as a heat engine, and examine the physical principles of its operation. From these physical principles the strengths and limitations of rocket motors can be understood and appreciated.

The thermal rocket motor is a heat engine: it converts the heat, generated by burning the propellants—fuel and oxidiser, in the combustion chamber—into kinetic energy of the emerging exhaust gas. The momentum carried away by the exhaust gas provides the thrust, which accelerates the rocket. As a heat engine, the rocket is no different in principle from other heat engines, such as the steam engine or the internal combustion engine. The conversion of heat into work is the same, whether the work is done on a piston, or on a stream of exhaust gas. It will be helpful if we look first at the basic form of the thermal rocket.

2.1 THE BASIC CONFIGURATION

A liquid-fuelled rocket engine (see Figure 2.1) consists of a combustion chamber into which fuel and oxidant are pumped, and an expansion nozzle which converts the high-pressure hot gas, produced by the combustion, into a high velocity exhaust stream. It is the expansion of the hot gas against the walls of the nozzle which does work and accelerates the rocket.

A solid-fuelled motor (Figure 2.2) operates in the same way, except that the fuel and oxidant are pre-mixed in solid form, and are contained within the combustion chamber. Normally the combustion takes place on the inner surface of the propellant charge. The exhaust nozzle is identical in form to that in the liquid-fuelled motor, and the principles of operation are the same. In this chapter we shall make little

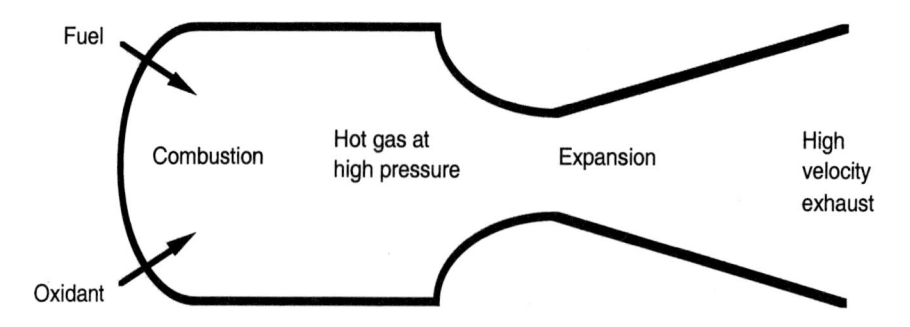

Figure 2.1. A liquid-fuelled rocket engine.

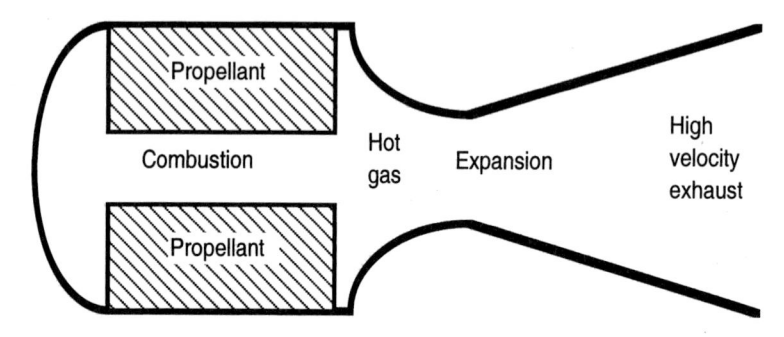

Figure 2.2. A solid-fuelled rocket motor.

distinction between the solid- and liquid-fuelled variants of the thermal rocket motor.

The combustion, which takes place in the chamber, can be any chemical reaction which produces heat. It may be simple oxidation of a fuel (hydrocarbon, or pure hydrogen, for example) by liquid oxygen; or it may be one of a number of other kinds of exothermic chemical reaction, as, for example, between fuming nitric acid and hydrazine. Solid propellants may contain an oxidiser such as potassium perchlorate together with finely divided aluminium and carbon, all bound together in a rubber-like material. Gunpowder is the classical solid propellant, and was used in the first Chinese rockets. A mixture of aluminium powder and sulphur is an example of a solid propellant with no oxidiser involved: the exothermic reaction produces aluminium sulphide. The main requirement for all propellant combinations is to maximise the energy release per kilogramme; as with any other rocket component, the lower the mass for a given energy release, the higher will be the ultimate velocity of the vehicle. The principles of the thermal rocket do not depend on specific types of propellant, so this aspect will only impact in a minor way on the following sections; solid and liquid propellants will later be dealt with in detail.

2.2 THE DEVELOPMENT OF THRUST AND THE EFFECT OF THE ATMOSPHERE

In Chapter 1 the discussion of the rocket equation and the application of Newton's third law to rocket propulsion ignored the effects of atmospheric pressure and the actual forces involved in producing the propulsive thrust. The concept of effective exhaust velocity enabled this simplification. The effective exhaust velocity is that velocity which, when combined with the actual mass flow in the exhaust stream, produces the measured thrust, $F = mv_e$, where m is the mass flow rate, and v_e is the effective exhaust velocity; v_e combines the true exhaust velocity, together with the effects of atmospheric pressure, and the pressure in the exhaust stream, into one parameter. The true exhaust velocity, however, is a function of these parameters, as well as the conditions of temperature and pressure in the combustion chamber. Here we shall look in more detail at the functioning of the rocket engine, and the development of exhaust velocity and thrust.

In the middle of the combustion chamber, the hot gas containing the energy released in the chemical reaction is virtually stationary. The energy—at this moment represented by the temperature and pressure of the gas—has to be converted into velocity. This occurs as the gas expands and cools while it passes through the nozzle. The velocity rises very rapidly, passing the speed of sound (for the local conditions) as it crosses the 'throat' or narrowest part of the nozzle. Thereafter it continues to accelerate until it leaves the nozzle. The accelerating force on the gas stream is the reaction of the nozzle wall to the gas pressure, as the gas expands against it. Thus the thrust is mostly developed by the nozzle itself, and is then transferred to the vehicle through the mounting struts. The accelerating force on the rocket is thus linked into the structure holding the rocket engine, and thereby to the base of the rocket itself.

The development of thrust, and the effect of atmosphere, can be examined through the derivation of the *thrust equation*, which relates the thrust of the rocket to the actual exhaust velocity, the pressure in the combustion chamber and the atmospheric pressure. It allows insight into some of the main issues in rocket motor design. The equation is derived by considering two separate applications of Newton's third law: once to the exhaust gases and once to the rocket motor, and the vehicle to which it is attached. It is important to recognise that the processes in a rocket engine result in two motions: the forward motion of the rocket and the backward motion of the exhaust stream, both of which require application of Newton's third law. There are two forces involved: the reaction of the internal surfaces of the rocket engine, which accelerates the gas, and the pressure force of the gas on those internal surfaces, which accelerates the rocket.

Figure 2.3 represents the action of the gas pressure on the combustion chamber and the exhaust nozzle, which is the force which accelerates the rocket. It also shows the reaction of the walls of the combustion chamber and of the exhaust nozzle, acting on the gas contained by them, which is the force that accelerates the exhaust gas.

The force accelerating the exhaust gas, the reaction of the walls, is equal to the surface integral of the pressure, taken over the whole inner surface of the chamber

38 The thermal rocket engine [Ch. 2

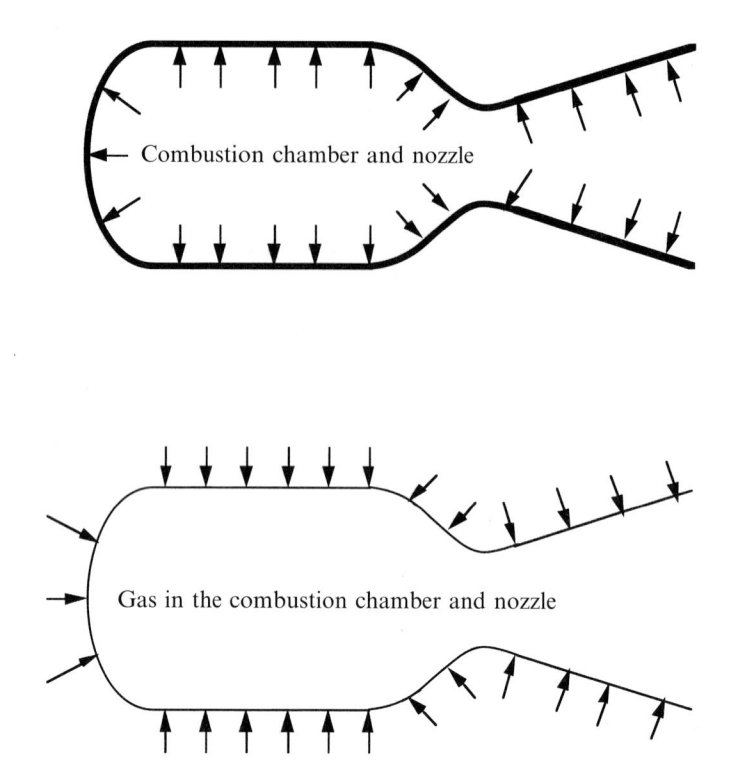

Figure 2.3. Forces in the combustion chamber and exhaust nozzle.

and nozzle: $F = \oint p\, dA$. This is not the only force acting on the gas: there is also a retarding force, which can best be appreciated by referring to Figure 2.4.

The gas flowing through the nozzle is impelled by the pressure gradient from the combustion chamber to the exit. At any point in the nozzle, the pressure upstream is greater than the pressure downstream. Considering the shaded portion of the exhaust stream represented in Figure 2.4, the net accelerating force acting on the shaded portion is

$$dF = pA - (p - dp)A$$

where A is the cross-sectional area at any given point, and the pressure gradient is dp/dx. This is the force that accelerates the gas through the nozzle. This formula applies at any point in the nozzle. For an element at the extreme end of the nozzle—the exit point shown in Figure 2.4—the outward force is pA, but the retarding force is the pressure at the exit plane, which can be denoted by p_e, multiplied by the area at the exit plane, A_e.

It is important to realise that the exhaust stream immediately beyond the end of the nozzle is not affected by ambient pressure: it is travelling at supersonic velocity, and hydrostatic effects can only travel at sound speed. Further downstream, effects

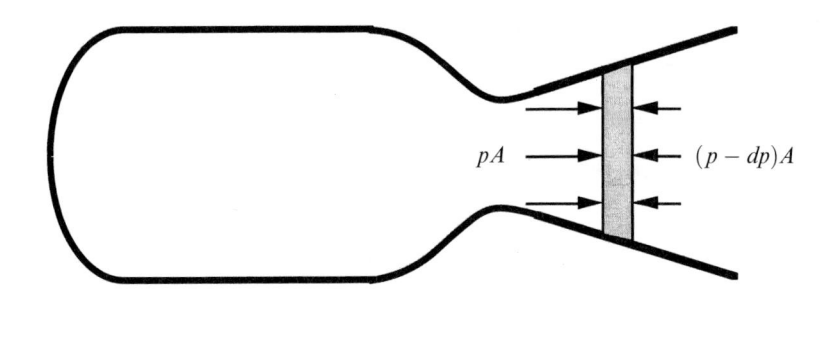

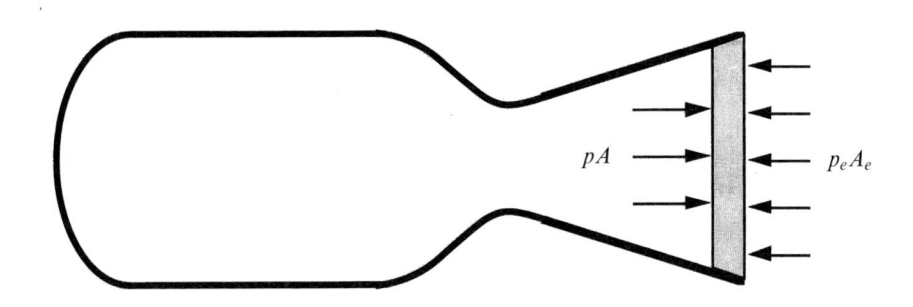

Figure 2.4. Gas flow through the nozzle.

of turbulence at the boundary between the exhaust and the atmosphere will make themselves felt, and under certain conditions shock waves can develop. But immediately beyond the exit plane the flow is undisturbed unless extreme conditions prevail. Thus for our purposes the above analysis holds.

Considering now the application of Newton's law to the exhaust gases, the accelerating force is represented by

$$F_G = \oint p \, dA - p_e A_e = m u_e$$

where m is the mass flow rate through the nozzle, and u_e is the exhaust velocity. This is the force that accelerates the exhaust stream in the nozzle; beyond the end of the nozzle the stream ceases to accelerate, and until turbulence starts to slow the stream down the exhaust velocity is a constant.

Turning now to the accelerating force on the rocket, this is represented by the surface integral of the pressure over the walls of the combustion chamber and nozzle:

$$F_R = \oint p \, dA$$

which is the force tending to accelerate the rocket.

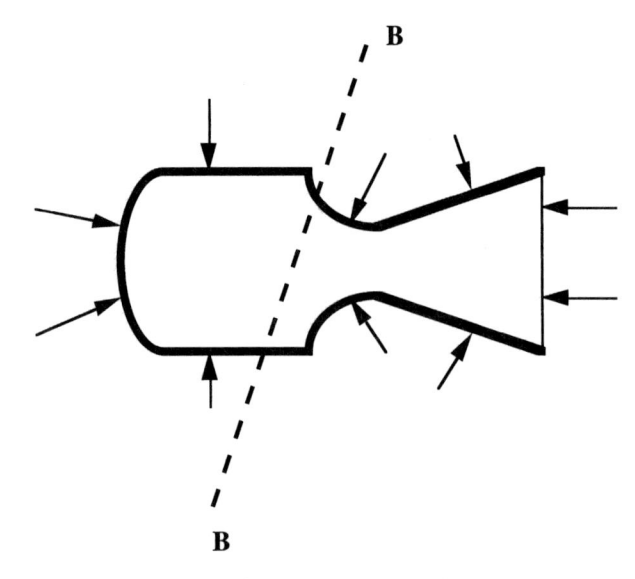

Figure 2.5. Static force due to atmospheric pressure.

Again there is a retarding force acting on the rocket due to the atmospheric pressure. This is a static force which operates whether or not the rocket is moving through the atmosphere. (There are of course, in addition, aerodynamic forces of drag and lift, developed through the motion of the rocket through the atmosphere; these are considered in Chapter 4.) To evaluate this static force, consider the cold rocket motor—not firing—shown in Figure 2.5.

As the rocket is stationary under the atmospheric pressure forces, they must balance across any arbitrarily chosen plane cutting the rocket, **BB**. When the rocket is active, and the supersonic exhaust stream occupies the region to the right of the nozzle exit plane, there is no longer a force due to atmospheric pressure acting on the exit plane (see the argument given above). Since the plane across which the atmospheric forces balance, in the cold case, can be chosen arbitrarily, **BB** can be moved to coincide with the exit plane of the nozzle without violating any physical principle. The unbalanced atmospheric force is then seen to be a retarding force, equal in magnitude to the atmospheric pressure integrated over the exit plane: $p_a A_e$, where p_a is the atmospheric pressure and A_e is the area of the exit plane.

So the net force accelerating the rocket is represented by:

$$F_R = \oint p\, dA - p_a A_e$$

This is the net thrust of the rocket. The surface integral, which appears in both equations, would be difficult to evaluate, but fortunately we have two expressions involving the same integral, and it can be cancelled. This arises because the magnitude of the force acting on the combustion chamber and nozzle is identical to that acting on the exhaust gases. Substituting for $\oint p\,dA$, from the equation for the acceleration of the exhaust gases, we find:

$$F_R = mu_e + p_e A_e - p_a A_e$$

This is the *thrust equation*.

The difference between this equation and the version given in Chapter 1 is that the true exhaust velocity u_e is used, together with the exit plane area of the nozzle A_e and the two pressures p_e and p_a. By using the real exhaust velocity, the various forces acting on the rocket are separated out. Using this equation, we can begin to examine performance parameters of a rocket, taking into account the ambient conditions.

An expression for the effective exhaust velocity may easily be derived from the above:

$$v_e = u_e + \left(\frac{p_e - p_a}{m}\right)$$

with the thrust written $F_R = mv_e$ (as in Chapter 1).

As formulated above, the thrust equation is incomplete: for a given true exhaust velocity the thrust can be derived, taking into account the ambient conditions; however, the true exhaust velocity u_e is not itself independent of the ambient conditions. Later in this chapter we shall derive an expression for u_e which includes the ambient conditions.

2.2.1 Optimising the exhaust nozzle

Among other parameters, the true exhaust velocity u_e depends on p_e, which in turn is related to the length of the nozzle. The pressure drops along the nozzle, and if the nozzle is lengthened the exit pressure decreases. For maximum exhaust velocity, and hence thrust, the design of the exhaust nozzle should be optimised so that the exit pressure p_e is equal to the ambient pressure p_a.

To understand this in a qualitative way, consider the force accelerating the exhaust gas, and the atmospheric retarding force; both are proportional to the magnitude of the surface integral of pressure over the area of the chamber and nozzle. If the nozzle is made longer, then the extra area will either add to the thrust or to the retarding force, depending on whether or not the internal pressure exceeds the atmospheric pressure. Thus, adding to the nozzle length will increase the thrust, provided $p_e \geq p_a$.

This is one of the most important issues in rocket motor design. Launches mostly begin at, or near, sea level, where the atmospheric pressure is high; however, the rocket rapidly gains altitude and the atmospheric pressure decreases. If the nozzle is the correct length for sea-level pressure, then at altitude the exit pressure p_e will be greater than ambient, and more thrust could have been developed if the nozzle had been made longer. Similarly a nozzle optimised for high altitude will have additional atmospheric retarding force at sea level, and will develop less thrust than a shorter nozzle. Thus, any nozzle is only optimal for a given ambient pressure. For maximum sea-level thrust, the nozzle should be short, with a high p_e; for maximum thrust at altitude, the nozzle should be longer, with a value of p_e equal to the local ambient pressure. For the vacuum of space, the nozzle should be of infinite length, so that $p_e = 0$. In practice, adding length to the nozzle adds mass to the rocket, and after a certain point there is no further benefit from additional length, because the extra thrust has less effect on the acceleration than does the extra mass.

2.3 THE THERMODYNAMICS OF THE ROCKET ENGINE

The thrust equation as developed above shows how the thrust depends on u_e and p_e, for a given atmospheric pressure p_a. Because the exhaust velocity is itself partially dependent on the performance of the nozzle, and hence on p_e, it is not an independent parameter. It must be expressed in terms of p_e and other independent parameters such as the temperature and pressure in the combustion chamber. To do this requires a thermodynamic argument which treats the rocket as a heat engine.

A heat engine converts the chaotic motion of the molecules in a heated fluid into the ordered motion of a piston, or in the case of a rocket, a high-velocity gas stream. In this process, the fluid expands and cools. The thermodynamic treatment will involve the equations relating internal and kinetic energy in the gas, and the equation of continuity for the gas flow through the nozzle. For our purposes it is sufficient to assume adiabatic expansion—strictly, isentropic expansion—with no exchange of energy between the rocket and the ambient. In a real rocket, heat loss to the walls is significant, and this will later be discussed qualitatively.

The first part of the derivation concentrates on the conversion of thermal energy into kinetic energy. This occurs in all heat engines, and there is an analogy between, say, the steam or internal combustion engine and the rocket engine. Figure 2.6 shows the familiar P–V diagram for a heat engine.

Referring to Figure 2.6(a) for the internal combustion engine, we see that from 1 to 2 the fuel–air mixture is being compressed; then after ignition, it expands at constant pressure from 2 to 3 as the piston moves downward. This is followed by adiabatic expansion, 3 to 4, as the gas does further work and cools. The final stroke, 4 to 1, shows the gas being exhausted at constant pressure.

In Figure 2.6(b) we see the rocket engine analogue to the internal combustion engine. There is no inlet stroke, as the rocket operates in a continuous manner. Instead, if we think of a fixed quantity of propellant introduced into the combustion chamber over a small time interval, we can follow a similar sequence. The propellant—fuel and oxidiser—enters the chamber at the combustion pressure (otherwise it would be blown back up the fuel line), so 1–2 is not appropriate, or may be thought of as the compression developed by the external turbo-pump. After the propellant ignites, the gas expands at constant pressure through the combustion chamber 2–3, until it enters the nozzle. At the nozzle it begins to expand adiabatically as it passes through the throat, and continues to expand until it exhausts at 4. There is no separate exhaust stroke, because the rocket is not reciprocating.

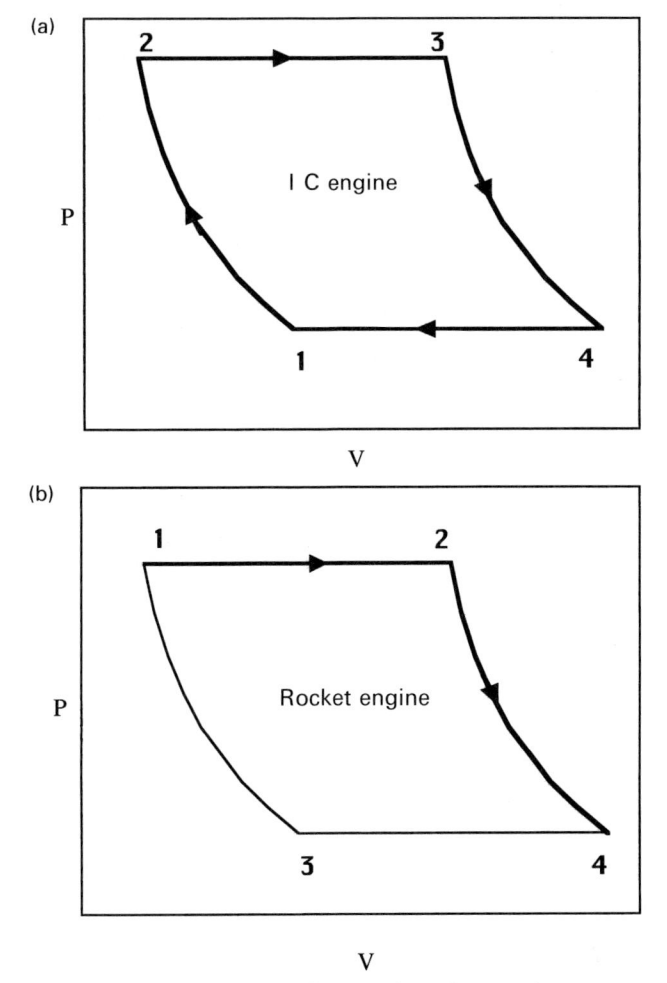

Figure 2.6. P–V diagram for a heat engine.

The thrust equation can now be elaborated in thermodynamic terms by considering the conversion of heat into kinetic energy, embodied in the exhaust velocity u_e, and the mass flow rate m:

$$F_R = mu_e + p_e A_e - p_a A_e$$

Referring to the thrust equation, repeated here for convenience, we shall first derive an expression for the exhaust velocity.

2.3.1 Exhaust velocity

The propellants enter the combustion chamber, mix, and are ignited. The gas produced is heated by the chemical energy of the combustion, and expands through the nozzle. This is a continuous process, but it can be analysed, and sensible results derived, by assuming that the heating and expansion are two successive processes. This can be done by imagining a small fixed mass of gas, and following it through the rocket. The exhaust velocity can be derived by setting the kinetic energy of the exhaust gas equal to the change in enthalpy (or internal energy) of the gas as it cools and expands through the nozzle. This is assumed to be under isentropic conditions—that no heat escapes from the gas to the nozzle walls, and the exhaust is assumed to behave like a perfect gas. Both are reasonable approximations for present purposes.

The change in internal energy for our assumed mass of gas is given by the well-known expression:

$$c_p M (T_c - T_e)$$

where c_p is the specific heat at constant pressure, and T_c and T_e are, respectively, the temperature of the gas in the combustion chamber (the initial temperature) and the temperature of the gas at the exit plane of the nozzle. The notional mass of our small 'packet' of gas is given by M. This change in energy is equal to the gain in kinetic energy of the exhaust gas, represented by $\frac{1}{2} M u_e^2$. Thus the square of the exhaust velocity is

$$u_e^2 = 2 c_p (T_c - T_e)$$

cancelling the mass.

The temperature in the combustion chamber is relatively easy to measure by means of a sensor placed in the chamber. As we shall see later, it is generally just a function of the propellant mixture, and does not really depend on the design of the combustion chamber. The temperature of the exhaust at the exit plane is more difficult to measure; it depends on the degree of expansion and hence on the nozzle design. Moreover, the thrust equation already contains the exhaust pressure, so it is more convenient to express the exhaust conditions in terms of pressure. Fortunately, this can easily be determined by using the well-known equations for adiabatic or isentropic expansion:

$$pV^\gamma = \text{constant}$$

$$T p^{\gamma/(\gamma-1)} = \text{constant}$$

The index γ is the ratio of the specific heat of the exhaust gases at constant pressure to that at constant volume; γ appears in the final equation for thrust, and its magnitude has a significant impact on the result. For air, it has a value of 1.3 at normal temperature and pressure (NTP). For rocket exhaust gases at high temperature, the value is generally smaller and is a function of the combustion conditions; a typical value would be about 1.2. γ is related to the specific heat, the gas constant, and the molecular weight of the exhaust gases by:

$$c_p = \frac{\gamma}{(\gamma-1)} \frac{R}{\mathfrak{M}}$$

where R is the universal gas constant, and $\mathfrak{M}$ is the molecular weight of the exhaust gases.

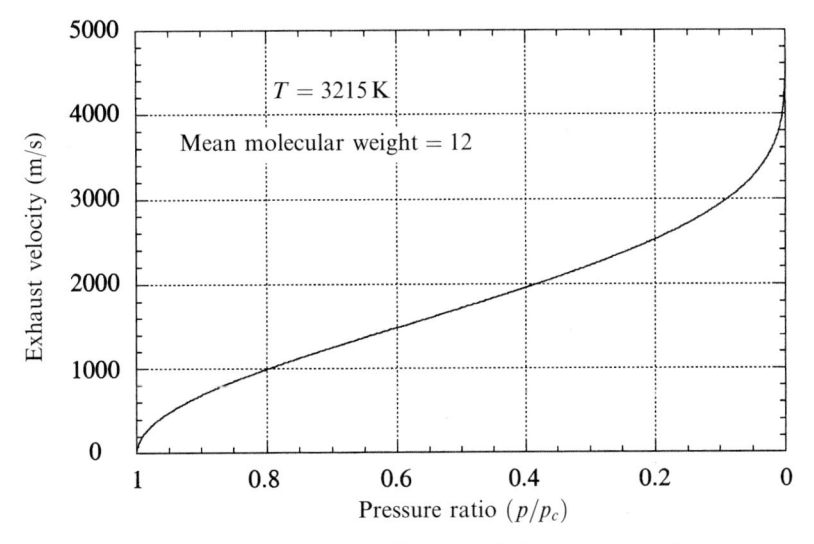

Sec. 2.3]   The thermodynamics of the rocket engine   45

Figure 2.7. Gas velocity as a function of the pressure ratio.

Substituting for T_e and c_p, the velocity can be expressed by

$$u_e^2 = \frac{2\gamma}{(\gamma-1)} \frac{RT_c}{\mathfrak{M}} \left[1 - \left(\frac{p_e}{p_c}\right)^{(\gamma-1)/\gamma}\right]$$

The ratio $(p_e/p_c)^{(\gamma-1)/\gamma}$ is, of course, the expression of the above temperature difference, in terms of the pressure difference between the combustion chamber (the entrance of the exhaust nozzle) and the exit plane.

This equation for the exhaust velocity immediately enables some insight into the physical factors which control its magnitude, and hence the performance of the rocket motor. The velocity is a function of the nozzle design, which determines the pressure ratio, p_c/p_e. Note that for p_e equal to zero—a perfect nozzle *in vacuo*—the exhaust velocity has a maximum value represented by

$$u_e^2 = \frac{2\gamma}{(\gamma-1)} \frac{RT_c}{\mathfrak{M}}$$

This demonstrates what is often stated, that a rocket is most efficient (delivers most thrust) in a vacuum. Thrust is, of course, proportional to exhaust velocity.

Figure 2.7 shows an example of the gas velocity as a function of the pressure ratio, as the gas expands down the nozzle. The velocity becomes hypersonic at the narrowest part of the nozzle (see below).

For the small values of γ which pertain in the exhaust, the velocity is a strong function of γ; as γ approaches unity the expression containing γ tends to infinity. The velocity can be seen to depend on the combustion temperature; this is intuitive. It also depends inversely on the molecular weight of the exhaust gases, which latter is a very important dependence, and plays a major part in the optimisation of

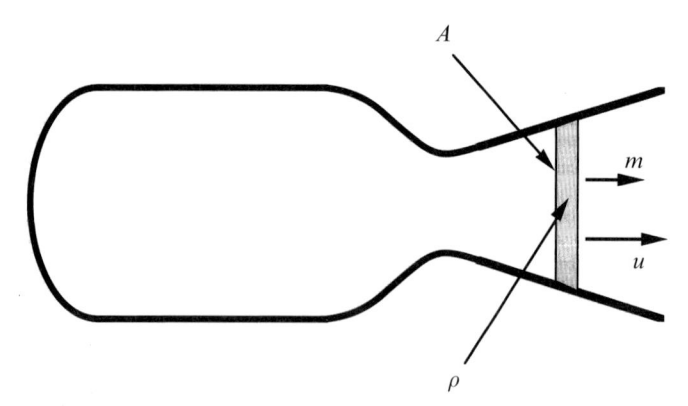

Figure 2.8. Mass flow in the nozzle.

propellant selection. Low molecular weight conveys a significant advantage if the combustion temperature can be kept high. The benefit of low molecular weight is so great that for liquid hydrogen–liquid oxygen engines, extra hydrogen is often added to the mixture simply to reduce the molecular weight of the exhaust; it plays no part in the combustion.

In general terms, the exhaust velocity is the most important performance indicator for a rocket engine. It determines the final velocity of the vehicle (as we have seen in Chapter 1), and it is a major contributor to the thrust development. The magnitude of the exhaust velocity is seen to be dependent on molecular and chemical properties of the propellant, and the expansion ratio of the engine. It has no component related to the actual size or dimensions of the engine. The exhaust velocity can be the same in a 1-mega-Newton thruster used on a heavy launcher, or a tiny micro-Newton thruster used for station keeping. In the next section, we shall deal with the parameter which *does* depend on the dimensions of the engine—the mass flow rate.

2.3.2 Mass flow rate

The remaining term in the thrust equation is the mass flow rate, m. This is determined by the conditions in the combustion chamber and in the nozzle. Once the exhaust velocity is defined, then the pressure difference between the combustion chamber and the exit plane of the nozzle, together with the cross-sectional area of the nozzle, will determine the mass flow rate.

The mass flow rate is constant throughout the nozzle, under steady flow conditions, because all the propellant entering the chamber has to pass through the nozzle and leave through the exit plane. The pressure decreases monotonically. The density of the gas varies dramatically: it is very high at the throat, and decreases to a low value at the exit plane. The velocity, on the other hand, will increase, reaching its maximum at the exit plane.

Sec. 2.3] The thermodynamics of the rocket engine 47

The mass flow rate can be expressed simply as

$$m = \rho u A$$

where m is the (constant) mass flow rate, ρ is the density at any particular point in the nozzle, and u and A are the velocity and the cross-sectional area, respectively, at that point.

The expression for the exhaust velocity has already been derived. The same formula can be used to give the velocity at any point in the nozzle, provided the pressure ratio is defined correctly. The velocity at any point is given by

$$u^2 = \frac{2\gamma}{(\gamma-1)} \frac{RT_c}{\mathfrak{M}} \left[1 - \left(\frac{p}{p_c}\right)^{(\gamma-1)/\gamma} \right]$$

where u and p, unsubscripted, represent the *local* pressure and velocity, rather than the exhaust values. Using this, the mass flow rate can be written as

$$m = \rho A \left\{ \frac{2\gamma}{(\gamma-1)} \frac{RT_c}{\mathfrak{M}} \left[1 - \left(\frac{p}{p_c}\right)^{(\gamma-1)/\gamma} \right] \right\}^{1/2}$$

In the above expression the density ρ is as yet unknown, and to proceed further we need to express it in terms of known parameters. In fact, the density of the gas is linked to the pressure and cross-sectional area of the nozzle by the gas laws for adiabatic expansion. It is this expansion through the nozzle which converts the energy contained in the hot dense gas in the combustion chamber into cooler high-velocity gas in the exhaust.

Using the gas laws

$$pV = nRT = \frac{\rho}{\mathfrak{M}} RT$$

$$pV^\gamma = \text{constant}$$

the density can be expressed in two ways:

$$\rho = p_c \frac{\mathfrak{M}}{RT}$$

$$\frac{\rho}{\rho_c} = \left(\frac{p}{p_c}\right)^{1/\gamma}$$

In this formulation, the density and pressure at the particular place in the nozzle under consideration are expressed in terms of the pressure and density in the combustion chamber, the expansion, defined by the gas laws, and γ. The density is therefore represented by

$$\rho = \frac{p_c \mathfrak{M}}{RT_c} \left(\frac{p}{p_c}\right)^{1/\gamma}$$

48 The thermal rocket engine [Ch. 2]

This can be substituted in the mass flow equation, which, after some cancellation and rearrangement, produces

$$m = p_c A \left\{ \frac{2\gamma}{(\gamma-1)} \frac{\mathfrak{M}}{RT_c} \left(\frac{p}{p_c}\right)^{2/\gamma} \left[1 - \left(\frac{p}{p_c}\right)^{(\gamma-1)/\gamma}\right]\right\}^{1/2}$$

Because of the continuity argument, the mass flow rate is constant; but A, the cross-sectional area, varies continuously, and is a free parameter in this equation. We can however look at the mass flow rate per unit cross-sectional area of the nozzle, which is not a constant:

$$\frac{m}{A} = p_c \left\{ \frac{2\gamma}{(\gamma-1)} \frac{\mathfrak{M}}{RT_c} \left(\frac{p}{p_c}\right)^{2/\gamma} \left[1 - \left(\frac{p}{p_c}\right)^{(\gamma-1)/\gamma}\right]\right\}^{1/2}$$

This is shown in Figure 2.9 as a function of pressure ratio.

Figure 2.9 depicts the way in which a rocket nozzle works. The flow density first increases as the pressure drops. When the pressure has reached about 60% of the value in the combustion chamber the flow density starts to decrease, and continues to decrease until the exhaust leaves the nozzle. The mass flow rate is constant, so this curve implies that for optimal expansion the cross-sectional area of the stream should first decrease and then increase. No assumptions about the profile of the nozzle are included; the requirement on the variation of cross section with pressure ratio has emerged simply from the thermodynamics. The convergent–divergent shape of the optimum rocket engine nozzle is therefore the result of a simple physical process.

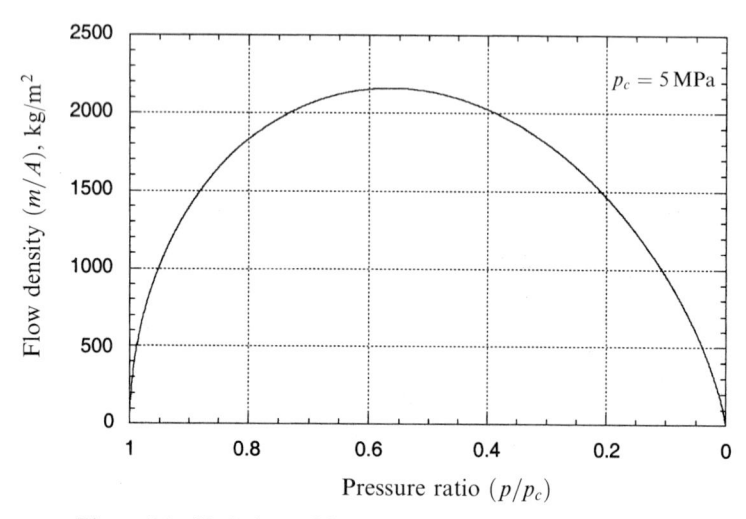

Figure 2.9. Variation of flow density through the nozzle.

From the same formula, the ideal cross-sectional area of the nozzle for any pressure is given by

$$A = \frac{m}{p_c} \left\{ \frac{2\gamma}{(\gamma-1)} \frac{\mathfrak{M}}{RT_c} \left(\frac{p}{p_c}\right)^{2/\gamma} \left[1 - \left(\frac{p}{p_c}\right)^{(\gamma-1)/\gamma}\right] \right\}^{-1/2}$$

This is just an inversion of the function shown in Figure 2.9. While this shows how the cross-sectional area and pressure ratio are related, it cannot show the shape of the nozzle, because this is dependent on the axial dimension, which is not represented in these formulae. The nozzle designer is therefore presented with a degree of freedom. The convergent–divergent nozzle can simply be composed of two truncated cones joined at the throat, which is the narrowest part. This would produce an appropriate expansion, the pressure and flow density adjusting themselves to the cross-sectional area through the above formulae. There would in fact be inefficiencies with this approach because the flow lines of the hypersonic gas stream would interact unfortunately with the sharp edge at the join, which would generate shocks. It is possible to calculate the proper shape of the throat region, and generate the appropriate smooth curve, but the 'method of characteristics' is beyond the scope of this book. Downstream of the throat, once the smooth contour is established, a simple cone shape is often used. The flow lines are divergent with such a shape, and some thrust is lost, although the nozzle is shorter than the correct shape, and the reduction in mass of the shorter nozzle may offset the loss in thrust. A shorter nozzle is also beneficial in a multistage rocket, in which the length, and hence dead mass, of the vehicle is reduced, leading to an improved mass ratio. The proper shape of a rocket nozzle—the so-called *de Laval* nozzle, has a smoothly curved converging part (the exact shape is not important)—joined to a bell-shaped diverging cone. With this shape the flow lines are constrained to be axial over the whole cross-sectional area, which ensures that all of the thrust is developed along the rocket axis and none is lost to divergence. The bell is longer than the equivalent cone for the same expansion ratio.

From the foregoing, the application of the expansion formula should be clear: it relates the pressure and the cross-sectional area. The distance down the axis of the nozzle, where the area and pressure have a particular value, is not defined.

Referring to Figure 2.9 it can be seen that the pressure drops to about half its initial value between the combustion chamber and the throat of the nozzle. The precise value can be derived by differentiating one of the above expressions. This leads to:

$$\frac{p}{p_c} = \left(\frac{2}{\gamma+1}\right)^{\gamma/(\gamma-1)}$$

The pressure ratio takes the value of 0.57 for $\gamma = 1.2$. Substituting in the expression for gas-stream velocity produces, after some manipulation,

$$u = \sqrt{\gamma \frac{RT}{\mathfrak{M}}}$$

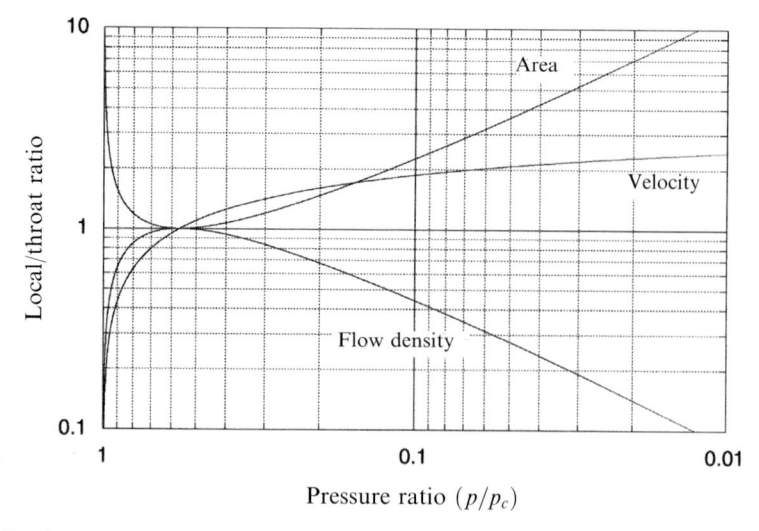

Figure 2.10. Area, velocity and flow density relative to the throat values as a function of the pressure ratio.

where T is the local temperature. Reference to gas physics shows this to be the expression for the local sound speed. This is quite general and occurs wherever hot confined gas is expanded through a throat; and it is even found in astrophysics, where jets of electrons emerge from quasars. Thus for nearly all conditions the accelerating gas reaches sound velocity at the throat, and the exhaust stream is hypersonic in the rest of the nozzle.

The cross-sectional area of the throat of the nozzle denoted by A^* is an important parameter of the rocket motor, being in effect a measure of the size. As we have already seen, the exit area A_e is chosen by the designer, depending on the expected ambient conditions, and can be different for motors of the same size. The conditions at the throat essentially form a basic set of values for a given rocket engine. Figure 2.10 shows the cross-sectional area, velocity and flow density normalised to the throat values, as a function of the pressure ratio.

By differentiating the expression for the mass flow rate per unit cross-sectional area, the peak value, which occurs at the throat, can be determined. This is:

$$\frac{m}{A^*} = p_c \left\{ \gamma \left(\frac{2}{\gamma + 1} \right)^{(\gamma+1)/(\gamma-1)} \frac{\mathfrak{M}}{RT_c} \right\}^{1/2}$$

Since the mass flow rate is everywhere constant, the following simpler expression can be used:

$$m = p_c A^* \left\{ \gamma \left(\frac{2}{\gamma + 1} \right)^{(\gamma+1)/(\gamma-1)} \frac{\mathfrak{M}}{RT_c} \right\}^{1/2}$$

Thus the mass flow rate can be seen to be determined mainly by the throat area, and the pressure and temperature in the combustion chamber. This is what we would expect intuitively: the product of throat area and pressure is the main factor, with the molecular weight and combustion temperature as modifiers. The mass flow rate depends positively on the molecular weight, which is in accord with intuition; but it decreases with temperature. This is because at a higher temperature a given mass of gas exerts a higher pressure; thus, the mass of gas flowing is less for a given pressure if the temperature increases. For multistage rockets in which the lower stages have the main task of developing high thrust rather than high velocity, it may be preferable to use a propellant with high molecular weight. This will allow a higher thrust to be developed through a smaller throat area in a physically smaller engine. Solid propellant boosters are in this class, and the use of aluminium—generating aluminium oxide in the exhaust—is beneficial.

This formula also gives insight into the problem of throttling the motor: the throat itself cannot be varied, so the pressure in the chamber must be changed by varying the supply rate of propellants. While the conditions for complete combustion pertain, a rocket engine is always propellant starved. Increasing the supply of propellant will increase the mass of hot gas produced, and it will result in a rise in chamber pressure and a corresponding increase in thrust. Complete combustion depends on there being sufficient time and volume in the combustion chamber for the evaporation and mixing of propellant droplets produced by the injector. If the injection rate becomes arbitrarily high, then this may not happen, and the combustion could become unstable. Similarly, for very low rates the thermal input to droplet evaporation from the combustion could be insufficient, leading again to unstable combustion. These difficulties explain why throttling is rarely used. The key example of the throttled motor is the Space Shuttle main engine, which has a thrust normally variable from 65% to 104%.

2.4 THE THERMODYNAMIC THRUST EQUATION

We are now in a position to substitute expressions for the velocity u_e and the mass flow rate m, derived from thermodynamics, into the thrust equation:

$$F_R = mu_e + p_e A_e - p_a A_e$$

After some manipulation and cancellation, this leads to

$$F_R = p_c A^* \left\{ \frac{2\gamma^2}{\gamma - 1} \left(\frac{2}{\gamma + 1} \right)^{(\gamma+1)/(\gamma-1)} \left[1 - \left(\frac{p_e}{p_c} \right)^{(\gamma-1)/\gamma} \right] \right\}^{1/2} + p_e A_e - p_a A_e$$

This is the full thermodynamic thrust equation, made up of three terms: the Newtonian thrust related to the mass ejection, the accelerating force of the static pressure in the exhaust stream as it leaves the nozzle, and the retarding force due to ambient atmospheric pressure.

It is interesting that this equation no longer contains the terms relating to combustion, the molecular weight and the temperature. These terms—which appear in the individual expressions for mass flow rate and exhaust velocity—cancel, and are now subsumed into the combustion chamber and exhaust pressures.

The mass flow rate depends on density, which is proportional to the square root of molecular weight over temperature; the velocity, on the other hand, depends on the energy contained in the hot gas, which is determined by the square root of the temperature over the molecular weight.

This does not mean that temperature and molecular weight are not important in the performance of rocket motors. The thrust is made up of the mass flow rate, which is predominantly determined by the throat area and the pressure in the combustion chamber, and the exhaust velocity, which depends on the temperature and molecular weight. The former factor is mainly determined by the size and shape of the rocket motor, and the latter is determined by the propellant combination and the combustion conditions. The product of throat area and chamber pressure, which appears at the beginning of the formula, is the fixed parameter which determines the size and other mechanical design properties of the rocket engine. The throat area determines the dimensions, while the pressure determines a whole host of requirements from the strength of the chamber walls to the power of the turbo-pumps.

The main parameter of rocket motor size is the throat area A^*, as we have seen. The defining property of the nozzle is the exit area A_e, and the shape of the nozzle can be expressed in a dimensionless way as the expansion ratio, A_e/A^*. This depends on the expected ambient pressure. For first-stage motors, intended for use low in the atmosphere, it takes a value of about 10; for high altitude motors, and for use in space, the area ratio can be as high as 80. We should recall that for maximum efficiency p_e should be equal to the ambient pressure, and that this value is controlled by the expansion ratio. An expression for the expansion ratio can be derived from the ratio of the above expressions for mass flow rate per unit area at the throat, and at the exit plane. The mass flow rates cancel, leaving, after some rearrangement, the following:

$$\frac{A_e}{A^*} = \left\{ \frac{\left(\frac{\gamma-1}{2}\right)\left(\frac{2}{\gamma+1}\right)^{(\gamma+1)/(\gamma-1)}}{\left(\frac{p_e}{p_c}\right)^{2/\gamma}\left[1-\left(\frac{p_e}{p_c}\right)^{(\gamma-1)/\gamma}\right]} \right\}^{1/2}$$

Figure 2.11 shows this ratio plotted as a function of the pressure ratio for $\gamma = 1.1$, 1.2 and 1.3. The value 1.2 is about average for many rocket motors. This figure can be used for reading off the pressure ratio for a given rocket motor nozzle, given the expansion ratio, or vice versa when trying to establish the correct expansion ratio for a given ambient pressure.

2.4.1 The thrust coefficient and the characteristic velocity

There are two other parameters of the rocket motor which can be defined, and which are helpful in calculating the performance. These are the *thrust coefficient*, denoted by C_F, and the *characteristic velocity*, denoted by c^*. The thrust coefficient represents the performance of the nozzle, and the characteristic velocity that of the propellants and combustion. The thrust coefficient is the ratio of the thrust to the notional force,

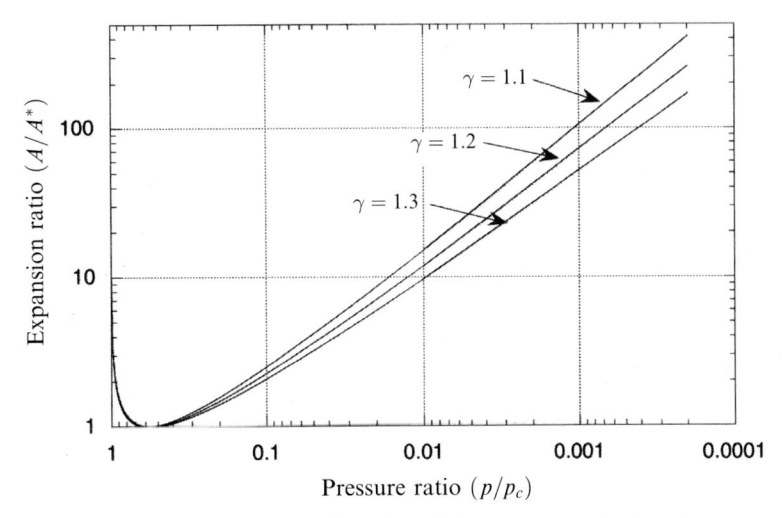

Figure 2.11. Expansion ratio as a function of the pressure ratio for changing γ.

defined by the product of combustion chamber pressure and the throat area. It is defined by the relationship

$$C_F = \frac{F_R}{p_c A^*}$$

Using the formulae derived above it can be written as

$$F_R = p_c A^* \left\{ \frac{2\gamma^2}{\gamma-1} \left(\frac{2}{\gamma+1}\right)^{(\gamma+1)/(\gamma-1)} \left[1 - \left(\frac{p_e}{p_c}\right)^{(\gamma-1)/\gamma}\right]\right\}^{1/2} + p_e A_e - p_a A_e$$

and having divided through by the notional force $p_c A^*$:

$$C_F = \left\{ \frac{2\gamma^2}{\gamma-1} \left(\frac{2}{\gamma+1}\right)^{(\gamma+1)/(\gamma-1)} \left[1 - \left(\frac{p_e}{p_c}\right)^{(\gamma-1)/\gamma}\right]\right\}^{1/2} + \left(\frac{p_e}{p_c} - \frac{p_a}{p_c}\right)\frac{A_e}{A^*}$$

The first expression is the ratio of quantities, which can be measured during the firing of the rocket motor. The combustion chamber pressure is measured by a sensor in the chamber, the thrust is measured directly during a static test bed firing, and the throat area is a dimension of the nozzle. The second expression is a theoretical calculation, based on thermodynamics. This immediately shows how the coefficient can be used to estimate the departure of an actual motor from its theoretical efficiency. In the same way, it can be used for a first estimate of the nozzle performance based on theory. Thus having selected a throat area and combustion chamber pressure, the effects of expansion ratio for particular ambient pressure conditions can be calculated.

The thrust coefficient depends on the ambient pressure, and is always larger for vacuum conditions than for a finite ambient pressure; for vacuum $p_a = 0$, the coefficient is given by

$$C_F = \left\{ \frac{2\gamma^2}{\gamma-1} \left(\frac{2}{\gamma+1}\right)^{(\gamma+1)/(\gamma-1)} \left[1 - \left(\frac{p_e}{p_c}\right)^{(\gamma-1)/\gamma}\right] \right\}^{1/2} + \left(\frac{p_e}{p_c}\right) \frac{A_e}{A^*}$$

Comparing the two expressions, we see that

$$C_F = C_{Fv} - \frac{A_e p_a}{A^* p_c}$$

Since the values of the two ratios are often given for rocket engines, this formula is useful.

To appreciate the thrust coefficient, we can set p_e equal to zero, in which case the expansion is assumed to be perfect, and the rocket is operating *in vacuo*. This gives

$$C_F = \left\{ \frac{2\gamma^2}{\gamma-1} \left(\frac{2}{\gamma+1}\right)^{(\gamma+1)/(\gamma-1)} \right\}^{1/2}$$

For $\gamma = 1.2$, it takes the value of 2.25.

This can now be compared with a notional motor, which does not have a nozzle. Here the exit pressure is the same as the throat pressure, and the exit area is the same as the throat area. From another differentiation of the mass flow equation the throat pressure can be determined to be

$$\frac{p^*}{p_c} = \left(\frac{2}{\gamma+1}\right)^{\gamma/(\gamma-1)}$$

and A_e/A^* is of course equal to unity. Substitution gives for the thrust coefficient, in the absence of a nozzle:

$$C_F = \left\{ \frac{2\gamma^2}{\gamma+1} \left(\frac{2}{\gamma+1}\right)^{(\gamma+1)/(\gamma-1)} \right\}^{1/2} + \left(\frac{2}{\gamma+1}\right)^{\gamma/(\gamma-1)}$$

The exit pressure is no longer zero, hence the presence of the second term. The coefficient, in the absence of a nozzle, takes a value of 1.24 for $\gamma = 1.2$. Thus, a rocket motor without a nozzle still develops thrust, as experience shows—for example, the behaviour of a toy balloon. However, the thrust developed will be nearly a factor of 2 lower, because much of the energy stored in the gas is wasted.

Using the thrust coefficient we can also re-examine the notion that the maximum efficiency is achieved for a nozzle when the expansion ratio is such that the exhaust pressure equals the ambient pressure. Figure 2.12 shows the thrust coefficient plotted against the expansion, or area ratio, for different values of the ratio of the combustion chamber pressure to the atmospheric pressure. It can be seen that for each atmospheric pressure value, the thrust coefficient peaks at a particular expansion ratio. Inspection of Figure 2.11 shows that the expansion ratio at the peak is that which produces the exhaust pressure equal to the atmospheric pressure. Thus, we have proved by thermodynamics the assertion at the beginning of this

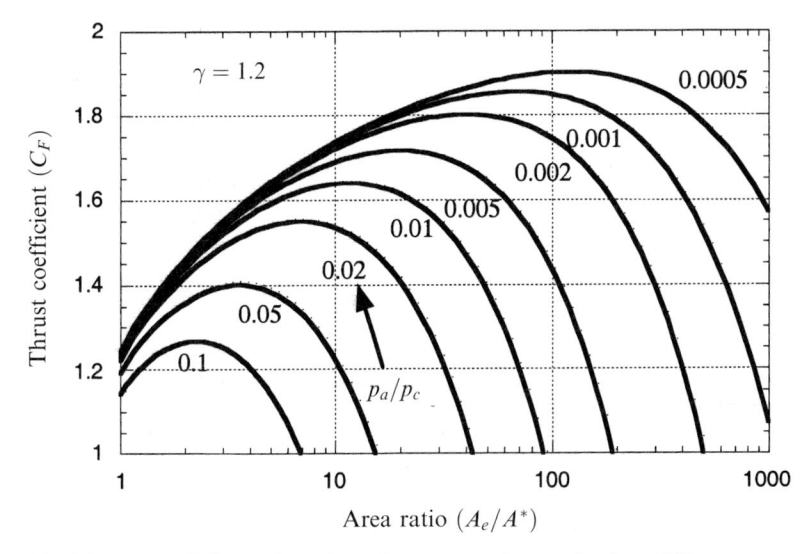

Figure 2.12. Thrust coefficient plotted against expansion ratio for different atmospheric pressures.

chapter—that the maximum thrust is developed when the exhaust pressure is equal to the ambient pressure.

The thrust coefficient is thus a measure of the efficiency with which the nozzle extracts energy from the hot gas in the combustion chamber.

The remaining parameter—the characteristic velocity—measures the efficiency of conversion of thermal energy in the combustion chamber into high-velocity exhaust gas. This is defined by the ratio of the notional force, given by the pressure in the combustion chamber, integrated over the throat area, now divided by the mass flow rate:

$$c^* = \frac{p_c A^*}{m}$$

It has the dimensions of a velocity, and is again based on measurable quantities. The thermodynamic form is given by

$$c^* = \left\{ \gamma \left(\frac{2}{\gamma+1} \right)^{(\gamma+1)/(\gamma-1)} \frac{\mathfrak{M}}{RT_c} \right\}^{-1/2}$$

The characteristic velocity depends on the temperature and, inversely, on the molecular weight. Again, the comparison between the expected and actual values can be used to assess the performance of the motor; or c^* can be used to estimate the expected performance of a new motor design.

Figure 2.13 shows the characteristic velocity as a function of the combustion temperature and the molecular weight of the exhaust gas. For plotting purposes, these parameters are combined in the combustion parameter $\sqrt{T_c/\mathfrak{M}}$. A typical

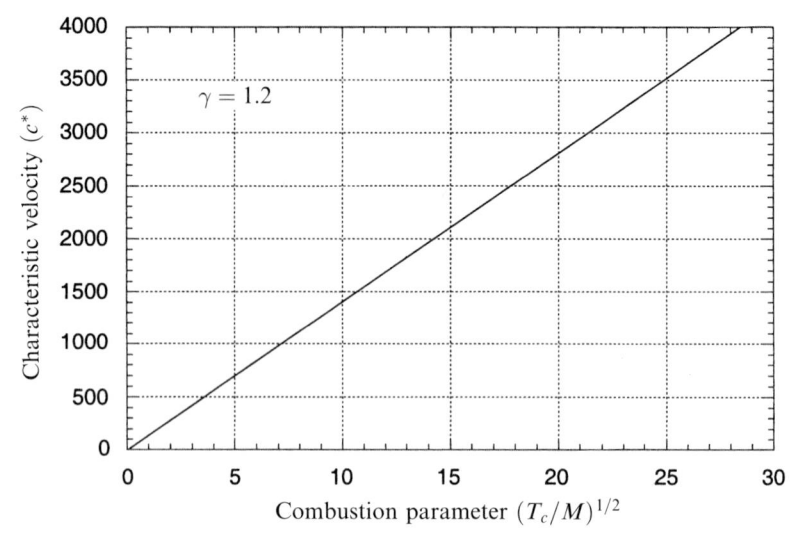

Figure 2.13. Characteristic velocity as a function of the combustion temperature and molecular weight.

value of characteristic velocity is around 2,000 m s^{-1}. Liquid oxygen–liquid hydrogen engines have a combustion parameter of approximately 16, while for solid propellants the value is about 10. The graph also shows what we have already deduced: high exhaust velocities are associated with high temperature and low molecular weight. In designing a rocket engine, choice of propellant combination will define the characteristic velocity. The effect of the nozzle and expansion can then be added by using the thrust coefficient. Choice of a high temperature combustion may be limited by the structural properties of the combustion chamber material.

2.5 COMPUTING ROCKET ENGINE PERFORMANCE

Having defined the thrust coefficient and the characteristic velocity, we can use them to compute the behaviour of specific motor designs. Before doing so, however, it is useful to summarise the relationships between them, and the performance of the rocket motor:

$$C_F = \frac{F_R}{p_c A^*}$$

$$c^* = \frac{p_c A^*}{m}$$

$$F_R = mc^* C_F$$

The last relationship derives from the previous two, and shows the real importance of C_F and c^*: together with the mass flow rate they define the thrust. Comparison with the thrust equation given in Chapter 1 shows that the product of C_F and c^* gives the *effective* exhaust velocity v_e—the parameter used earlier to define rocket performance in the rocket equation.

We have therefore returned to the essential property of a rocket, the effective exhaust velocity, which is now defined in terms of thermodynamics, and in terms of parameters measured in the engine itself. Combustion chamber pressure, mass flow rate, the throat area and the measured thrust combine to give the effective exhaust velocity. This can be inserted in the rocket equation to reveal the capability of a rocket to launch a payload or make an orbital manoeuvre.

2.5.1 Specific impulse

The exhaust velocity of a rocket engine is quoted and calculated in this book in S.I. units of metres/second. In rocket engineering the exhaust velocity is almost universally quoted in terms of the *specific impulse*. The apparent units of specific impulse are seconds, and the equation relating specific impulse to exhaust velocity is

$$v_e = g I_{sp}$$

where g is the acceleration of gravity. In metric units, g is close to $10\,\mathrm{m\,s^{-1}\,s^{-1}}$, so it is sufficient, when faced with a velocity quoted as specific impulse, to multiply by 10 to express it in $\mathrm{m\,s^{-1}}$.

There are three reasons for this usage. The first is historical, and as astronomers know, it is difficult to change historical usage. The second reason is that, the apparent units of specific impulse being seconds, it takes the same value in the metric and imperial systems. The actual exhaust velocity emerges in the same units in which the acceleration of gravity is quoted. Remembering that until recently NASA has used imperial units, this is a useful convention. There is also a plausible physical argument, as follows.

'Impulse' is the term used to describe the effect of a force applied for a very short time to an object, as in, for example, the contact between a bat and a ball. The product $F\,dt$ is equal to the momentum given to the object. The 'specific impulse' is defined as the impulse given to the rocket by unit *weight* of propellant:

$$I = F\,dt = m v_e\,dt$$

$$I_{sp} = \frac{F\,dt}{mg\,dt} = \frac{m v_e\,dt}{mg\,dt} = \frac{v_e}{g}$$

where m, in our convention, is the mass flow rate in $\mathrm{kg\,s^{-1}}$. The real units of specific impulse are Newton-seconds/Newton, or kilogramme-metre/second per kilogramme-metre/second/second; cancellation of dimensions produces the familiar unit of seconds.

The specific impulse can be most usefully considered as a measure of the fuel efficiency of the rocket; that is, the momentum imparted to the rocket per kilogramme of propellant expelled. If the exhaust velocity is high then the propellant efficiency is high, which is clear from the equation for thrust. Specific impulse is proportional to exhaust velocity, and is a direct measure of the propellant efficiency.

It is convenient to express some of the other formulae in terms of specific impulse, remembering that gI_{sp} is equal to the *effective* exhaust velocity v_e:

$$F_R = mc^*C_F = mv_e = mgI_{sp}$$

$$C_F = \frac{mgI_{sp}}{p_c A^*}$$

$$C_F c^* = gI_{sp}$$

It would perhaps be an advantage if the confusion of units could be removed from the business of rocket engineering, and with the universal adoption of S.I. this might be achieved. Use of mixed metric and imperial units can occasionally cause problems, as recent events have shown. In this book the units of exhaust velocity will be metres per second, except where unavoidable. Performance tables will use the convention of specific impulse.

2.5.2 Example calculations

As an example calculation we can consider a real engine design: the Viking series used on Ariane 4. This is a storable propellant motor using nitrogen tetroxide and UDMH25 (unsymmetrical dimethyl hydrazine with 25% hydrazine hydrate) as propellants. This mixture is self-igniting, and both propellants are liquid at NTP. The information provided by the manufacturers is as follows

Property	Viking 5C	Viking 4B
Vacuum thrust	752 kN	805 kN
Sea-level thrust	678 kN	n/a
Specific impulse	278.4 s	295.5 s
Pressure p_c	58 bar	58.5 bar
Area ratio	10.5	30.8
Mass flow	275.2 kg s^{-1}	278.0 kg s^{-1}
Exit diameter	0.990 m	1.700 m

The thrust coefficient can be determined from the tabulated data. The throat area can be calculated from the exit diameter, and the expansion ratio

$$A^* = \pi \left(\frac{(0.99/2)^2}{10.5} \right)$$

The thrust coefficient, C_F, is then 1.751. Referring to Figure 2.11, for the lowest atmospheric pressure (approximating a vacuum) and the expansion ratio 10.5, we see that this value of thrust coefficient is far from optimum. This is because the motor is designed for sea-level use, while we have computed the coefficient using the vacuum thrust. This is an example of an under-expanded motor. Using the sea-level thrust, the coefficient becomes 1.579, which is at the peak of the curve for sea-level pressure. Thus, this motor is optimised for sea-level use on the first stage of the Ariane 4.

A similar calculation for the Viking 4B high-altitude motor—which has the same throat area but a much larger exit area—yields a value of 1.86 for the vacuum thrust coefficient. Referring to Figure 2.11 for the lowest atmospheric pressure, the expansion ratio of 30.8 shows that the thrust coefficient is not quite optimum for vacuum. The exhaust is still under-expanded, and the value of the thrust coefficient corresponds to a position to the left of the appropriate peak. This is deliberate, and is the result of a compromise: to make the expansion perfect would entail a much longer nozzle and a consequent increase in weight. The adverse effect on the mass ratio of the rocket would eliminate the advantage gained in thrust.

These simple calculations show how the performance of the nozzles can be estimated and compared with the intended use. The characteristic velocity is the same for both rocket engines: $1,560 \text{ m s}^{-1}$. This is as would be expected, since this parameter is independent of the shape of the nozzle, and depends only on the conditions in the combustion chamber. These are almost the same in both cases. The effective exhaust velocity of the engines is the product of the thrust coefficient and the characteristic velocity. It is equal to $2,732 \text{ m s}^{-1}$ for the Viking 5C, and $2,899 \text{ m s}^{-1}$ for the high-altitude Viking 4B. This latter value shows the effect of the longer nozzle in increasing the exhaust velocity. These calculated values are slightly less than the quoted values derived from the specific impulse.

Another way to use the thrust coefficient and the characteristic velocity is to determine the theoretical values from the thermodynamics of the motor, and to compare these with the values derived from measured motor parameters. We are on less secure ground here in respect of actual data, because not all the parameters are quoted in the data sheet. In particular, the combustion temperature is not given, and the value of γ and of the mean molecular weight in the exhaust stream will have to be estimated.

These parameters are dependent on the conditions in the combustion chamber and exhaust, and a proper estimation is complex. The exhaust products are not simply the compounds predicted from chemistry at normal temperatures; at high temperatures partial dissociation takes place, there are many different molecules present, and γ is also affected. The estimation of the temperature is also complex, and depends to some extent on the combustion chamber pressure, as well as on the propellants. For present purposes a value of 3,350 K would be a reasonable value to assume. A suitable mean value of the molecular weight for this mixture is $\mathfrak{M} = 23$, while $\gamma = 1.2$, which we have assumed throughout, should be fairly close to the real value. These topics are dealt with in more detail in a subsequent chapter.

Substituting the above values in the thermodynamic formula produces $c^* = 1,697$. The accuracy of the estimated temperature, and so on will have an effect, but most of the discrepancy will be due to inefficiencies in the real motor, which are not dealt with in our simple theory. The most important of these is the loss of thermal energy from the system due to conduction and radiation—effects which have been neglected in our adiabatic assumption. Some of this lost heat can be returned to the combustion process by circulating part of the propellant around the nozzle and combustion chamber and so pre-heating the propellant before it enters the chamber. The Viking engine does not do this, and the losses are larger than would otherwise be the case.

Unlike the characteristic velocity, the thrust coefficient depends on the ambient conditions; that is, whether the motor is operating at sea level, or in a vacuum. The pressure ratio, however, is independent of the ambient, and depends only on the nozzle shape. For the Viking 5C operated at sea level, the thermodynamic argument predicts $C_F = 1.57$, compared with the actual value 1.58. This is very close, showing a nozzle well optimised for sea-level operation. For the vacuum case the theoretical value for the highly expanded Viking 4B (1.749) is again close to the measured value of 1.751.

2.6 SUMMARY

In this largely theoretical chapter we have considered the thermal rocket as a heat engine, and derived the thrust from thermodynamic considerations. In doing so, some insight into the physics of the rocket motor has been gained, and the parameters important to the performance of the rocket have been identified. The use of the thrust coefficient and the characteristic velocity enabled us to compare the performance of a real motor with theoretical predictions. It is a tribute to the efficiency of modern rocket design that these predictions are so close to the actual performance.

In Chapter 1 the rocket vehicle was examined in terms of its effective exhaust velocity and mass ratio. Here the effective exhaust velocity has been explained in terms of simple design parameters of the rocket engine, and the chemical energy available in the propellants. This is the basic science of the rocket engine. In the next two chapters we shall examine the practicalities of liquid-fuelled and solid-fuelled engines.

3

Liquid propellant rocket engines

In the previous chapter the thermal rocket engine was considered in isolation from the way the hot gas, which is the working fluid, is produced. The basic operation of both liquid- and solid-fuelled engines is the same, but behind the broad principles, technical issues have a significant impact on efficiency and performance. In this chapter we shall examine the technical issues pertaining to liquid-fuelled rocket engines, and see how they affect the performance; several practical examples will be used.

3.1 THE BASIC CONFIGURATION OF THE LIQUID PROPELLANT ENGINE

A liquid propellant rocket engine system comprises the combustion chamber, nozzle, and propellant tanks, together with the means to deliver the propellants to the combustion chamber.

In the simplest system, the propellant is fed to the combustion chamber by static pressure in the tanks. High-pressure gas is introduced to the tank, or is generated by evaporation of the propellant, and this forces the fuel and oxidiser into the combustion chamber. As we have seen in Chapter 2, the thrust of the engine depends on the combustion chamber pressure and, of course, on the mass flow rate. It is difficult to deliver a high flow rate at high pressure using static tank pressure alone, so this system is limited to low-thrust engines for vehicle upper stages. There is a further penalty, because the tanks need to have strong walls to resist the high static pressure, and this reduces the mass ratio. The majority of large liquid propellant engine systems use some kind of turbo-pump to deliver propellants to the combustion chamber. The most common makes use of hot gas, generated by burning some of the propellant, to drive the turbine.

Since high combustion temperature is needed for high thrust, cooling is an important consideration in order to avoid thermal degradation of the combustion chamber and nozzle. The design of combustion chambers and nozzles has to take

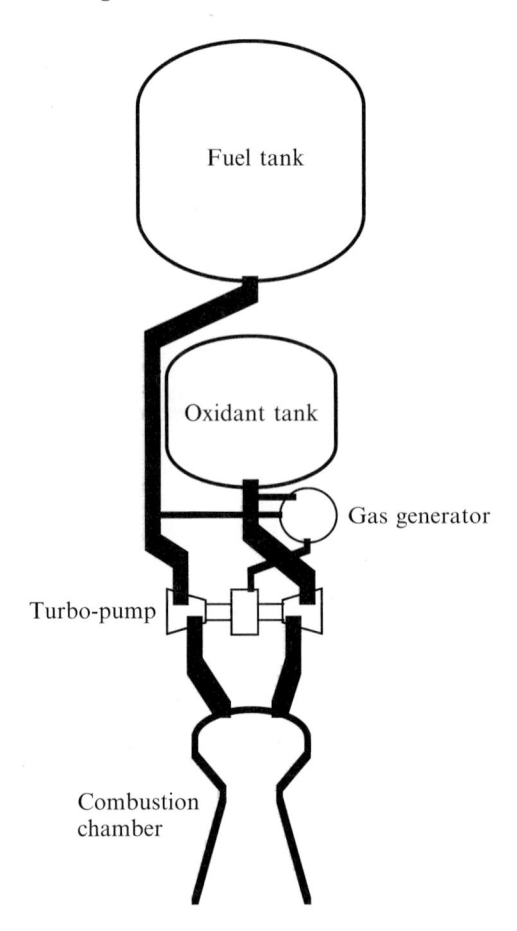

Figure 3.1. Schematic of a liquid-propellant engine.

this into account. In addition, safe ignition and smooth burning of the propellants is vital to the correct performance of the rocket engine.

3.2 THE COMBUSTION CHAMBER AND NOZZLE

The combustion chamber and the nozzle form the main part of the engine, wherein the thrust is developed. The combustion chamber comprises the injector through which the propellants enter, the vaporisation, mixing, and combustion zones, and the restriction leading to the nozzle. The throat is properly part of the nozzle. The combustion chamber has to be designed so that the propellants vaporise and mix efficiently, and so that the combustion is smooth. It must also withstand the high temperature and pressure of combustion, and in some cases cooling of the chamber walls is arranged. The combustion chamber joins smoothly on its inner surface to the nozzle, and the restriction in the combustion chamber and the nozzle together form the contraction–expansion or de Laval nozzle (discussed in the previous chapter). The shape is defined by the thermodynamic and fluid flow laws (also mentioned in the previous chapter) together with the design requirements.

3.2.1 Injection

The injector has to fulfil three functions: it should ensure that the fuel and oxidiser enter the chamber in a fine spray, so that evaporation is fast; it should enable rapid mixing of the fuel and oxidiser, in the liquid or gaseous phase; and it should deliver the propellants to the chamber at high pressure, with a high flow rate. Figure 3.2 shows schematically the vaporisation, mixing and combustion zones in the combustion chamber.

The specific injector design has to take into account the nature of the propellants. For cryogenic propellants such as liquid oxygen and liquid hydrogen, evaporation into the gaseous phase is necessary before ignition and combustion. In this case a fine spray of each component is needed. The spray breaks up into small droplets which evaporate, and mixing then occurs between parallel streams of oxygen and hydrogen. For hypergolic or self-igniting propellants such as nitrogen tetroxide and UDMH, the two components, which react as liquids at room temperature, should come into contact early, and impinging sprays or jets of the two liquids are arranged. In some cases pre-mixing of the propellants in the liquid form is needed, and here the swirl injector is used, in which the propellants are introduced together into a mixing tube. They enter the chamber pre-mixed, and are exposed to the heat of combustion. In all cases, the heat of the gases undergoing combustion is used to evaporate the

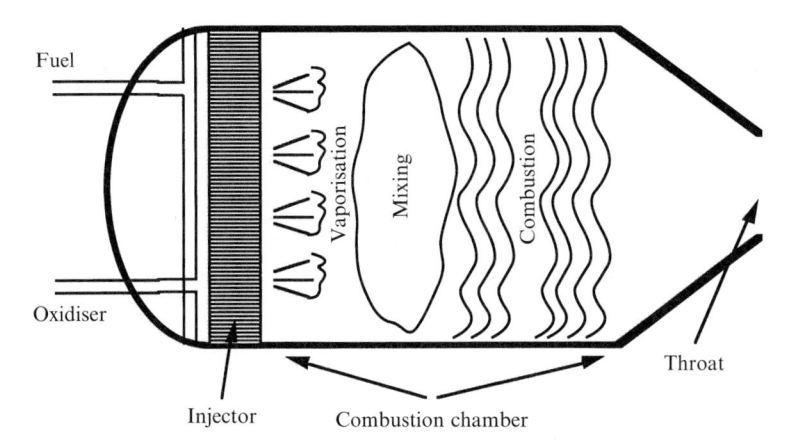

Figure 3.2. Injection and combustion.

propellant droplets. The heat is transferred to the droplets by radiation, and conduction through the gas. The propellant passing through the combustion chamber has a low velocity, and does not speed up until it reaches the nozzle.

The requirement for a fine spray, together with a high flow rate, is contradictory, and can be realised only by making up the injector of many hundreds of separate fine orifices. Good mixing requires that adjacent jets consist of fuel and oxidiser. Thus, the hundreds of orifices have to be fed by complex plumbing, with the piping for two components interwoven. The design of the injector is a major issue of combustion chamber design.

Types of injector

The simplest type of injector is rather like a shower head, except that adjacent holes inject fuel and oxidant so that the propellants can mix. Improved mixing can be achieved with the use of a coaxial injector. Here each orifice has the fuel injected through an annular aperture which surrounds the circular oxidant aperture, and this is repeated many times to cover the area of the injector. These types of injector are shown in Figure 3.3.

The above injectors are used for propellants which react in the vapour phase. The fine sprays quickly form tiny droplets, which also evaporate quickly. The *impinging jet* injector is shown in two forms in Figure 3.4. The first is designed to make sure that propellants mix as early as possible, while still in the liquid phase, and is useful for hypergolic propellant combinations. In the second form, jets of the *same* propellant impinge on one another. This is useful where fine holes are not suitable. The cross section of the jets can be larger, while the impinging streams cause the jets to break up into droplets.

The injector can be located across the back of the combustion chamber, as indicated in Figure 3.2, or it can be located around the cylindrical wall of the rear end of the combustion chamber. The choice depends on convenience of plumbing, and the location of the igniter, where used. For example, the HM7-B cryogenic engine, used to power the third stage of Ariane 4, uses a frontal injector unit with 90 coaxial injector sets which feed the liquid oxygen and liquid hydrogen into the combustion chamber at a pressure of 35 bar. In contrast, the Viking engine used to power the first stage of Ariane 4 uses 216 parallel injector pairs set in six rows around the wall of the combustion chamber, and these feed the hypergolic propellants UMDH and nitrogen tetroxide into the chamber. The number of injectors controls the flow rate and for high thrust engines many more are used. For the Vulcan engine, used as the single motive power for the Ariane 5 main stage, 516 coaxial injectors are used, delivering liquid hydrogen and liquid oxygen at 100 bar. This engine generates more than 1 mega-Newton of thrust.

3.2.2 Ignition

Secure and positive ignition of the engine is essential in respect of both safety and controllability. The majority of engines are used only once during a mission, but the

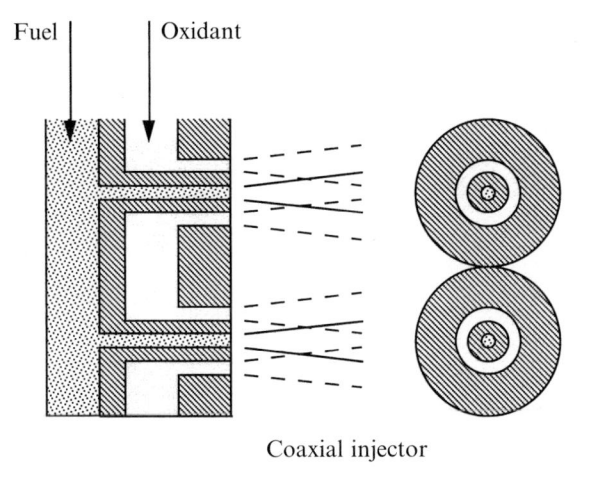

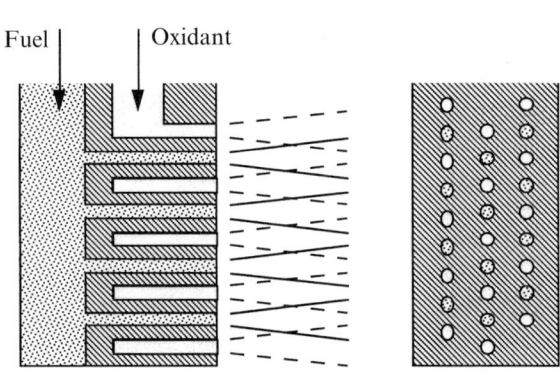

Figure 3.3. Types of injector.

ability to restart is vital to manned missions, and contributes greatly to the flexibility of modern launch vehicles. A typical requirement is to restart the upper-stage engine after an orbital or sub-orbital coast phase, which enables the correct perigee of a transfer orbit to be selected, for example, irrespective of the launch site. The restart capability is therefore becoming a more common requirement.

For single-use engines, including all solid propellant engines, starting is usually accomplished by means of a pyrotechnic device. The device is set off by means of an electric current, which heats a wire set in the pyrotechnic material. The material ignites, and a shower of sparks and hot gas from the chemical reaction ignites the gaseous or solid propellant mixture. Pyrotechnic igniters are safe and reliable. They have redundant electrical heaters and connections, and similar devices have a long history, as single-use actuators, for many applications in space. For this reason, they are often the preferred method of starting rockets. They are clearly one-shot devices, and cannot be used for restarting a rocket engine.

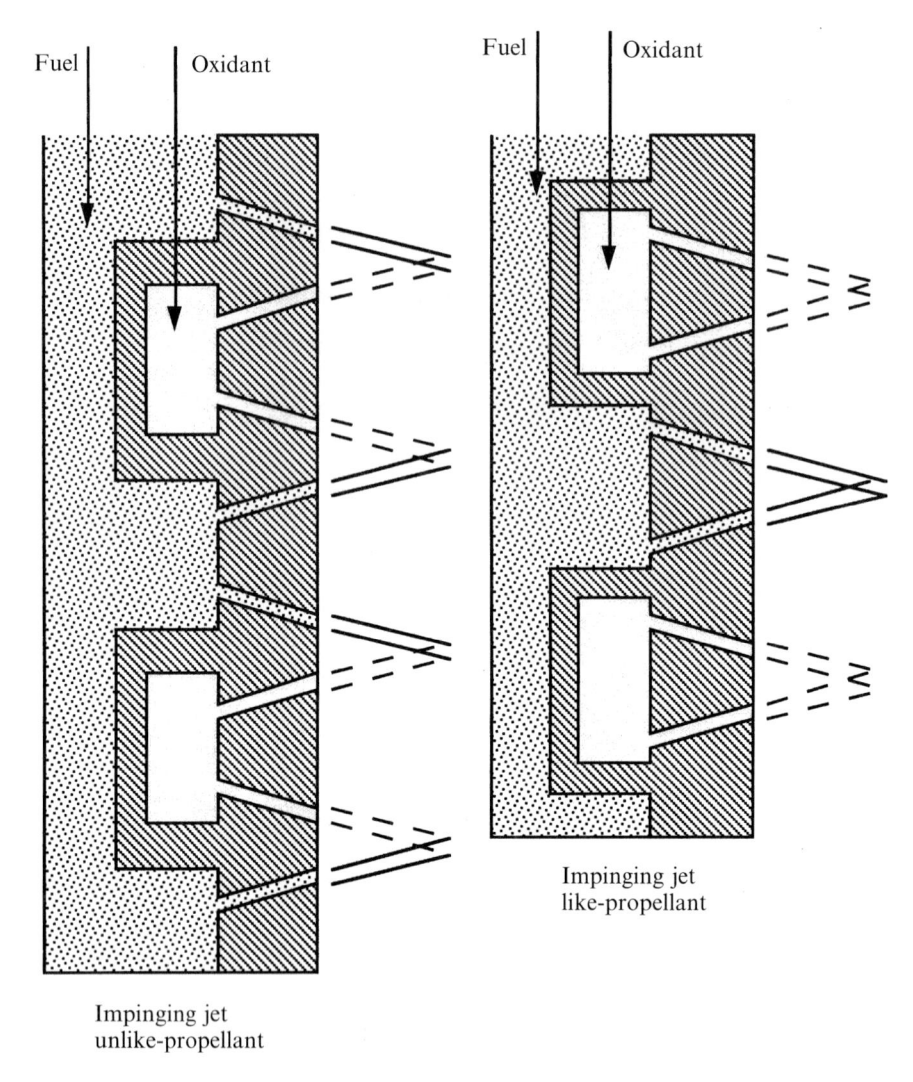

Figure 3.4. The impinging jet injector.

An electrical spark igniter, analogous to a sparking plug is generally used to ignite LH2/LO2 engines, which in principle provides the possibility of a restart. However, there is a difficulty in that the electric spark releases less energy than a pyrotechnic device, and there is also the possibility of fouling during the first period of operation of the engine, which may then put the restart at risk. Much design effort has been put into reusable igniters, and this will continue as restart capability becomes more desirable. For a single use, the Space Shuttle main engine has electric ignition for both the main combustion chamber and for the turbo-pump gas generators. In this case the spark is continuous for the period during which the igniter is switched on, and the system is contained in a small tube which forms part of the injector. The gaseous hydrogen and oxygen in the tube ignite first, and then the flame spreads to the rest of the chamber. By confining the initial gas volume to that in the tube, the risk of the flame being quenched by a large volume of cool gas is reduced. There is sufficient heat in the flame, once established in the tube, to prevent quenching.

For a secure restart capability on manned missions, hypergolic propellants must be used. These have the property that they ignite on mixing, and so starting the engine is simply a matter of starting the flow of propellants into the combustion chamber. This process is used for all manned flight critical engines. It was used for the Apollo lunar transfer vehicle, and is used for the de-orbiting of the Space Shuttle. The most common combination of propellants is nitrogen tetroxide and UDMH. As mentioned before, these are liquid at room temperature and can be stored safely on board for a long time, with no special precautions. The disadvantage of these propellants is their rather low specific impulse, which is a little more than half of that achievable with liquid hydrogen and liquid oxygen. Safe and secure restartable engines using more powerful propellants would be a major advance, but these are yet to be produced. Restartable engines are preferred for upper stages, particularly for injecting spacecraft into elliptical transfer orbits. The use of this facility means that the argument of perigee can be selected correctly, independent of the launch site and time of launch. The higher exhaust velocity of cryogenic propellants combined with such a facility would convey a much greater advantage.

The starting sequence for cryogenic engines is complicated, and will be dealt with after the propellant supply and distribution have been considered. Before this, the steering of rocket vehicles using thrust vector control will be discussed.

3.2.3 Thrust vector control

As we have seen, the thrust of the rocket engine is developed mostly on the exhaust nozzle, and is transferred to the vehicle itself through the mounting struts of the rocket engine. In effect, the rocket vehicle is being accelerated by a force applied at its lower extremity. This is a very unstable dynamical system, and a brief look at the history of rocketry shows that failure to control the thrust vector has been a major cause of loss. Indeed the development of modern control systems and on-board computers has significantly contributed to the success of the space programme.

Control systems and their theory are beyond the scope of this book, but some useful ideas emerge from simple considerations. Consider any portion (see Chapter 5) of the rocket's trajectory: the early part in which the ascent is vertical, the flight at constant pitch angle, or any controlled path while the rocket is firing. The requirement is to keep the thrust vector parallel to the rocket axis, unless the pitch angle is being changed. To keep the thrust vector parallel to the axis there must be an attitude measurement system, a computer to calculate the attitude error and the necessary correction, and a steerable exhaust stream to bring the rocket back on course. The earliest control system used gyroscopes (the A4 rocket), and these are still used, in a more sophisticated form, to generate the error signal. The A4 gyroscopes used a direct electrical connection to the rocket engine. When the rocket went off course, the relative motion of the gyroscope was transferred electromagnetically to the engine to generate a corrective transverse thrust.

The A4 used four graphite vanes, set in the exhaust stream of the single rocket engine to divert the thrust vector, in order to correct errors in pitch and yaw. This was the first successful use of thrust vector control and was an essential step in the development of the modern rocket vehicle. The vanes, being set in the hypersonic exhaust stream, generated shocks, which reduced the thrust. The benefit of a simple and robust thrust vector control far outweighed the loss of thrust. An uncontrollable rocket is useless as a vehicle.

Many modern engines have either a gimballed mounting for the whole engine, or a flexible nozzle joint so that the nozzle itself can be moved. The latter is often used on solid boosters. Use of a gimballed engine requires that the propellant supply lines are to some extent flexible. These flexible joints are usually in the lines from the tanks to the turbo-pump inlets; the turbo-pump and propellant distribution system are all mounted on the engine itself, to avoid flexible joints operating at high pressures. If there is more than one motor to a stage, then separate control of the motors can be used to correct roll errors as well as those in pitch and yaw. If there is only a single motor then roll must be controlled by separate small rocket motors mounted, typically, on the side of the vehicle and directed tangentially.

The Japanese Mu rocket is solid fuelled, and has a fixed nozzle. For thrust vector control, cold liquid is injected into the exhaust stream through a ring of injectors set around the throat of the nozzle. Transverse thrust is generated by injecting liquid at the side toward which the rocket axis should be tilted. The liquid flows down the inside of the nozzle, and cools the exhaust stream as it evaporates. The pressure at that side of the nozzle is thereby reduced, and the stream diverts towards the cooler gas. By carefully selecting which injectors are opened, accurate control of the thrust vector can be achieved without resorting to a flexible joint on the nozzle.

This difficult engineering problem has had to be solved for the large solid-fuelled boosters employed on the Space Shuttle and Ariane 5. Here the thrust is so great that failure to steer the booster thrust vector would render the vehicle uncontrollable. At the same time it is not possible by the use of cooling techniques to generate sufficient transverse thrust. The flexible joints on the Space Shuttle and Ariane 5 boosters are a major engineering achievement. They allow the whole thrust of the boosters to be steered, and at the same time they survive the heat and pressure existing at the entrance to the nozzle.

Because the thrust of the engine is so high, only a very small deflection of the stream or nozzle is needed. On the other hand, the mechanism for deflection has to deal with large forces, extreme conditions, or both. Gimballed engines can be steered by hydraulic or gas-powered pistons attached to the nozzle, and if the thrust is not too large then electromechanical actuators can also be used.

3.3 LIQUID PROPELLANT DISTRIBUTION SYSTEMS

The most commonly used distribution system employs turbo-pumps to deliver the propellants to the injectors at high pressure and flow rate. The turbo-pumps are driven by hot gas, generated in a separate combustion chamber or gas generator. This basic idea has many variants which seek to confer improvements in efficiency, but here we shall examine only the basic concept, leaving variants until later in this chapter.

There are a number of design problems to be overcome. Above all, such pumps have to be reliable over the life of the engine, and they have to work under extreme conditions and at maximum efficiency. The mass of turbo-machinery is part of the dead weight of the rocket and limits the achievable mass ratio. The total mass of a rocket engine seems small compared with that of the vehicle and propellant at lift-off. However, once the propellant is exhausted, the rocket engine mass becomes a significant part of the total dry mass. There is therefore a strong incentive to reduce the mass of the engine as far as possible, and this has an affect on the design of the turbo-pumps.

The propellants will be of different densities, and the mixture ratio is generally quite far from 50:50, so the pump for each component has to be individually sized. Changes in mixture ratio during flight, or minor adjustments to keep the ratio constant, require the pumps to be controllable individually. The propellants may be corrosive or cryogenic, or they may have other properties not compatible with simple engineering: for example, hydrogen leaks very easily through gaskets and seals. Mass limitations prevent the use of redundant delivery systems, and so reliability is of paramount importance.

The propellant tanks should be thin-walled to reduce dead weight, but have to be stiff to transmit the thrust up the rocket without the need for additional structure. The propellants also have to be delivered to the turbo-pump at quite a high pressure to prevent *cavitation* (see below). For both of these reasons it is convenient to pressurise the propellant tanks to 5–10 bar, although this is far too small to deliver the propellant to the combustion chamber of a high-thrust engine. Sometimes this pressure is provided by a separate compressed gas supply—generally helium or nitrogen—stored on board, or one of the propellants is converted to gaseous form by the heat of the combustion chamber and used to pressurise the tanks. This pressure can also be supplied by bleeding off into the propellant tanks some of the gas used to drive the turbines. The temperature of this gas needs to be kept low, so some cooling may be needed before introduction to the tanks.

The gas generator which provides the high-pressure gas to drive the pumps is a miniature combustion chamber, burning part of the propellant supply. It needs a separate igniter, and it has to be supplied with propellant, usually by a branch line from the turbo-pump itself. The propellant burned in the gas generator represents a loss of thrust, and is included in the mass ratio. The mass decreases as a result of the gas generator operation, but no thrust is produced. Some of this loss can be recovered by exhausting the gas, after it has driven the turbine, through a proper miniature nozzle in order to develop some additional thrust. The natural temperature of the burning propellant in the gas generator would be close to that in the main combustion chamber, rather too high for the turbine blades. Sometimes water is injected into the gas generator to reduce the temperature of the emerging gas, or a very fuel-rich mixture is used, which achieves the same result. A fuel-rich mixture is also less corrosive. Basically, the former measure requires a water tank on board, and the latter implies a waste of propellant; both reduce the efficiency of the rocket. Some rocket engines with turbo-pumps make use of propellant evaporated in the cooling of the combustion chamber to drive the pump. This saves the mass of the gas generator, but generally results in a lower inlet pressure, and is suitable for low thrust engines. For modern high-thrust engines, the inlet pressure needs to be of the order of 50 bar, and this requires a gas generator powered turbo-pump.

Turbines are most efficient when the hot gas inlet and exhaust pressures are very similar. When used for electricity generation or on ships, for example, many stages are used with different sized turbines, each with a small pressure drop to make the most efficient use of the energy. If the turbine exhaust of a rocket turbine were to go directly to the ambient, then the pressure ratio would be too large, and the efficiency would be low. This can be overcome by utilising a multi-stage turbine, but the extra stages add weight. It is therefore necessary to reach a compromise. An important variant of the gas generator system is the *staged combustion* system. The exhaust from the turbine enters the combustion chamber, instead of the ambient. This has two advantages: the pressure ratio for the turbine is more compatible with high efficiency, and the remaining energy in the turbine exhaust contributes to the main combustion chamber energy and ultimately to thrust.

Where the exhaust from the turbines is used directly to generate thrust the efficiency is low, because the temperature at the nozzle inlet is much lower than that in the main combustion chamber. As we have seen, the exhaust velocity depends on the square root of the combustion chamber temperature. If the exhaust from the turbines is allowed to enter the combustion chamber, the residual heat contained in the gases contributes to the heating of the combustion products in the chamber. This provides a way of using the waste heat that is thermodynamically much more efficient. Engines which do this in general produce a higher exhaust velocity for a given thrust.

While turbo-pumps driven by gas generators are widely used, there are other methods of providing the hot gas that can save in complexity by making use of the gas created by regenerative cooling. Here the propellant, often liquid hydrogen, is passed through the cooling channels of the combustion chamber and nozzle, emerging as a hot gas, which is then diverted to drive the turbine. In some cases, there is only one turbine and the oxidant pump is driven by a gear chain. In others, there are two turbines, in series, with the hot hydrogen emerging from one turbine and flowing to the second. The exhaust is sometimes then diverted into the combustion chamber to recover any remaining heat. This system, because of its simplicity, was used on early engines, and is now being specified again for upper-stage engines to reduce the cost and mass associated with a separate gas generator. This system is sometimes described as the *topping cycle* or *expander cycle*, and has many variants.

3.3.1 Cavitation

This is a well-known problem which occurs when a liquid is in contact with a rapidly moving vane on, for instance, a ship's propeller. The pressure in the liquid at the retreating surface of the vane is reduced, and it can be low enough to allow local boiling to take place. Bubbles of vapour are produced, and they then collapse when they enter a region of normal pressure. The tiny shock waves produced damage the surface of the vane. Severe cavitation can produce significant quantities of vapour at the inlet of the turbo-pump. This very quickly reduces the efficiency of liquid transfer as the rotation speed increases and larger regions of vapour appear.

For rocket engine turbo-pumps the rotation speed is very high indeed—10,000–30,000 rpm is typical—and the liquids are quite likely to be cryogenic. These are ideal conditions for cavitation to take place. Damage to the vane surface may not be too serious for a single-use pump, but if a significant amount of vapour forms then the turbo-pump will 'race' due to the reduced load and damage the bearings and other components. The flow rate of propellant will also suddenly decrease, with a consequent drop in thrust, which in most cases would lead to disaster. This is therefore a very serious problem.

To avoid cavitation, the pressure at the inlet to the pump must be kept high enough to prevent local evaporation of the liquid. This can be realised in several ways. Static pressure in the propellant tanks may be sufficient, and the acceleration of the rocket can generate additional pressure at the pump inlet, if the propellant is reasonably dense and the supply lines are axial and sufficiently long. This is unlikely to be the case with liquid hydrogen, which has a specific gravity of 0.071. Where other measures cannot succeed, and particularly where the pump speeds are very high, a two-stage pump is required. The low-pressure pump—often called an impeller—simply has the task of raising the inlet pressure at the main pump to an acceptable level of, say 10–20 bar. It can therefore be rather simple in design. If it is mounted directly at the inlet and on the same shaft as the main pump, it may not be possible to avoid cavitation at the impeller blades if the shaft speed is very high. Some improvement can be realised by correct shaping of the impeller blades, although they may need to be driven by a gear train. The Space Shuttle main engine uses separate low-pressure turbo-pumps driven by a small fraction of the propellant flow, diverted from the outlets of the high-pressure pumps. This allows the low and high-pressure pumps to be individually optimised.

3.3.2 Pogo

This humorously named phenomenon is nevertheless a serious problem. The pressure at the inlet to the combustion chamber should be constant for a steady flow of propellant to the combustion chamber and hence a steady thrust. As mentioned above, the acceleration of the rocket raises the pressure at the pump inlet, and it is possible to develop a feedback loop between the instantaneous thrust and the pump inlet pressure. If this happens, then a small natural fluctuation in thrust will result in a fluctuation in flow rate to the combustion chamber. The fluctuation will make itself felt, with a slight delay (the time taken for the propellant to flow from the pump to the chamber), as a further fluctuation in thrust. This, in turn, changes the inlet pressure at the pump, which causes another thrust fluctuation, and so on. The time delay is usually in the 10-ms region, and the reinforcement mechanism can result in the build-up of an oscillation in thrust with a period of about 100 Hz. This is very damaging to the rocket and the payload, as a small fluctuation in a mega-Newton of thrust is a large force. For this reason, pogo correction systems are fitted to liquid-propellant rocket engines.

The basic principle is to introduce some capacitance to the system in order to smooth out fluctuations in inlet pressure. A small sealed volume is connected to the propellant line, adjacent to the combustion chamber inlet, and is filled with propellant. It is pressurised using gas from the tank pressurisation system. If the line pressure falls momentarily, additional propellant is very quickly injected from the storage volume, to raise the pressure to its original value. If the pressure rises then some of the excess propellant flows into the sealed volume, again restoring the line pressure to normal. It is usually only necessary to fit pogo correction to one propellant line, which in most cases is the oxidant line. This system can be passive, or it can be actively controlled, to deal with, for example, the much greater pressure fluctuations which occur when an engine is shut down or is started. In such cases the pogo correction system can also protect the turbine from cavitation. For a single-use engine, damage to the turbines at shut-down is of no consequence, but for a reusable engine a turbine damaged due to 'racing' in a heavily cavitated fluid is a serious matter.

3.4 COOLING OF LIQUID-FUELLED ROCKET ENGINES

Before considering examples of actual rocket engines it is convenient to consider the cooling of the combustion chamber and nozzle. High combustion temperature produces a high exhaust velocity. A typical temperature is 3,000 K, but the melting point of most metals is below 2,000 K and so the combustion chamber and nozzle must be cooled. This is done by allowing part of the cool unburnt propellant to carry away the heat conducted and radiated to the walls of the chamber and nozzle. This can be done in a number of ways.

Technically, the simplest method is *film cooling*. Part of the liquid propellant is caused to flow along the inside surface of the combustion chamber and down the inside surface of the nozzle. The evaporation of this liquid film has a certain cooling effect, and results in a layer of cool gas between the wall and the hot gases passing from the chamber and through the nozzle. The cooling film is introduced through part of the injector next to the wall. This type of cooling works best with lower combustion temperatures such as are encountered in storable propellant engines. The Ariane 4 Viking engine uses film cooling, which results in the simplest configuration of the combustion chamber and nozzle. In this engine the injector is mounted on the cylindrical wall of the chamber rather than at the end, and it is therefore simple to inject part of the UDMH parallel to the wall. This method is suitable for cooling the combustion chamber and throat, because the efficiency of cooling decreases with distance from the injector. The nozzle is less well cooled and may glow red hot, cooling by radiation. The use of a refractory cobalt alloy enables the nozzle to retain its structural strength at this high temperature.

Cryogenic propellants, liquid hydrogen and liquid oxygen generate much higher combustion temperatures, and the cold liquid lends itself to efficient cooling. In such cases the walls of the combustion chamber and the nozzle are made hollow, and one of the propellants—usually hydrogen—is passed through the cavity. This cools the chamber and nozzle walls effectively, at the expense of additional complication and cost in construction. The gas resulting from the waste heat carried away from the walls can be used in various ways. The simplest approach is to exhaust the gas through many small nozzles round the rim of the main nozzle, to generate a little additional thrust. This is called *dump cooling*, and it can be used to cool the long nozzle of an engine designed for use in a vacuum, where it may be inconvenient to pipe the gas back into the top of the engine. As mentioned above, the gas may also be used to drive the turbine or to pressurise the propellant tanks. The most efficient way of using this gas is to feed it back into the combustion chamber and burn it to contribute to the main thrust. This has two advantages: the chemical energy of the gas—part of the propellant load of the rocket—is not wasted, and the waste heat conducted and radiated out of the combustion chamber is returned to the main combustion. This latter point is very important. Fundamental thermodynamics tells us that extraction of energy from a hot gas depends on the temperature difference between the source and the sink. After cooling the walls, the temperature of the propellant is far below that in the combustion chamber, so not much energy or thrust can be extracted from it. On the other hand, if it is passed into the combustion chamber and heated to the combustion temperature, then much more of the energy acquired during cooling is released. This technique is called *regenerative cooling,* and results in the most efficient engines. Of course, it leads to further complications and results in a heavier engine, and as always there must be a correct balance between extra thrust and extra weight.

If hot spots on the chamber and nozzle walls are to be avoided, the propellant must be in contact with the wall everywhere, and the flow must be smooth and continuous. Moreover, there is a large quantity of heat to carry away. Most engines therefore have the nozzle and lower part of the combustion chamber made from metal tubes welded together, wall to wall, to form a continuous surface. The propellant flows through this multiplicity of tubes freely and is, at the same time, constrained to cover the entire inner wall. In some cases the tubes are parallel to the axis of the thrust chamber, and in others a spiral form is used to produce a longer flow path. The two may be combined, with the spiral form being used on the nozzle and the axial form of the combustion chamber. The design of such a complicated structure is very demanding both on the materials and on the function. The operating temperature and pressure are very high, and any interruption of the flow during operation would be fatal. Nevertheless the advantage to be gained in terms of exhaust velocity is significant. The Saturn V engine developed an exhaust velocity of around $4,200 \, \text{m s}^{-1}$, while the SSME develops a velocity of $4,550 \, \text{m s}^{-1}$.

As mentioned above, these apparently small gains have a major impact on the performance of the rocket, in terms of payload and achievable velocity increment.

3.5 EXAMPLES OF ROCKET ENGINE PROPELLANT FLOW

For most modern launchers, gas-pressure-fed systems are not sufficiently powerful for use in first or second stages. This is just a matter of the required thrust, as pressure-fed systems cannot deliver propellant at a very high flow rate without prohibitively high tank pressures. Pressure-fed systems are advantageous for upper stages, because the reduction in weight helps to produce a high mass ratio, and the thrust and propellant flow requirements are less demanding. Before considering examples of gas generator and turbo-pump systems, a modern pressure-fed system used on the Ariane 5 upper stage will be described.

3.5.1 The Aestus engine on Ariane 5

This is the restartable engine used on the upper stage of the Ariane 5 rocket (Figure 3.5). The propellants are hypergolic: monomethyl-hydrazine (MMH) and nitrogen tetroxide. Both of these are liquid at normal temperature and pressure (NTP) and can be stored safely. Ignition of the rocket results simply from the chemical reaction that occurs spontaneously when the propellants meet in the combustion chamber. The propellant delivery scheme is shown in Plate 1.

There is a single combustion chamber gimballed to allow $\pm 6°$ of thrust vector control through two actuators. The nozzle is bell-shaped with an expansion ratio of 30 to develop an exhaust velocity of $3,240 \, m \, s^{-1}$ *in vacuo*; as an upper stage, it operates only *in vacuo*. Regenerative cooling is employed for the combustion chamber walls and the inboard part of the nozzle, for which the MMH is used. It flows from the tank into the lower part of the hollow walls, and having extracted heat it enters the combustion chamber through the injector. This is a multi-element coaxial injector with which the swirl technique is used to mix the MMH with the nitrogen tetroxide. While the combustion chamber and the inboard part of the nozzle are regeneratively cooled the nozzle extension is not; it is allowed to glow red-hot in use, dissipating heat by radiation.

There are two fuel tanks and two oxidiser tanks. The fuel (MMH) tanks are spherical, while the oxidiser tanks are slightly elongated, reflecting the differing volumes of fuel and oxidant. The oxidant–fuel ratio is 2.05. Both types of tank are made of aluminium alloy. The spherical shape uses the minimum volume of aluminium to contain the propellant, and also produces the minimum wall thickness to safely contain a given pressure. Thus the propellant tanks are optimised for a pressure delivery system. This can be employed for an upper stage in which the quantity of propellant is relatively modest, but the huge amounts of propellant needed for the first stage cannot be contained in spherical tanks. This approach for an upper stage also minimises the length, and hence the structural mass required. The tanks are pressurised with helium from a pair of high-pressure tanks;

Sec. 3.5] Examples of rocket engine propellant flow 75

Figure 3.5. The Aestus engine on Ariane 5. The high-expansion ratio nozzle and two of the four propellant tanks can be seen. Upper-stage engines should be short to reduce the overall length of the vehicle; here the propellant tanks cluster round the engine.
Courtesy ESA.

the gas pressure being moderated by a reducing valve to around 18 bar to pressurise the propellant tanks. The propellant is delivered to the engine at 17.8 bar, and the combustion chamber itself operates at 11 bar. There is a considerable pressure reduction across the injector. The passive anti-pogo system is fitted to the oxidant line.

Before the engine is started, the system is purged with helium to remove propellant residues from test firings. The oxidiser valve is then opened, followed, after a short delay, by the fuel valve. The full thrust of 29 kilo-Newton is developed 0.3 seconds after the start signal. Shutdown is initiated by closing the MMH valve, followed shortly by the closure of the oxidiser valve. The engine is then purged with helium to prepare it for the restart. The total burn time of the engine is 1,100 seconds, and the vacuum exhaust velocity is $3,240 \, \text{m s}^{-1}$. This engine has been used successfully for the upper stage of Ariane 5 since 1999. The restart capability has been demonstrated for an improved range of orbit options. A pump-fed version has been tested for higher thrust applications (Figure 3.6).

3.5.2 The Ariane Viking engines

This series of rocket engines is used to power the first and second stages of the Ariane 4 launch vehicle. There are three variants. The short nozzle version—Viking 5C—is used in groups of four to power the first stage; the Viking 6—more or less identical to

Figure 3.6. The pump-fed variant Aestus engine firing. In this test the long nozzle extension has been removed.
Courtesy ESA.

the 4C—is used for the strap-on boosters; and the Viking 4B powers the second stage and has a long nozzle to produce greater efficiency at high altitude. There is little difference in the propellant delivery systems. The general scheme is shown in Plate 2.

The Viking engine uses the storable hypergolic propellants nitrogen tetroxide and UDMH25 (unsymmetrical dimethyl hydrazine with a 25% admixture of hydrazine hydrate). There is no ignition system because the propellants ignite on contact, which, as mentioned before, is convenient for restartable engines and is also a very reliable system even when the engine is not restartable. In addition to the tanks of propellant, water is also carried to act as a combustion coolant, and high-pressure nitrogen to operate the valves. (High flow rates demand large-diameter pipes and large valves, which are difficult to operate purely electrically). There are two valves to control the flow of the individual propellants to the turbo-pump. This is a single turbine, developing 2,500 kW at 10,000 rpm and driving two pumps on the same shaft; the different flow rates are accommodated by having different sized pumps. A separate pump, driven through a reduction gear, distributes the water. Part of the propellant flow (about 0.5%) is diverted to the gas generator, where the propellants react to produce the hot gas which powers the turbine. Water is injected to cool the combustion products. The hot gases pass to the turbine and then to the turbine exhaust, which is nozzle shaped to add to the thrust. Part of the hot gas is diverted to pressurise the propellant tanks. This static pressure is quite high—about 6 bar—and is enough to prevent cavitation at the pump blades with these room-temperature liquids.

The thrust is stabilised by two control loops. One controls the temperature of the hot gases from the gas generator by varying the amount of injected water, and the other uses the combustion chamber pressure to control the flow of propellant into the gas generator and thus the turbo-pump speed. In this way the thrust is kept constant. A third balancing system controls the relative pressures of the two propellants at the injector to keep the mixture ratio correct. The pogo corrector is a small cylindrical chamber surrounding the main oxidiser pipe and linked to it by small holes; it is pressurised from the nitrogen supply used to operate the valves. The combustion chamber and nozzle are cooled with a film of UDMH from the lower part of the injector. The 5C develops 678 kN of thrust at sea level, with an exhaust velocity of 2,780 m s^{-1}. The high altitude 4B variant develops 805 kN of thrust with a higher exhaust velocity of 2,950 m s^{-1}.

This engine has been used successfully for the upper stage of Ariane 5 since 1999. The restart capability has been demonstrated for an improved range of orbit options. A pump-fed version has been tested for higher thrust applications.

3.5.3 The Ariane HM7 B engine

The HM7 B liquid hydrogen–liquid oxygen engine is used to power the third stage of the Ariane 4 series of launchers and a version is presently used as a cryogenic upper stage for the Ariane 5 while the Vinci engine is being developed. The schematic is shown in Plate 3. It uses a single gas generator and turbine driving two pumps on different shafts. The high-speed pump driven directly by the turbine at 60,000 rpm delivers the liquid hydrogen at 55 bar, while the low speed pump driven through a gear chain at 13,000 rpm delivers the liquid oxygen. The static pressure in the gas lines is raised by coaxial impellers to a level sufficient to prevent cavitation The gas from the turbine is exhausted through a shaped nozzle to generate additional thrust. The nozzle throat and combustion chamber are cooled regeneratively by passing most of the hydrogen through 128 axial tubes forming the wall, before it enters the combustion chamber itself. The rest of the nozzle is dump cooled by routing a fraction of the hydrogen through 242 spiral tubes and then through micro-nozzles at the end of the main nozzle. The gas generator is fed a hydrogen rich mixture, which keeps the temperature down and reduces oxidation of the turbine blades. The gas generation rate—and therefore the propellant flow rate—is stabilised by controlling the oxygen flow into the gas generator. The valves which control the flow of propellant are operated by helium at high pressure, switched by electro-magnetic valves. A pogo corrector is fitted to the liquid oxygen line, pressurised by helium.

A particular requirement of cryogenic engines is to purge the system before ignition, and to deal with the boil-off of the cryogenic propellants. Neither liquid oxygen nor liquid hydrogen can remain liquid under achievable pressures, and so the tanks have to vent continuously to the atmosphere until a few minutes before launch. The need for purging is twofold. Firstly, all the components—the valves, pumps and combustion chambers—need to be brought down to the temperature of the propellants to avoid localised boiling of the cryogen. This would generate back pressure and interrupt flow. Secondly, the entire system must be freed of atmospheric gases which would freeze and block the system on coming into contact with the cryogenic liquids. For this reason, purging valves are provided to enable a free flow of cold gas from the boiling cryogens through the system before the main valves are opened.

The pre-launch sequence includes the chilling and purging of the system. The gas generator is then started—in this case by a pyrotechnic igniter. When the turbines are delivering full power, the main propellant valves are opened and the main combustion chamber is started by another pyrotechnic igniter.

3.5.4 The Vinci cryogenic upper-stage engine for Ariane 5

Further increase in the payload mass to geostationary transfer orbit (GTO) with the Ariane 5 makes use of a cryogenic upper stage, to replace the Aestus storable propellant engine. This engine, called Vinci will power the upper stage from 2006 (Figure 3.7). For an upper stage, mass ratio is very important, and the system does not use a gas generator to power the turbo-pumps, instead the turbines are driven by hot hydrogen emerging from the cooling channels of the combustion chamber and upper nozzle. This *expander cycle* can be used when the propellant delivery rates and chamber pressure are not too high. The two turbines are connected, in series, on the hot hydrogen line, the gas being routed first to the hydrogen turbine. On emerging from the oxygen turbine, the gas enters the combustion chamber; all the hydrogen follows this route while the oxygen is delivered in liquid form to the combustion chamber straight from the turbo-pump. The exhaust velocity is 4,650 m/s, thanks to this efficient regenerative cooling and an expansion ratio of 240 (achieved by a deployable nozzle extension). The thrust is 180 kN. These values are to be compared with the thrust of the Aestus, 29 kN, and its exhaust velocity, 3,240 m/s. This is an example of the modern trend to reduce the complexity of rocket engines, and to address all the factors that make the vehicle efficient. This engine only weighs 550 kg, which helps to keep the mass ratio of the upper stage high.

3.5.5 The Ariane 5 Vulcain cryogenic engine

The Vulcain cryogenic engine used for the main propulsion stage of Ariane 5 develops 1.13 MN of thrust and operates at 110 bar combustion chamber pressure. It is similar to the HM7 B in design, but uses full regenerative cooling of the combustion chamber and nozzle. The single gas generator drives two separate turbo-pumps, with nozzle exhausts. The propellants enter through 516 coaxial

Sec. 3.5] **Examples of rocket engine propellant flow** 79

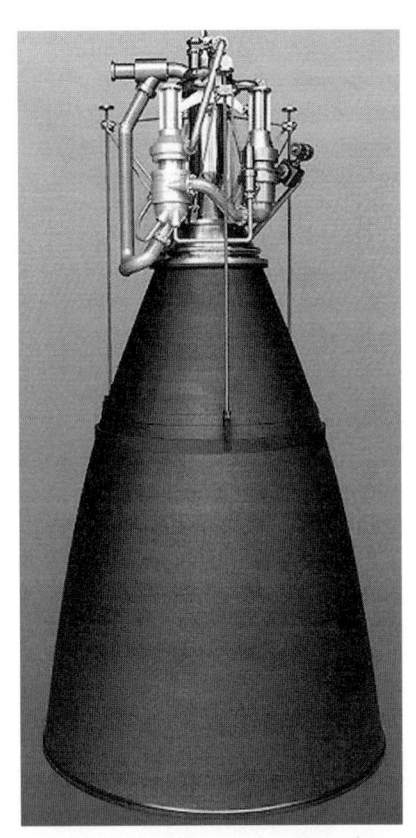

Figure 3.7. The Vinci cryogenic upper-stage engine. Note the very long nozzle extension to give the high exhaust velocity; it is deployed after separation of the main stage. This new cryogenic engine is specified for the updated Ariane 5, which will have a 10 tonne capacity to GTO.
Courtesy ESA.

injectors and generate an exhaust velocity of $4,300 \, \mathrm{m \, s^{-1}}$. The schematic is shown in Plate 5.

The propellants are stored in a cylindrical tank 24 metres long, which also provides the main structural element of the stage. Combining the functions of fuel tank and rocket structure reduces the dead weight. The 25.5 tonnes of liquid hydrogen occupies most of the volume of the tank, the 130 tonnes of oxygen being stored in the upper portion, separated by a hemispherical bulkhead. The density of liquid oxygen is much higher than that of liquid hydrogen. The hydrogen tank is pressurised by gaseous hydrogen produced by the regenerative cooling circuit—that is, heated by the combustion chamber. The oxygen tank is pressurised by helium stored in a spherical tank containing 140 kg of liquid helium. The helium is heated by the turbo-pump exhaust. A separate gaseous helium supply is used to operate the propellant valves and the pogo corrector, and to pressurise the liquid helium tank. This is stored in separate spherical tanks.

80　Liquid propellant rocket engines　　　　　　　　　　　　　　　　　　　　[Ch. 3

The gas generator and the combustion chamber are both fitted with pyrotechnic igniters. A separate solid propellant cartridge provides the gas pressure to start the turbo-pumps. The hydrogen and oxygen then enter the gas generator and the combustion chamber and are ignited. The engine is started 8 seconds before firing the boosters. This allows it to be checked out before the irrevocable booster ignition. The engine is stopped by closing the propellant valves.

The Vulcain 2 engine (Figure 3.8) specified for Ariane 5 launchers after 2002, to give an additional tonne of payload into GTO, is an updated version of the Vulcain engine used before 2002. The new engine incorporates a number of improvements, the most notable being an increase of 10% in the mass of propellant available, as a result of changing the fuel–oxidiser ratio of the engine in favour of more oxygen; the ratio was changed from 5.3 to 6.15. Because of the higher density of liquid oxygen, this can be accomplished without increasing the total volume of the propellant tanks. More oxygen increases both the mass ratio and the thrust. Normally, this would be expected to decrease the exhaust velocity because the mean molecular weight of the

Figure 3.8. The Vulcain 2 under test. This is the new version of the Vulcain specified for Ariane 5. It uses a more oxygen-rich mixture to improve the mass ratio, and a longer nozzle to restore the exhaust velocity.
Courtesy ESA.

exhaust increases, however other improvements mitigate this effect and in fact the exhaust velocity is some 30 m/s faster. The exhaust velocity is maintained by a higher expansion ratio—60 compared with 45. The cooling of the longer nozzle is accomplished by routing the turbo-pump exhaust into the nozzle extension to create a film of cooler gas, protecting the walls from the hot exhaust. The quantity of oxygen carried is increased by 23% and a re-designed two-stage turbo-pump for the oxygen line gives a 40% higher delivery rate. This combined with an increase in throat area gives a higher thrust of 1,350 kN, compared with 1,140 kN for the Vulcain.

3.5.6 The Space Shuttle main engine

The SSME uses the same cryogenic propellants as the Ariane engines, but is different in concept. It is intended to be reused many times, and to be highly efficient. It uses the *staged combustion* system to drive the turbo-pumps, and has full regenerative cooling. The vacuum exhaust velocity is 4,550 m s^{-1}, and the thrust is controllable from 67% to 109% of nominal. The propellant distribution system is shown in Plate 4.

The propellants are stored in the external tank. The hydrogen tank is pressurised by gas from the regenerative cooling of the combustion chamber, and the oxygen tank by gas resulting from regenerative cooling of the oxidiser gas generator. The propellants are delivered to the combustion chamber by separate turbo-pumps, with individual gas generators. These are called 'pre-burners' because the exhaust from the turbo-pumps passes to the combustion chamber for further burning. The propellants are raised from tank pressure to combustion chamber pressure in two stages, using separate low-pressure and high-pressure turbo-pumps.

The most important aspect of the SSME design, for our purposes, is the fact that all the exhaust from the fuel delivery system passes into the combustion chamber so that all the energy stored in the exhaust contributes to the thrust. This recovery of energy is much more efficient if enabled at high temperature in the combustion chamber than by venting the gas at the turbine exhaust temperature as in, for example, the Vulcain engine. Since the propellant flow is rather complicated, we shall examine each propellant system in turn.

The unique aspect of the SSME is that nearly *all* of the hydrogen from the fuel tank passes through the pre-burners or gas generators, and only a small fraction passes directly to the main combustion chamber after driving the low-pressure fuel pump; as the exhaust from the pre-burners will eventually enter the combustion chamber, this does not matter. It has the further advantage that a fuel-rich mixture—to keep the pre-burner exhaust temperature low enough for the turbine blades—is automatically achieved.

Liquid hydrogen arrives at the inlet of the low-pressure pump at the static pressure of about 2 bar. The pump raises this to 18 bar. It is powered by hot hydrogen gas emerging from the cooling channels in the combustion chamber. The liquid hydrogen is then pressurised to 440 bar by the high-pressure turbo-pump. It then follows three separate paths. Part of the flow enters the cooling channels in the

Figure 3.9. The SSME on a test stand. Note the long bell-shaped nozzle to extract the maximum exhaust velocity from the hot gas, and the complexity of the propellant feed system above. The hydrogen turbo-pump is visible on the left and the (smaller) oxygen pump on the right of the engine. The pipes to feed liquid hydrogen into the cooling channels of the nozzle are visible.
Courtesy NASA.

combustion chamber and emerges as hot gas, which is routed to the low-pressure fuel pump turbine to drive it. Emerging (now cool) from the turbine some of it goes to pressurise the fuel tank, and the rest cools the hot gas manifold before entering the combustion chamber. The second path passes through the cooling channels of the nozzle (Figure 3.9) before joining the third path, which routes most of the hydrogen to both of the pre-burners.

The exhaust from the pre-burners is effectively hydrogen-rich steam, at quite a high temperature (850 K). This is the fuel supply for the main combustion chamber. Consequently the hydrogen 'injector' is handling hot gas rather than cold liquid, and is called the 'hot gas manifold'. This takes the exhaust from both turbo-pumps and feeds it into the combustion chamber, where it burns with the liquid oxygen, to generate a thrust of 2 MN.

All of the liquid hydrogen is routed through the chamber or nozzle cooling channels, and afterwards becomes gaseous. In contrast, most of the oxygen remains in the liquid state right up to the combustion chamber injector. The static pressure in the oxygen tank is higher than in the hydrogen tank—about 6 bar—and the low-pressure oxygen turbo-pump raises this pressure to about 30 bar for the inlet to the high-pressure oxygen turbo-pump. After this turbo-pump the pressure of the liquid oxygen is 300 bar. The flow now divides into four separate paths. The first path carries some of the liquid oxygen to the low-pressure turbo-pump to drive the turbine, and on leaving the turbine it re-enters the main flow to the high-pressure turbo-pump. In the second path the liquid oxygen cools the high-pressure pre-burner and is converted into gas, which is used to pressurise the main oxygen tank and the pogo corrector. The third path carries most of the oxygen to the main combustion chamber injector. The fourth and final path takes liquid oxygen to an additional turbo-pump attached to the main pump shaft, which boosts the pressure to 500 bar for injection into the two pre-burners. This oxygen is burned with part of the hydrogen and forms the hot steam in the pre-burner exhaust, which then enters the combustion chamber. These routes can be followed in Plate 4. The thrust and the mixture ratio are controlled by the fuel and oxidant pre-burner valves which regulate the flow of oxygen to the pre-burners, and hence the turbine speed. Since the mixture is fuel rich, it is only necessary to vary the oxygen flow to the pre-burners to control the speed.

As with all highly developed devices the SSME (Plates 8 and 9) seems complicated in its propellant distribution. The main aim is, however, simple: to run each element of the system at its maximum efficiency, and then convert all the energy released from the burnt propellants into thrust, at a high exhaust velocity. In previous chapters we have seen that high exhaust velocity is the ultimate determinant of the success of a rocket as a launcher.

3.5.7 The RS 68 engine

From 1990 onwards the United States has been developing the Evolved Expendable Launch Vehicle, a complementary vehicle to the Shuttle. The Delta family of launchers is one manifestation of this programme, and amongst its technological innovations has been the RS 68 engine (Figure 3.10), claimed to be the first new large rocket engine to be developed in the United States since the SSME. Its main features, compared with the SSME, are its simplicity and low cost. The number of separate components has been reduced by 80%, compared with the SSME, and the amount of manual manufacture has been cut to the minimum, most components being made by digitally controlled machines. The RS 68 has now been flight qualified on launches of the Delta IV vehicle. It is the United States counterpart to the Vulcain 2 engine on Ariane 5. The vacuum thrust is about twice that of the SSME, being 3.13 MN, while the exhaust velocity is relatively low for a liquid hydrogen–liquid oxygen engine, at 4,100 m/s. This is because of the low expansion ratio; this engine is intended to

Figure 3.10. The RS 68 engine firing. This is the expendable equivalent to the SSME, it is much cheaper to build, and has twice the thrust, all useful cost saving properties—only one engine needed rather than two.
Courtesy NASA.

operate on the main stage of the Delta IV, and so is not optimised for vacuum. The sea-level thrust is relatively high at 2.89 MN, reflecting its purpose as an all-altitude booster. The weight of the engine is 6.6 tonnes, heavier than the SSME, but the thrust-to-weight ratio is about the same. Like the SSME, it can be throttled from 100% down to 60%. An engine of this thrust needs to make use of the gas generator–turbo-pump propellant delivery system to provide the necessary mass flow rates, and this contributes to the lower exhaust velocity; the hydrogen emerging from the turbo-pump exhaust is used for the roll-control thrusters of the Delta vehicle. Fundamentally, this is a low-cost expendable engine designed to provide high thrust for a heavy launcher.

3.5.8 The RL 10 engine

This engine, still a workhorse of the United States programme, has a heritage going back to the earliest liquid hydrogen–liquid oxygen engines designed in the United States (Figure 3.11); the first RL 10 was built in 1959. A pair of RL 10s power the Centaur upper stage, used on Atlas and Titan launchers. In its latest manifestation, the RL 10A-4-1, it has a vacuum thrust of 99 kN, weighs only 168 kg, and develops an exhaust velocity of 4,510 m/s. It is the archetypal upper-stage engine, optimised for vacuum use. It uses the expander cycle, with hydrogen heated in the cooling channels of the combustion chamber and upper nozzle powering the turbine of the liquid hydrogen pump, before entering the combustion chamber as gas. The liquid-oxygen pump is driven by a gear chain, from the hydrogen turbine; it delivers oxygen, as a liquid, to the injector. The engine is re-startable, giving a greater range of potential orbits.

Figure 3.11. An early photograph the RL 10 engine. The nozzle extension has been removed here. This engine is used in pairs to power the Centaur cryogenic upper stage, and has a heritage going back to the earliest use of liquid hydrogen and liquid oxygen in the United States.
Courtesy NASA.

The RL 10 engine has recently been considered as a potential chemical engine for Mars exploration, because it can be adapted to run using methane, instead of hydrogen, with the liquid oxygen. It is thought possible to produce methane on Mars from the carbon dioxide in the atmosphere, and this could be used for a return journey. Methane has other useful properties in that it is easy to store and has a high density as a liquid. It may therefore be the propellant of choice for long chemically propelled voyages. The exhaust velocity is of course smaller because of the presence of carbon dioxide in the combustion products; values as high as 3,700 m/s are predicted.

3.6 COMBUSTION AND THE CHOICE OF PROPELLANTS

Having examined the practicalities of propellant distribution in the liquid-fuelled engine, we shall now discuss the different types of propellant and the combustion process. Referring for the moment to Chapter 2, we recall that the exhaust velocity and thrust are related to the two coefficients c^*, the characteristic velocity, and C_F, the thrust coefficient. The thrust coefficient is dependent on the properties of the nozzle, while the characteristic velocity depends on the properties of the propellant and the combustion. It is defined by

$$c^* = \left\{ \gamma \left(\frac{2}{\gamma+1} \right)^{(\gamma+1)/(\gamma-1)} \frac{\mathfrak{M}}{RT_c} \right\}^{-1/2}$$

The exhaust velocity and thrust defined by

$$v_e = C_F c^*$$
$$F = m C_F c^*$$

For a given rocket engine the performance depends on the value of c^*, defined above in terms of the molecular weight, the combustion temperature, and the ratio of specific heats, all referring to the exhaust gas. Different propellant combinations will produce different combustion temperatures and molecular weights. The exhaust velocity will also depend on the nozzle and ambient properties, but the primary factor is the propellant combination.

3.6.1 Combustion temperature

The exhaust velocity and thrust depend on the square root of the combustion temperature. The temperature itself varies a little depending on the expansion conditions, but the main dependence is on the chemical energy released by the reaction: the more energetic the reaction, the higher the temperature. Table 3.1 shows the combustion temperature under standard conditions for a number of propellant combinations.

The data in Table 3.1, which are calculated for adiabatic conditions, provide an insight into the effects of chemical energy. The combustion temperatures directly

Table 3.1. Combustion temperature and exhaust velocity for different propellants.

Oxidant	Fuel	Ratio[4] (O/F)	T_c (K)	Density (mean)	c^* (m s^{-1})	v_e (m s^{-1})
O_2	H_2	4.83	3,251	0.32	2,386	4,550
O_2	RP1[1]	2.77	3,701	1.03	1,783	3,580
F_2	H_2	9.74	4,258	0.52	2,530	4,790
N_2O_4	MMH[2]	2.37	3,398	1.20	1,724	3,420
N_2O_4	N_2H_4 + UDMH[3]	2.15	3,369	1.20	1,731	3,420

(1) RP1 is a hydrocarbon fuel with hydrogen/carbon ratio 1.96, and density 0.81.
(2) MMH is monomethyl hydrazine.
(3) UDMH is unsymmetrical dimethyl hydrazine.
(4) The mixture ratios are optimised for expansion from 6.8 bar to vacuum.

reflect the chemical energy in the reaction. With oxygen as the oxidant, hydrogen produces a lower temperature than the hydrocarbon fuel RP1, the molecules of which contain more chemical energy. Fluorine and hydrogen produce a still higher temperature. This combination produces the highest temperature of any bi-propellant system. The corrosive nature of fluorine has prevented its use except with experimental rockets.

If the temperature is calculated theoretically for the complete reaction—for example, the combustion $2H_2 + O_2 = 2H_2O$—then a much higher value of about 5,000 K is predicted. In fact at this temperature, and for pressures prevalent in combustion chambers, much of the water formed by the reaction dissociates and absorbs some energy, lowering the temperature to the values shown in Table 3.1. If we deliberately introduce additional fuel, which cannot be burned without additional oxygen, then these atoms have to be heated by the same amount of chemical energy, and the temperature will be lowered further. This was discussed in the section on gas generators. Dissociation is an important phenomenon because it alters the molecular weight of the exhaust gases and the value of γ. For the oxygen–hydrogen combination the composition of the exhaust at 3,429 K is roughly 57% water and 36% hydrogen, with 3% monatomic hydrogen, 2% OH and 1% monatomic oxygen. The ratio of specific heats (γ) for this mixture is about 1.25.

3.6.2 Molecular weight

The expression for exhaust velocity also shows that v_e depends inversely on the square root of molecular weight: lower molecular weight produces a higher velocity. This is very obvious in Table 3.1, comparing hydrogen and the hydrocarbon RP1 as fuels, with oxygen as oxidant. Although the RP1, with its greater chemical energy, produces a much higher combustion temperature, the carbon atoms produce heavy carbon dioxide molecules which raise the mean molecular weight of the exhaust gases. The net result is a significantly lower exhaust velocity for the RP1 fuel. In fact, except for fluorine—which has very high chemical energy—the $H_2:O_2$ combination

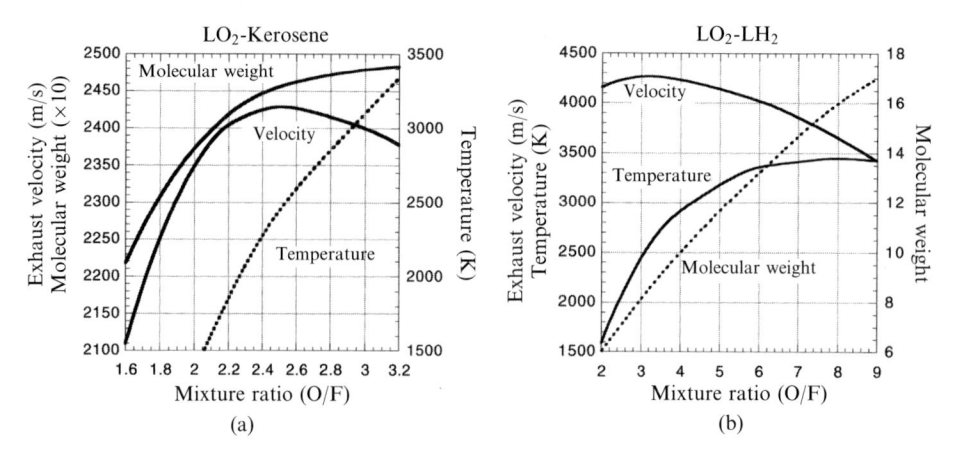

Figure 3.12. The variation of exhaust velocity, temperature and molecular weight for different propellant combinations.

produces the highest exhaust velocity, largely due to the low molecular weight of the exhaust gases. Additional hydrogen can be added to the mixture, in which case the exhaust velocity is actually *raised*, although the mean chemical energy and therefore the combustion temperature is reduced by the addition.

Figure 3.12 shows, for different propellant combinations, how the exhaust velocity (represented as specific impulse v_e/g) varies, together with temperature, molecular weight, and γ, as the mixture ratio is changed. It is remarkable how the maximum exhaust velocity is shifted away from the stoichiometric value, in the direction of lower molecular weight for each mixture. This fact is made use of when choosing the mixture ratio for maximum exhaust velocity; the fuel rich mixture in the SSME contributes directly to the high exhaust velocity.

It might be asked: Why use a propellant with a high molecular weight? If exhaust velocity were the only criterion, then this is a valid question. We should not forget, however, that thrust also depends on mass flow rate, and a heavy propellant may give a higher mass flow rate, for an engine of a given throat area, than a low-mass propellant. The ultimate velocity may be lower in this case, because the exhaust velocity is lower, but the overall thrust will be increased. This may be appropriate for the first stage of a rocket where the main objective is to raise it off the launch pad and gain some altitude; it is particularly applicable to strap-on boosters.

3.6.3 Propellant physical properties

In addition to the chemical energy and molecular weight, there are other propellant properties which affect their application. Most obviously, cryogenic propellants require special tanks and venting arrangements, as well as careful design of the distribution system and combustion chamber. Provision also has to be made for the shrinkage of the pipe-work when the cold liquid first enters it. Over tens of metres this shrinkage is significant, and flexible joints have to be included in the system. Hydrogen is a small molecule and is notorious for leaking through materials—especially organic materials used in seals. In all liquid hydrogen systems, the oxygen and hydrogen have to be kept separate until combustion, in order to avoid the formation of explosive mixtures. In the SSME turbo-pumps, for example, helium is used to purge the space between the double shaft seals to prevent hydrogen leaking through and mixing with the oxygen in the gas generator. A further problem with cryogenic liquids is the need to purge away all atmospheric gases to avoid their freezing and blocking the pipes. The constituent gases of air all freeze solid at liquid hydrogen temperatures. Liquid oxygen has a vapour pressure of 1 bar at 90 K, and liquid hydrogen has the same vapour pressure at 20 K. This means that the liquids boil under atmospheric pressure at these temperatures. It is not possible to keep such liquids under pressure at temperatures above their boiling point. The normal way of dealing with such cryogenic liquids is to allow a fraction of the liquid to boil off. The latent heat of vapourisation taken from the remaining liquid keeps it cold, and in the liquid state. Ultimately, all the liquid will have boiled away. This is familiar from the use of liquid nitrogen for cooling purposes in many laboratories. For cryogenic propellants the same procedure has to be used. Thus, the rocket is fuelled only a day or two before the launch. The liquids continue to boil away, and are topped up until just before the launch. Ice forms around the vents, and is a familiar sight when breaking off and dropping down in the first moments after lift-off. The use of cryogenic propellants adds all these problems to the design of a rocket vehicle and its ground support equipment. In some cases it may be better to use other propellants which do not involve such complications. In particular, so-called *storable* propellants are indicated for many applications. They are essential for long-duration missions such as the Space Shuttle and lunar and interplanetary transfer. Fortunately, cryogenic propellants are well adapted to the most energetically demanding role as fuels for launchers.

Hydrogen as a fuel is energetic and provides low molecular weight, as we have seen. It also has a very low density ($SG = 0.071$), and—remembering that it is the mass of propellant that determines mass ratio and hence the ultimate velocity—a large volume of hydrogen has to be carried. This is reflected in the need for large tanks, which add to the dead weight of the rocket: the mass of empty tanks may be as much as four times that needed for other fuels. This may, of course, be counterbalanced by the higher exhaust velocity, but another propellant, which may have a higher density and therefore require smaller tanks, may fit a particular application. This applies particularly to first stages where very high thrust is needed.

The whole design and structure of a rocket is simplified if room-temperature liquids are used, and there are many applications for these. Even using a room-temperature liquid fuel with liquid oxygen provides significant simplification; in 1944 the A4 rocket used alcohol and liquid oxygen. There is still a significant use of petroleum derivatives such as RP1 with liquid oxygen for first stages. The real difficulty with liquified gases is the need to vent the tanks to avoid a dangerous pressure build-up. Such a procedure is virtually impossible for long-duration flights, and so room-temperature liquids are essential.

Hydrazine and its derivatives are very commonly used. Hydrazine itself, N_2H_4, is a useful *mono-propellant*—it dissociates exothermically on a catalyst (iridium or platinum) to produce a hot mixture of hydrogen (66%) and nitrogen (33%) together with a little ammonia, without the need for another propellant. This provides a very simple system. The value of T_c is quite low—about 880 K—but the mean molecular weight is also low at about 11, so the exhaust velocity is significant at 1,700 m s^{-1}. It is not suitable for use in launchers, but hydrazine is a very important propellant for use in attitude and orbital control of spacecraft. The simplicity of a single propellant, the absence of ignition and restarting problems, and the relatively easy handling properties of the liquid, contribute to a safe and reliable system.

Hydrazine can be used in a conventional bipropellant system with liquid oxygen or with the nitrogen-based oxidisers such as nitrogen tetroxide. Hydrazine has a relatively high room-temperature vapour pressure and a relatively high freezing point—274 K—and for some applications monomethyl hydrazine, CH_3NHNH_2, is safer and easier to use than hydrazine. It is used in bipropellant systems for the Space Shuttle orbital control engines, and on the Apollo lunar transfer vehicle, both with nitrogen tetroxide. This is a self-igniting mixture and is safe for restartable engines. Unsymmetrical dimethyl hydrazine (UDMH), $(CH_3)_2NNH_2$, is very commonly used with nitrogen tetroxide and other oxidisers in bipropellant systems of all kinds. It is again easier to handle than hydrazine: it is liquid between 216 K and 336 K, and MMH is often added to it in 50% or 25% concentration to improve the performance. The additional carbon atoms in the molecule cause an increase in mean molecular weight of the exhaust, and so adding a fraction of MMH or indeed hydrazine itself to the mixture improves the exhaust velocity. Pure hydrazine systems need to be kept warm to avoid freezing, but the mixed propellants have lower freezing points.

The oxidiser nitrogen tetroxide has replaced fuming nitric acid. Red fuming nitric acid (RFNA) was commonly used in the early years of the space programme. This is pure nitric acid with dissolved NO_2, which produces the colour and the fumes. As its name suggests it is very corrosive, and dangerous to handle, but was a readily available room-temperature oxidant. Even nitrogen tetroxide has a high vapour pressure (1 bar at 294 K), and requires careful handling.

In general, all propellants are more or less difficult and/or dangerous to handle. A large quantity of stored chemical energy, such as is found in a fuelled rocket, always has the potential for disaster. As has been proved so many times, only the most meticulous attention to detail, and correct procedure, reduces the risk of accident on the ground, or during launch, to an acceptable level.

3.7 THE PERFORMANCE OF LIQUID-FUELLED ROCKET ENGINES

Having discussed the different types of rocket engine and how they work, it is useful to complete this chapter by examining the performance of some examples and how

Sec. 3.7] **The performance of liquid-fuelled rocket engines** 91

they are adapted to particular requirements. It is worth recalling from Chapters 1 and 2 that the requirements for thrust and exhaust velocity (or specific impulse) are different, as they determine different properties of the vehicle. The thrust of an engine determines the mass that can be accelerated, while the specific impulse determines the ultimate velocity to which that mass can be accelerated, for a given quantity of propellant. In general, first-stage engines should have high thrust, while second and third-stage engines should have high specific impulse or exhaust velocity. For orbital manoeuvres, high specific impulse is again the important parameter, but the need for safe and storable propellant systems is paramount.

The thrust of a rocket engine is mainly determined, by the product of chamber pressure and throat area (see Chapter 2), and so high-thrust engines will tend to have large values of this product. Since the mass of the engine will depend roughly on the throat area, improvements in efficiency will tend towards high chamber pressures, which allow the same thrust with a smaller engine. Of course, high pressures require high propellant inlet pressure and hence more elaborate propellant delivery systems, and the walls of the combustion chamber and nozzle also need to be stronger.

The exhaust velocity, or specific impulse, depends on the temperature of combustion and the molecular weight of the exhaust gases. Improvements will tend towards lighter exhaust gases and higher combustion temperatures. Again, this will place additional stress on the combustion chamber and nozzle.

3.7.1 Liquid oxygen–liquid hydrogen engines

In Table 3.2, several liquid oxygen–liquid hydrogen engines are compared. In general the propellant combination determines the combustion chamber temperature, and hence the vacuum specific impulse for a well-designed nozzle. The first two—the Vulcain and the SSME—are unique because they each have to work efficiently throughout the launch, from sea level to vacuum. This requirement cannot easily be met; the sea-level thrust is about 20% lower than the vacuum thrust, and the exhaust velocity is 25% lower. Both engines are designed to have high thrust, as they form the main propulsion system from sea level. (The boosters provide the main thrust at lift-off, but burn only for the early part of the ascent.) Both use very high combustion chamber pressures to achieve this high thrust. These engines represent the current state of the art as engines for high performance launch vehicles: the Space Shuttle, and the Ariane 5 launcher.

The throat diameter, expansion ratio and exit diameter are different for the two engines, although the thrust differs only by a factor of 2. The very high pressure in the SSME means that the throat can be quite small—only 27 cm in diameter—and a high expansion ratio can be used without producing a huge nozzle. The high expansion ratio, high pressure, and the efficient use of regenerative cooling, all result in a very high specific impulse, which contributes to the high thrust. The high weight of the SSME partly reflects its use on a manned vehicle (reliability) and partly the fact that it is reusable. The Vulcain is used as a single engine (the SSME is used in a cluster of three), and it can therefore have a large-diameter nozzle. The lower expansion ratio helps with the sea-level thrust, and the throat diameter is 80 cm,

Table 3.2. Liquid oxygen engines.

Engine	Propellants	O/F	Thrust v. (kN)	Thrust sl	I_{sp} (v.) (s)	I_{sp} (sl) (s)	Mass (kg)	D_E (m)	P_C (bar)	R_{exp}	C_F (v.)	C_F (sl)
Vulcain	LO_2/LH_2	5.2	1,075	815 kN	431	310	1,300	2.0	105	45	1.87	1.44
Vulcain 2	LO_2/LH_2	6.1	1,350		434	318	2,040	2.15	116	58.5		
SSME	LO_2/LH_2	6.0	2,323	1,853 kN	455	363	3,177	2.4	204	78	1.91	1.53
RS 68	LO_2/LH_2	6.0	3,312		420	365	6,597	2.46	96	21.5		
HM7 B	LO_2/LH_2	5.14	62		445		155	0.99	36	83		
Vinci	LO_2/LH_2	5.8	180		465		550	2.15	60	240		
RL 10	LO_2/LH_2	5.0	68	0.16k	410	10	131	0.90	24	40	1.76	0.09
RL 10A-4-1	LO_2/LH_2	5.5	99		451		168	1.53	39	84		
J-2	LO_2/LH_2	5.5	1,052		425	200	1,438	2.1	30	28		
F-1	LO_2/Kerosene	2.27	7,893	6,880 kN	304	265	8,391	2.0	70	16	1.82	1.59
RS 27	LO_2/Kerosene	2.25	1,043	934 kN	295	264	1,027	1.1	48	8	1.60	
XLR 105-5	LO_2/Kerosene		370	250 kN	309	215	460	3.1	48	25	1.74	1.22
11D-58	LO_2/Kerosene		850		348		300	1.2	78	189	1.82	
RD 170	LO_2/Kerosene	2.63	8,060	1,925 kN	337	309	9,750	4.2	245	37		

which partly compensates for the lower chamber pressure and combustion temperature.

The other three engines are intended for use in second and third stages. Here the thrust requirement is less, as much of the original launch vehicle mass has been lost either as expelled propellant or as the jettisoned empty stages and boosters. Exhaust velocity is the most important property, together with a low mass for the engine. Chamber pressures are low, and this contributes to a low mass for the engine; for third stages the engine mass is a significant part of the dead mass of the vehicle. The degree to which these engines are optimised for vacuum use can be seen in the extremely low thrust and thrust coefficient of the RL 10 at sea level (Plate 7).

3.7.2 Liquid hydrocarbon–liquid oxygen engines

Historically this propellant combination has played a major role in the space programme in the US and Russia. The combustion chamber temperature is higher than for hydrogen (3,700 K as against 3,500 K), but the specific impulse is lower because of the heavy oxides in the exhaust (higher mean molecular weight). Having a room-temperature liquid as a propellant simplifies the design, and this combination has evolved from the A4 rocket, which used liquid oxygen and alcohol. The fuel is variously described as *kerosene, gasoline* or *RP1*, but all are very similar in performance. Typically, this propellant set is used for high thrust engines, the largest engine used by NASA being the F1 on the Saturn V launcher. The chamber pressure at 70 bar is high, and this, combined with the throat area, produces the high thrust. Five of these engines were used on the first stage of the Saturn V.

The equivalent Russian engine is the RD 170, using the same propellants, and developing the same thrust but through four combustion chambers and exhaust nozzles with a common propellant distribution system. The chamber pressure is very high and the mixture is unusual in being oxygen rich. The RD 170 is used on the Energia rocket. The engine is throttleable over a wide range of thrust from 100% down to 56%, and is re-useable.

Smaller engines are in use on other launchers: the XLR 105-5 used for Atlas (NASA), the 11D-58 used on the Proton (Russia), and the RS 27 used on the Delta launchers. The general robustness and reliability of this propellant combination has contributed to its worldwide use in first-stage engines. The low exhaust velocity is less of a problem here, and high thrust is important.

3.7.3 Storable propellant engines

This combination is used on the Ariane Viking series and on many other launchers, and has the advantages of comprising room-temperature liquids and being self-igniting. Table 3.3 compares some typical engines.

Storable propellants have many advantages. They are much easier to handle on the ground, and so found favour for use with the Ariane 1–4 series launchers. They also do not need vented tanks, and are for that reason convenient for upper stages;

Table 3.3. Storable propellant engines.

Engine	Propellants	O/F	Thrust v. (kN)	Thrust sl (kN)	I_{sp} v. (s)	I_{sp} sl (s)	Mass (kg)	D_E (m)	P_C (bar)	E_{Exp}	C_F v.	C_F sl
Viking 5C	N_2O_4/UDMH	1.70	725	678	278	248	826	0.99	58	10.5	1.67	1.56
RD 253	N_2O_4/UDMH		1,670	1,410	316	267	1,280	1.5	147	26	1.62	1.37
Aestus	N_2O_4/MMH	2.05	29	–	324	–	1,200	1	11	80	1.87	–

although some propellant is inevitably lost during ascent before the upper stage is ignited. They have been particularly used for upper stages in the Russian programme, but they have also been used in Chinese and Western launchers.

Storable propellants are essential for interplanetary stages and for orbital correction, station keeping, and attitude control in satellites. The Space Shuttle uses storable propellants for orbital manoeuvres, and to initiate re-entry, and the Apollo lunar transfer vehicle used them for the Space Propulsion System (SPS; the engine of the Apollo Lunar Module), to journey to and from the Moon and for the lunar landing and take-off engines.

Nitrogen tetroxide is the oxidiser of choice, while various hydrazine derivatives are used for fuels. As shown in Table 3.3, UDMH is used, mixed with 25% of hydrazine hydrate, in the Viking engine. The Aestus engine uses MMH, which has a somewhat better performance and a higher density, which is useful in improving the mass ratio of an upper stage. The Apollo engines used MMH for the same reason.

The Viking engine is intended for first-stage use, and has an expansion ratio of 10, which is typical. The thrust coefficient of 1.57 at sea level is again optimal for the early part of the flight when the atmospheric pressure is still substantial. The engine does not operate at high altitudes, and the vacuum thrust coefficient is far from ideal. On the other hand the vacuum and sea-level specific impulse are not very different, showing that this engine is indeed optimised for low-altitude work. The value of the exhaust velocity of around $2,500 \text{ m s}^{-1}$ is not high, reflecting the nature of storable propellants, although it is quite suitable for a first stage, where high thrust is the most important parameter.

The Russian RD 253 engine is a typical high-thrust Russian engine used on the Proton series of launchers. This version is a high-altitude engine with an expansion ratio of 29. This expansion ratio—which is rather conservative, and may reflect anxiety about cooling a much longer nozzle—is offset by a very high combustion chamber pressure of 147 bar, compared with the 58 bar of the Viking. This contributes to the high thrust of 1.6 MN, and to the high specific impulse, producing a respectable exhaust velocity for an upper stage of $3,168 \text{ m s}^{-1}$. The thrust coefficients are not very high. The low thrust coefficients reflect the small expansion ratio for an upper stage engine, although the vacuum coefficient is reasonable for this use. This engine is very compact for its mega-Newton thrust. The exit diameter is only 1.5 metres, and the throat diameter is only 30 cm. The high combustion chamber pressure allows this compactness. This high-thrust engine places the Proton launcher competitively in the present-day launcher portfolio.

The Aestus on Ariane 5 is the latest example of the storable propellant engine. It uses pressure-fed propellants as we have already described, and so its chamber pressure is low. It uses MMH as the fuel, and this, together with efficient design of the nozzle produces a high exhaust velocity. The thrust coefficient is high, and so the Aestus is the archetypal upper-stage engine. The thrust is not very large, but great attention has been paid to the optimisation of the mass ratio. The restart capability provides the Ariane 5 with great flexibility in launching spacecraft of different requirements.

The liquid-fuelled rocket engine produces the highest performance, but it is very complicated and requires both high quality engineering in the combustion chamber and in the propellant delivery systems. The cost is in general very high, and for most applications the engine is used only once and then discarded. In the next chapter we shall see how the solid-fuelled rocket motor has evolved from the simple gunpowder rocket into a very efficient propulsive unit for many space applications—in particular, for the high-thrust strap-on boosters used in association with many modern launchers.

4

Solid propellant rocket motors

Considering the complexities of the liquid propellant rocket engine, it does not seem remarkable that so much attention has been given to the design and development of the much simpler solid propellant motor. This has a range of applications: the main propulsion system for small and medium launchers; as a simple and reliable third stage for orbital injection; and most of all as a strap-on booster for many modern heavy launchers. The solid propellant is storable, and is relatively safe to handle; no propellant delivery system is required, and this produces a huge improvement in reliability and cost. There are two main disadvantages: the motor cannot be controlled once ignited (although the thrust profile can be preset), and the specific impulse is rather low because of the low chemical energy of the solid propellant.

4.1 BASIC CONFIGURATION

Thermodynamically a solid-fuelled rocket motor is identical to a liquid-fuelled engine. The hot gas produced by combustion is converted to a high-speed exhaust stream in exactly the same way, and so the nozzle, the throat and the restriction in the combustion chamber leading to the throat are all identical in form and function. The thrust coefficient is calculated in the same way as for a liquid-fuelled engine, as is the characteristic velocity. The theoretical treatment in Chapter 2 serves for both.

The hot gas is produced by combustion on the hollow surface of the solid fuel block, known as the *charge*, or *grain*. In most cases the grain is bonded to the wall of the combustion chamber to prevent access of the hot combustion gases to any surface of the grain not intended to burn, and to prevent heat damage to the combustion chamber walls. The grain contains both fuel and oxidant in a finely divided powder form, mixed together and held by a binder material.

Figure 4.1 shows a typical solid-motor configuration. In comparison with the liquid rocket combustion chamber it is very simple. It consists of a casing for the propellant, which joins to a nozzle of identical geometry to that of a liquid-fuelled

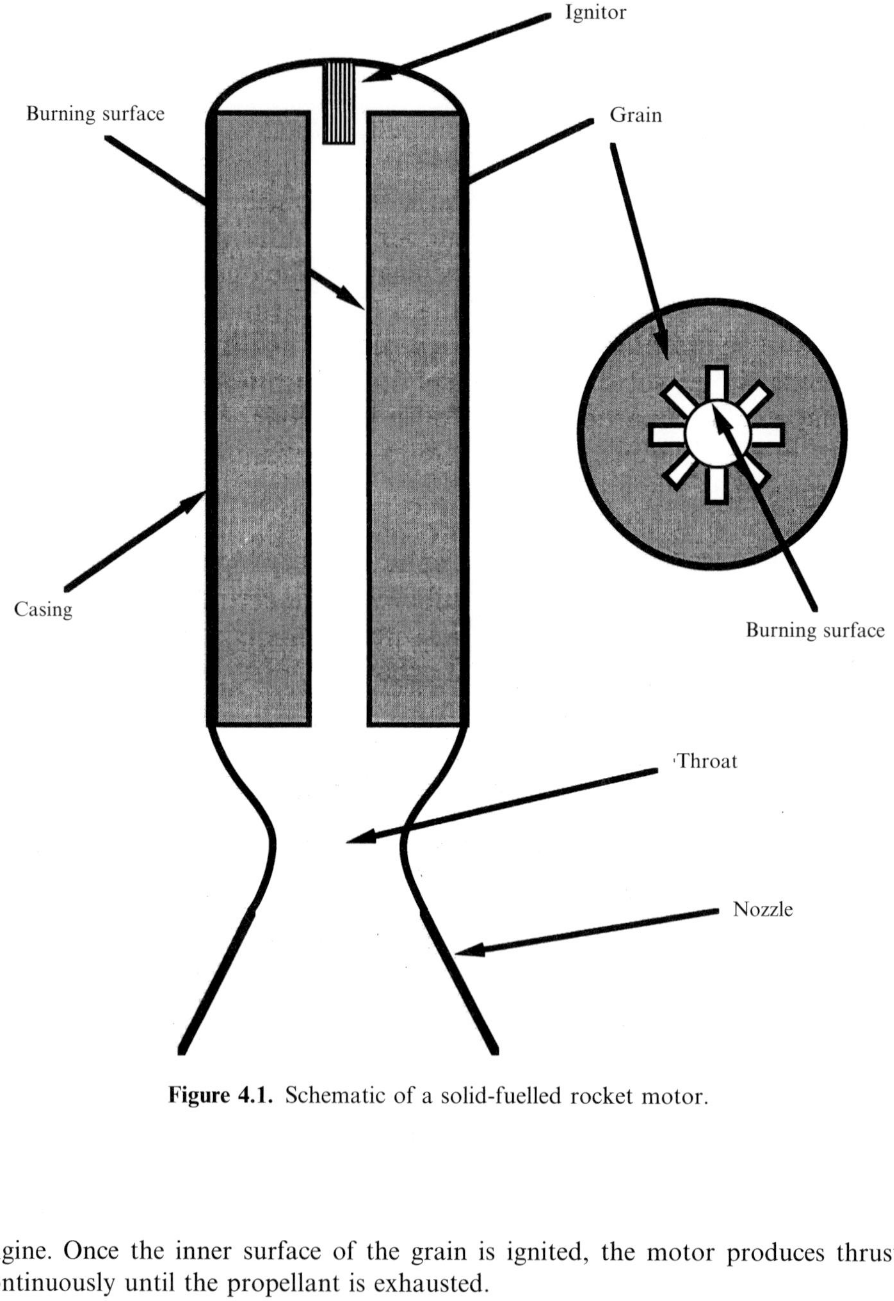

Figure 4.1. Schematic of a solid-fuelled rocket motor.

engine. Once the inner surface of the grain is ignited, the motor produces thrust continuously until the propellant is exhausted.

The fundamental simplicity of the solid propellant rocket enables wide application. The exhaust velocity is not very high—the most advanced types can produce about 2,700 m s^{-1}—but the absence of turbo-pumps and separate propellant tanks, and the complete absence of complicated valves and pipelines, can produce a high mass ratio, low cost, or both. In addition, the reliability is very high, due to the small number of individual components compared with a liquid-fuelled engine. The one big disadvantage is that the device cannot be test fired, and so the reliability has to be established by analogy and by quality control. The two areas in which solid motors excel are as strap-on boosters and as upper stages, particularly for orbit insertion or for circularisation of elliptical transfer orbits. Solid propellants are, by definition, storable.

As a booster, a solid motor can have a very high mass-flow rate and therefore high thrust, while the engineering complexity and cost can be low in a single use item. This is ideal for the early stages of a launch where high exhaust velocity is not an issue. To produce the same thrust with a liquid-fuelled rocket would not require such a large engine, because of the higher specific chemical energy of some liquid propellants, but it would be much more costly and less reliable. Very large solid boosters can be made and fuelled in sections which are then bolted together, which again makes for simplicity of construction and storage of what would otherwise be a very large unit.

As a final stage the solid motor is again reliable, and is well adapted to high mass-ratio. While the dead weight of a liquid stage includes turbo-pumps and empty tanks for two separate propellants, the dead weight of a solid stage is just the casing and the nozzle. The casing for upper stages is often made of composite materials, reducing the mass even further. It is also convenient to make such a stage with a spherical or quasispherical form, so as to minimise the mass of containing walls.

4.2 THE PROPERTIES AND THE DESIGN OF SOLID MOTORS

In comparison with a liquid-fuelled engine, the solid motor is very simple, and the design issues are therefore fewer. There is no injector, and no propellant distribution system. Design issues related to the propellant are mostly concerned with selection of the propellant type and the mounting and protection of the propellant in the casing, and ignition is similar to that of a liquid-fuelled engine. There are no propellant tanks, but the casing has to contain the propellant and also behave as a combustion chamber. For boosters the casing is large, and to combine large size with resistance to high combustion pressure is very different from the same issue in a liquid system where the requirements are separated. Cooling is totally different, because there are no liquids involved and heat dissipation has to be entirely passive.

Combustion stability—which for a liquid-fuelled rocket is dependent only on a steady supply of propellant once the chamber and injector have been optimised—is very complicated for a solid propellant. Here the supply of combustible material is dependent on conditions in the combustion chamber, and there are increased chances for instabilities to arise and propagate. Associated with stability is thrust control. For a liquid rocket the thrust is actively controlled by the rate of supply of propellants, and in the majority of cases it is stabilised at a constant value. For a solid rocket the thrust depends on the rate of supply of combustible propellant; this depends on the pressure and temperature at the burning surface, and it cannot actively be controlled. In the same way, a liquid-fuelled engine can be shut down by closing valves, whereas the solid motor continues to thrust until all the propellant is exhausted. These design problems are central to the correct performance of a solid propellant rocket motor.

While the solid-fuelled rocket is essentially a single-use item, the cost of large boosters is very high, and the necessary engineering quality of some components—specifically the casing—may make them suitable for reuse. This was a design feature for both the Space Shuttle and the Ariane 5 solid boosters. The Space Shuttle boosters are recovered, and the segments are reused. The Ariane 5 boosters are also recovered, but only for post-flight inspection.

4.3 PROPELLANT COMPOSITION

While there is a wide choice of propellant composition for liquid-fuelled rocket engines, the choice is considerably more narrow for solid propellants. Rather than selecting a particular propellant for a particular purpose, each manufacturer has its own optimised propellant mixture. The basic sold propellant consists of two or more chemical components which react together to produce heat and gaseous products. Solid propellants have been used since the earliest times, and until this twentieth century were based on gunpowder—a mixture of charcoal, sulphur and saltpetre. Modern propellants do not differ in fundamentals from these early mixtures. The oxidant is usually one of the inorganic salts such as potassium nitrate (saltpetre) although chlorates and perchlorates are now more commonly used. The fuels sometimes include sulphur, and carbon is present in the form of the organic binder.

As with any other type of rocket, the aim is to achieve the highest combustion temperature together with the lowest molecular weight in the exhaust. The difficulty with solid oxidants is that they are mostly inorganic and contain metal atoms. These lead to higher molecular weight molecules in the exhaust. Similarly the solid fuels generally have a higher atomic weight than hydrogen, and so again the molecular weight in the exhaust is driven up. The chemical energy, per unit mass of propellant, can be the same as for the main liquid propellants, and so the combustion temperature is similar. A particular problem is that some of the combustion products may form solid particles at exhaust temperatures. This affects the performance of the nozzle in converting heat energy into gas flow. All of these properties affect the performance of solid motors.

The charge of propellant in a solid rocket motor is often called the *grain*. The basic components of the grain are fuel, oxidant, binder, and additives to achieve burning stability and stability in storage. The finished charge must also be strong enough to resist the forces induced by vehicle motion and thrust. It must also be thermally insulating to prevent parts of the grain—other than the burning surface—from reaching ignition temperature.

In the past, two different kinds of solid propellant have been used. The first kind is the mixture of inorganic oxidants with fuels, as described above. This is the most commonly used today. The other type is based on nitrated organic substances such as nitroglycerine and nitrocellulose. These came into use as gun propellants after gunpowder, and it was natural that they should be considered as rocket propellants. These materials have the property that they contain the oxidant and fuel together in a single molecule or group of molecules. Heat induces a reaction in which the complex organic molecule breaks down, which produces heat and gaseous oxides of nitrogen, carbon and hydrogen. The molecular weight of such gas mixtures is rather low, giving an advantage in terms of exhaust velocity. These propellants are termed *homogeneous* propellants, for obvious reasons. They are not used for launcher boosters and most orbital change motors, because they have been superseded by more advanced mixture propellants.

The fundamental requirement is to develop high thrust per unit mass. As discussed in Chapter 2, this requires a high combustion temperature and low molecular weight of the combustion products. In general a relatively high temperature of combustion is easy to achieve, but it is impossible to have the same low molecular weight of the products achievable with liquid hydrogen and liquid oxygen. The presence of carbon and the byproducts of the inorganic oxidants, potassium and sodium salts, produces a higher molecular weight and hence a lower exhaust velocity. Referring back to Chapters 1 and 2, we can see that high molecular weight does not prevent the solid motor from developing high thrust, which is just a matter of high mass flow and throat area. High ultimate vehicle velocity is harder to achieve with a solid motor because of the low exhaust velocity. A typical value would be about $2,700 \, \text{m s}^{-1}$. For final stages, optimisation is directed towards improving the mass ratio rather than the exhaust velocity.

In modern propellants metallic powders are often added to increase the energy release and hence the combustion temperature. Aluminium is usual, and in this case the exhaust products will contain aluminium oxide, which has a high molecular weight and is refractory, and so is in the form of small solid particles. Particles in the exhaust stream reduce efficiency: they travel more slowly than the surrounding high-velocity gas, and they radiate heat more effectively (as black bodies) and therefore reduce the energy in the stream. The loss of exhaust velocity may be balanced by the higher combustion temperature and an increase in effective density of the exhaust gases. This increases the mass flow and hence the thrust. High thrust is applicable for a first-stage booster where ultimate velocity is not as important as the thrust at lift-off. In designing a motor for high thrust, increasing the exhaust density may be preferable to an increase in throat diameter and hence in overall size of the booster; the mass ratio is also increased if the grain density is higher. The presence of particles in the exhaust produces the characteristic dense white 'smoke' seen when the boosters ignite. The exhaust from a liquid-fuelled engine is usually transparent.

The most commonly used modern solid propellant is based on a polybutadiene synthetic rubber binder, with ammonium perchlorate as the oxidiser, and some 12–16% of aluminium powder. The boosters for the Space Shuttle use this type of propellant, as do the boosters for Ariane 5 and many upper stages. The combustion temperature without the aluminium is about 3,000 K with 90% of ammonium perchlorate. The addition of 16–18% aluminium increases the temperature to

3,600 K for the Ariane 5 booster, and the oxidiser concentration is reduced correspondingly.

The chemical composition of the exhaust is approximately 32% aluminium oxide, 20% carbon monoxide, 16% water, 12% hydrogen chloride, 10% nitrogen, 7% carbon dioxide and 3% chlorine and hydrogen. A major part of the aluminium oxide condenses into solid particles, but fortunately this does not contribute to the molecular weight in the expanding gases: Al_2O_3 has a molecular weight of 102. The combined effect of the gaseous components is to produce an average molecular weight of about 25. The combustion parameter is 12, giving a characteristic velocity of $1,700 \text{ m s}^{-1}$. The particles will reduce the mean exhaust velocity because of the effects mentioned above. The quoted vacuum exhaust velocity is $2,700 \text{ m s}^{-1}$, which is fairly close to the theoretical value if we assume a reasonable thrust coefficient. So this *two-phase flow*—in which the exhaust gases follow the normal expansion, cooling and acceleration, alongside particles which are accelerated by the gas—does not reduce the exhaust velocity very much. If the particles were to evaporate then a very high molecular weight gas would result, producing a very low exhaust velocity. This solid propellant is therefore rather efficient in producing high thrust and a reasonable exhaust velocity.

4.3.1 Additives

In a heterogeneous propellant the oxidant is the main constituent by mass, and the binder—usually polybutadiene rubber, a hydrocarbon—is the fuel. Aluminium is also present at 16–18%, and other materials are added to improve performance or safety. Carbon is present to render the propellant opaque to infrared radiation, so that the propellant cannot be internally ignited by heat radiated through the bulk material from the burning surface; it produces the characteristic black colour. Plasticisers are added to improve moulding and extrusion of the material. Other materials, such as inorganic salts, are added to control burning and to achieve the desirable value for the pressure-burning rate index. This is necessary for the so-called 'double base propellants'—those consisting mainly of nitroglycerine and nitrocellulose. For the heterogeneous propellants the oxidants themselves act in this way. Iron oxide is added at about the 1% level to assist smooth combustion, and waxes are also added to some propellants to lubricate extrusion.

Having arrived at an optimum composition for the main constituents, the additives are included to produce stability, storage qualities and mechanical strength. The latter is an important property. The whole mass of propellant—which is sometimes the biggest single mass in the whole vehicle (a single Ariane 5 booster weighs 260 tonnes)—has to be accelerated by the thrust. The propellant has to support this acceleration, without rupture or significant distortion, and also has to transfer the combustion pressure to the casing and maintain its integrity. The development of large boosters depends to some extent on the physical strength of the propellant.

4.3.2 Toxic exhaust

Launch vehicle boosters are fired close to the ground, and most of the exhaust is dispersed over a wide area of the launch site. While the products of liquid engines are mostly harmless, the chlorine in the oxidants of solid boosters produces hydrogen chloride, and particulates can also be dangerous. Thus it is important that during lift-off the booster exhaust is channelled away safely by water-cooled open ducts. Of course, once the rocket is in flight this is beyond control, and dilution of the exhaust products by the atmosphere, as they fall to earth, has to be relied upon.

4.3.3 Thrust stability

The overall thrust profile can be controlled by the shape of the charge, but other factors are important in understanding the way a solid motor performs, the most important of which is the stability of the thrust. In the liquid-fuelled engine the chamber pressure is usually constant and, with the mass flow rate, is determined by the rate at which the propellants are delivered through the injectors. On the other hand, in the solid motor the mass flow rate is not determined by external supply but by the rate at which the surface of the burning charge is consumed, which itself is a function of the pressure in the combustion chamber.

Because of this peculiarity of solid propellant systems the rate of supply of combustible propellant increases with pressure, and stable burning is not necessarily a given. If we arbitrarily assume that the rate of consumption of the grain depends on pressure as $m = ap^\beta$, then the value of β controls the stability, as follows. From Chapter 2 we see that the mass flow rate out of the chamber depends linearly on the pressure. Thus if $\beta > 1$, the supply of gas from the burning grain increases faster with pressure than the rate of exhaust, and an uncontrolled rise in burning rate and pressure could result from a small initial increase. Similarly a small initial decrease in pressure could result in a catastrophic drop in burning rate. Home-made rockets tend to exhibit one or other of these distressing tendencies. If on the other hand, $\beta < 1$ then the rate of change of burning rate is always less than the (linear) rate of change of mass flow through the exhaust, and the pressure in the chamber will stabilise after any positive or negative change in burning rate. This problem—which does not occur with liquid propellant engines—is a primary consideration in the design of the solid motor and in grain composition and configuration. Some additives are used to achieve the correct dependency, where this does not arise naturally. Typical values of β range from 0.4 to 0.7.

The rate of burning of the propellant, expressed as a linear recession rate of the burning surface, depends on the rate of heat supply to the surface from the hot gas. This heat evaporates the propellant. The recession rate should be constant, under constant pressure conditions, with $\beta < 1$. There is also another effect which can change the rate at which the surface recedes: *erosive* burning, which occurs because of the velocity of the gas over the surface. With a liquid-fuelled engine, it is a fair assumption that the velocity of the gas in the combustion chamber is small and constant; it is finite because the gas has to leave the chamber. Because of the length of a solid propellant combustion chamber the gas accelerates down the void; the velocity near the nozzle can be quite large. The conditions of burning at the upper and lower portions of the charge are then different. At the top, the hot gas is fairly stagnant, while near the bottom it is moving fast, and constantly supplying energy to the burning surface. The result of this is a faster evaporation and a faster recession rate near the nozzle. If this is not checked or allowed for, then the charge can burn through at the nozzle end before the upper portion is exhausted. This may lead to failure of the casing or an unforeseen decrease in thrust, neither of which is pleasant. Paradoxically, it can be ameliorated by designing the hollow void within the grain to have an increasing cross-sectional area towards the nozzle. For constant flow rate the increased cross section requires a corresponding decrease in the velocity, and in this way the effects of erosive burning are counterbalanced.

4.3.4 Thrust profile and grain shape

The pressure in the chamber, and hence the thrust, depends on the rate at which the grain is consumed. The pressure depends on the recession rate and on the area of the burning surface, and the mass flow rate depends on the volume of propellant consumed per second. The shape of the charge can be used to preset the way the area of the burning surface evolves with time, and hence the temporal thrust profile of the motor. The pressure and the thrust are independent of the increase in chamber volume as the charge burns away, and depend only on the recession rate and the area of the burning surface.

The simplest thrust profile comes from linear burning of a cylindrical grain (as with a cigarette): a constant burning area produces constant thrust. This shape, however, has disadvantages: the burning area is limited to the cylinder cross section, and the burning rim would be in contact with the wall of the motor. Active cooling of the wall is of course not possible with a solid motor, and this type of charge shape can be used only for low thrust and for a short duration because of thermal damage to the casing.

The most popular configuration involves a charge in the form of a hollow cylinder, which burns on its inner surface. This has two practical advantages: the area of the burning surface can be much larger, producing higher thrust, and the unburned grain insulates the motor wall from the hot gases. In the case of a simple hollow cylinder, the area of the burning surface increases with time, as do the pressure and the thrust. If a constant thrust is desired, the inner cross section of the grain should be formed like a cog, the teeth of which penetrate part way towards the outer surface. The area of burning is thus initially higher, and the evolving surface profile corresponds roughly to constant area and hence constant thrust. Other shapes for the grain produce different thrust profiles, depending on the design. Figure 4.2 illustrates some examples. It is common in large boosters to mix the profiles; for example, the forward segment may have a star or cog profile, while the aft segments may have a circular profile. In this way the thrust profile can be fine-tuned. It is also common to have at least one segment with a tapered profile to ameliorate erosive burning and to modify the thrust profile.

Sec. 4.3] Propellant composition 105

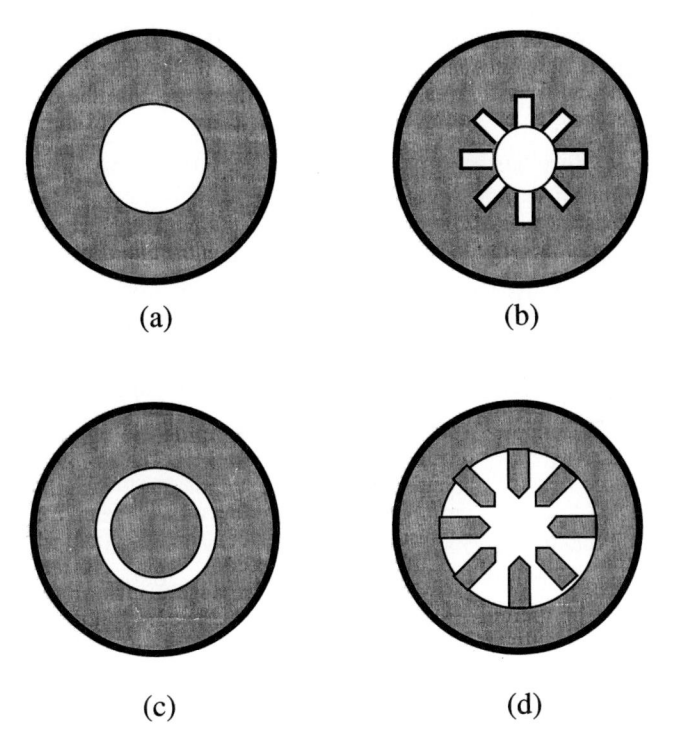

Figure 4.2. Cross sections of grains.

The thrust profiles associated with the shapes in Figure 4.2 can be understood from simple geometric arguments. The recession rate is assumed to be constant over the whole exposed area, which in the diagrams may be assumed to be proportional to the length of the perimeter of the burning surface. Type (a), called 'progressive', is the simplest to understand. Here the circumference of the circular cross section increases linearly with time, as does the area of the burning surface, and there is a linear increase in mass flow rate and hence in thrust. Type (b)—which is perhaps the most commonly used—produces a quasi-constant thrust, because the initial burning area is quite large due to the convolutions of the cog shape; as the cog 'teeth' burn away the loss of burning area is compensated by the increasing area of the cylindrical part. This profile is simple to cast, and is effective in producing an almost constant mass flow rate. Type (c) produces a perfectly flat thrust profile, because burning takes place both on the outer surface of the inner rod and on the inner surface of the outer cylinder. The decrease in burning area of the outer surface of the rod is exactly compensated by the increasing burning area on the inner surface of the cylinder. This type of grain profile is difficult to manufacture and sustain because of the need to support the rod through the hot gas stream. It is not used for space vehicles. The final example (d) shows how an exotic profile can be used to tailor the thrust profile for a particular purpose. The narrow fins of propellant initially produce a very high surface area, and so the thrust is initially very high. Once they have burned away then a low and slowly increasing thrust is produced by the cylindrical section. When the diameter of the burning cross section is large, the area changes more slowly than in the initial stages. Such a profile may be useful for strong acceleration followed by sustained flight.

Ambient temperature has a significant effect on the rate of burning and hence on the thrust profile. At first it may seem surprising that this has not arisen for liquid-fuelled rockets. However, they are much less sensitive to ambient temperature, because the temperature of the propellant and the supply rate are determined by the conditions in the combustion chamber and not by outside effects. For the solid propellant rocket this is not the case. The rate of evaporation of the combustible material from the burning surface of the grain depends on the rate at which the material is heated. This depends both on the rate of supply of heat from the combustion (which we have already dealt with) and on the temperature of the grain itself. If it is cold then more heat has to be supplied to reach evaporation point. The grain is massive, and is itself a good insulator, which means that during waiting time on the launch pad, or in space, it can slowly take up the temperature of its surroundings. This will not change appreciably over the short time of the burn because of the large heat capacity of the mass, and its good insulating properties. The burning rate will therefore shift, depending on the temperature of the grain. Variations of as much as a factor of two between $-15°C$ and $20°C$ have been reported. This affects the thrust profile, which could be a serious matter. It appears that the same factor which affects pressure sensitivity—β in the pressure index—also affects temperature dependence. Small values of β are beneficial here, and specific additives can also reduce the temperature effect. Even so, solid motors should not be used outside their specified temperature limits—particularly for launchers, for which a predictable thrust profile is very important.

We recall from Chapter 1 that, for orbital manoeuvres, the ultimate velocity of the vehicle depends on the exhaust velocity and mass ratio, and not on the thrust profile. Provided that the total impulse produced by the motor is predictable, the exact thrust profile is not important. Active temperature control of a solid motor in space would require far too much electrical power. But given the above argument, variation in thrust profile due to temperature changes is less important for this application.

4.4 INTEGRITY OF THE COMBUSTION CHAMBER

The combustion chamber of a liquid-fuelled engine is rather small. It is just big enough in diameter to allow proper mixing, and long enough to allow evaporation of propellant droplets. The combustion chamber of a solid motor is also the fuel store, and is large. In addition, since high thrust is usually the main requirement, the throat diameter is larger. The pressures experienced by each of them are about the same in modern rockets—about 50 bar. However, designing a large vessel to accommodate high pressure and high temperature is much more difficult than the equivalent task of designing a smaller vessel. The skin has to take the pressure, and as the diameter

Sec. 4.4] **Integrity of the combustion chamber** 107

increases the thickness has to increase; and because of the large surface area this has a major effect on the mass. In general, high-tensile steels are used. 4SCDN-4-10 high-strength low alloy steel is used for the Ariane 5 boosters.

4.4.1 Thermal protection

The walls of the vessel cannot be cooled by the propellant as in the liquid-fuelled engine, and this imposes a considerable difficulty. As in the case of liquid-fuelled combustion chambers, the temperature of combustion is much higher than the softening point of most metals. The combustion products cannot be allowed to contact the walls for any extended period, or disaster will result. The best solution is to bond the propellant to the walls and to cover the remaining inside surfaces with a refractory insulating layer. This technique is known as *case bonding*, and is used in most modern solid motors. The grain burns only on its inside surface, so the propellant acts as an insulator. Boosters are normally used only once, and so any residual damage caused to the walls when the propellant is exhausted is not important. In fact, a thin layer of propellant usually remains after burn-out, due to the sudden drop in pressure, which extinguishes the combustion. Where particular care is required on manned missions, and for potentially reusable casings, a layer of insulating material is also placed between the grain and the casing before it is bonded in.

The lack of any active means of cooling for solid rocket components would make them unusable if the time factor were not important. The motor has only to operate for a short time, and after this time it does not matter if components exceed their service temperature, although it is, of course, important if they are to be reused. So, provided the fatal rise in temperature of the casing or the throat of the nozzle can be delayed till after burn-out, then the motor is perfectly safe to use. The means for doing this were developed and used in the early space programme, for atmospheric re-entry. The conditions and requirements are the same: to keep the important parts cool, for a limited time, against a surface temperature higher than the melting point of metals. The method used is called *ablative* cooling. The surface is, in fact, a composite structure. Furthest removed from the source of heat is the metallic structural component, which provides the strength and stiffness if it is kept cool. This is covered with many layers of non-metallic material, which have a dual purpose. Undisturbed, they provide good heat insulation, but when exposed to the full effects of the hot gases they evaporate slowly, or *ablate*. This process extracts heat of vaporisation from the gas layers nearest the surface, and forms an insulating cool gas layer analogous to that provided by film cooling in a liquid rocket engine. The materials used are combinations of silica fibres, phenolic resins and carbon fibres. The material is refractory, but because it is fibrous and flexible it does not crack, and retains its strength and integrity even while being eaten away by the hot gases. Needless to say, a crack in the insulator would allow hot gases to penetrate to the wall. The development of these ablative insulators was a vital step in the development both of solid motors and re-entry capsules. The time factor is of course important. The process works only for a certain length of time, after which

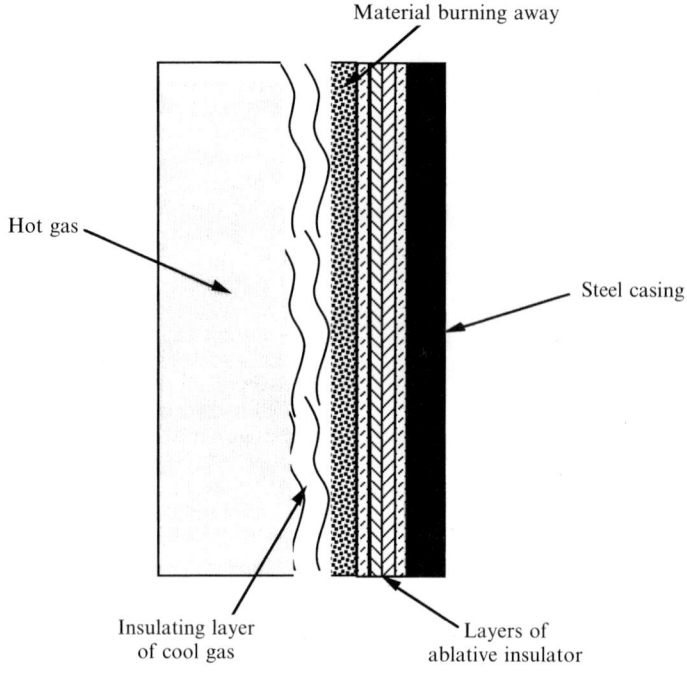

Figure 4.3. Thermal protection.

the heat reaches the structural material and the component fails. The time can be extended by including one or more *heat sinks* in the construction. These are thick pieces of metal with a high thermal capacity, which are used locally to slow down the rise in temperature of a sensitive component. They cannot be used for casings, but are often used in nozzle throats in which the sensitive structural components are smaller.

The propellant is generally cast or extruded into the required shape, and is then inserted into the chamber and bonded to the wall. The end faces of the charge—which should not burn—are coated with an insulating inhibitor, and the other surfaces of the chamber are insulated as described.

These activities are much easier if the motor is made in sections, and large ones may consist of several identical cylindrical sections as well as the top cap and the rear section containing the nozzle. The grain is cast in identical forms to match the sections, and each form is separately case-bonded into its section. For very large boosters the grain is cast directly into the casing section after the insulation is installed. The sections are then joined together to make the complete booster. This technique is not used for third stages or orbital change boosters in which the mass ratio advantage of a quasi-spherical shape for the motor is paramount; here the grain is formed as a unit and bonded into the case. These types of motor often have carbon fibre reinforced plastic (CFRP; called graphite reinforced plastic (GRP) in the United States) walls to improve the mass ratio.

4.4.2 Inter-section joints

The joints between sections have to be gas tight, and they also have to transmit the forces arising from the high thrust of the boosters. The combustion pressure of 50 bar is sufficient to cause some deformation of the cylindrical sections, and the joints must be proof against this. The forces involved are testified to by the large number of fasteners obvious when looking at a booster. Each section case is a cylinder with the wall as thin as possible (about 12 mm) to minimise mass. At each end there has to be a sturdy flange to take the fasteners and to properly transmit the forces to the cylindrical wall. Turning the whole section from solid material is a safe approach, but is costly, and other methods of forming the flanges—such as flow turning—can be employed.

There are two kinds of joint between sections: the *factory* joint and the *field* joint. The factory joint is assembled before the charge is installed, and results from the need to make up large booster casings from steel elements of a manageable size. These joints can be protected by insulation before the grain is installed, and are relatively safe. The field joint is so called because it is made 'in the field'—that is, at the launch site—and is used because of the impossibility of transporting and handling a complete booster. Field joints allow the booster to be assembled from ready-charged sections, more or less at the launch pad. They have two safety issues: they are made under field conditions away from the factory; and they cannot be protected with insulation in the same way as a factory joint, because the two faces of the propellant charge come together on assembly, and access to the inner surface of the joint is impossible

The simplest pressure seal is an 'o'-ring located in a groove in one flange, and clamped by the surface of the mating flange. This is a well tried and reliable seal, but requires very stiff and heavy flanges otherwise the flexing of the structure under thrust could open the seal. Organic seals like 'o'-rings are generally not resistant to high temperature, and should be protected. This is all the more difficult because at the junction of two sections in a large motor, such as a field joint, the grain is not continuous, so that it is possible for hot gases to reach the intersection joint in the casing. In general, for a simple solid-fuelled rocket the thrust acts axially, and so the loads on the joints are even and the effect is to close the joint even more firmly. For strap-on boosters there is the possibility of a bending load caused by the asymmetry of the structure. Simple face joints are used quite safely in solid-fuelled rockets, although with strap-on boosters a different joint is required.

To reduce mass and to give some protection from flexing, overlap joints are used between sections. By overlapping the joints the pressure in the motor compresses the 'o'-ring seals, while flexure can cause only small transverse movements of the joint faces.

It is essential that the casing is gas-tight at the operating pressure of 50–60 bar, as any leakage would allow hot gas to reach sensitive components. Being organic, 'o'-rings cannot withstand even moderately high temperatures, and should be kept within a reasonable range around room temperature. They cannot be exposed to the hot combustion products, and a thermal barrier has to be placed between the hot gases and the seal—it does not form a seal itself and is simply there to protect the 'o'-ring. Typically this thermal barrier is a flexible silicone sealant or thermal 'putty'. Provided the seal is intact, then a small amount of gas leaks past the thermal protection and equalises the pressure. This small amount of gas cools and does no harm to the pressure seal. While the thermal protection remains intact, the 'o'-rings are safe, but if either the sealant leaks or the 'o'-ring leaks then the joint can fail. If the sealant leaks then hot gas can reach the 'o'-ring and cause it to scorch or melt. If the 'o'-ring leaks, then the cool gas supporting the sealant leaks away, and the sealant flows, which again may allow hot gas to reach the 'o'-ring.

This kind of failure caused the loss of *Challenger*. An external temperature a little above 0° C may have caused the 'o'-ring in the aft field joint to become stiff, and unable to follow the movements of the opposing flanges so as to remain leak-tight. Hot gas leaked past the thermal putty and destroyed part of the 'o'-ring. High-temperature combustion products then escaped and damaged the main propellant tank containing liquid hydrogen and liquid oxygen. The loss of the Space Shuttle and of seven lives then became inevitable.

The joint was redesigned after the disaster, both to eliminate the thermal putty by replacing it with a rubber 'J' seal, and to use three 'o'-rings in series. The 'J' seal is configured to seal more tightly when the internal pressure rises. It is interesting to contemplate that rubber and silicone rubber—both of which are used domestically—can, under the right circumstances, resist the most undomestic temperatures and pressures.

4.4.3 Nozzle thermal protection

The nozzle and throat are protected from the heat of the exhaust by using similar techniques to those used to protect the casing. Here the problem is more severe because of the high velocity of the exhaust gases. The main structure of the throat and nozzle is made of steel, but many layers of ablative insulator are applied to the inside. A heat sink is also used at the throat to reduce the transfer of heat to the steel structure. Most of the thrust is developed on the walls of the nozzle, and so the structure needs to remain within its service temperature until burn-out. Ablation, heat diffusion into the heat sink, and the thermally insulating properties of the throat lining keep the steel cool long enough to do its job. After burn-out it does not matter if the outer structure becomes too hot. It is worth mentioning that without such a lining the steel would reach its melting point in less than one second, but the lining prolongs this by a factor of about 200.

4.5 IGNITION

Solid motors are used for two applications, both of which require ignition which is as stable and reliable as for a liquid-fuelled engine. The main propulsion unit for a rocket stage or satellite orbital injection system requires timely ignition in order to achieve the eventual orbit. A booster forms one of a group of motors which must develop thrust together. In all cases a pyrotechnic igniter is used. Pyrotechnic devices have an extensive and reliable heritage for space use, in a variety of different applications. To ignite a solid motor, a significant charge of pyrotechnic material is needed to ensure that the entire inner surface of the grain is simultaneously brought to the ignition temperature: 25 kg of pyrotechnic is used in the Ariane 5 solid boosters. It is itself ignited by a redundant electrical system.

4.6 THRUST VECTOR CONTROL

For orbital injection, thrust vector control is not normally needed, as the burn is too short to require the spacecraft to change its course while the motor is firing. Thrust vector control is essential for solid boosters because their thrust dominates the thrust of a launcher for the first few minutes, and so course corrections require the booster thrust to be diverted.

The technique of liquid injection applicable to small solid propellant launchers cannot produce sufficient transverse thrust to manoeuvre a large launch vehicle like the Ariane or the Space Shuttle. It requires a moveable nozzle, mounted on a gimballed flexible bearing so that it can be traversed by about 6°C in two orthogonal directions. The large forces are needed to move the nozzle quickly, and the motion is contrived using hydraulic rams controlled by electrical signals from the vehicle's attitude control system. The flexible joint has to be protected from the heat of the combustion products by flaps of material similar to that used to insulate the joints and casing.

4.7 TWO MODERN SOLID BOOSTERS

To complete this chapter we shall describe two important solid boosters: those of the Space Shuttle and of Ariane 5. Their similarities reflect both the similarity in application and the relative maturity of the technology.

4.7.1 The Space Shuttle SRB

Table 4.1 shows that the SRB is about twice the size of the MPS. It develops a thrust of 10 MN. The casing consists of eight steel segments flow-turned with the appropriate flanges. The fore and aft sections are fitted with the igniter and nozzle respectively. The casing sections are joined in pairs by factory joints which are then thermally protected by thick rubber seals, and the inner walls of the casing are protected by insulating material to which the propellant will be bonded. The propellant is mixed and cast into each pair of segments, with a mandrel of the appropriate shape to form the hollow core. The booster pairs are thrust-matched by filling the appropriate segments of each pair together, from the same batch of propellant. Insulating material and inhibitor are applied to the faces, which are not

Table 4.1. Two modern solid boosters.

	SRB (Space Shuttle)	MPS (Ariane 5)
Thrust (individual)	10.89 MN	5.87 MN
Thrust (fractional)	71% at lift-off	90% at lift-off
Expansion ratio	11.3	≈10
Exhaust velocity	2,690 m s^{-1}	2,690 m s^{-1}
Temperature	3,450	3,600
Pressure	65 bar	60 bar
Total mass	591 T	267 T
Propellant mass	500 T	237 T
Dry mass	87.3 T	30 T
Burn-time	124 s	123 s
Charge shape		
Upper section	11 point cog	23 point cog
Middle section	Truncated cone	Cylinder
Lower section	Truncated cone	Truncated cone
Propellant	Ammonium perchlorate 69.6%	Ammonium perchlorate 68%
	Aluminium powder 16%	Aluminium powder 18%
	Polymeric binder 14%	Polymeric binder 14%
	Additive iron oxide 0.4%	
Factory joints	4	6
Field joints	3	2
Length	45.4 m	27 m
Diameter	3.7 m	3.0 m
Casing	Steel 12 mm	4SCDN-4-10 steel

to burn. The filled sections are then transported to the launch site where they are assembled, using the field joints, into the complete booster.

The nozzle is made of layers of glass and carbon fibre material bonded together to form a tough composite structure which can survive temperatures up to 4,200 K. This composite is then bonded onto the inside of the steel outer cone, which provides the structural support. Rings attached to the cone provide the anchorage for the hydraulic actuators. Inboard is the flexible joint which allows the nozzle to be tilted.

In addition to the propellant in its casing, each booster has a redundant hydraulic system to displace the nozzles by ±8°. There are two actuators, one for each orthogonal direction of displacement. The nose cap of the booster also contains avionics and recovery beacons.

The casing segments are reused. After recovery from the sea the casings are cleaned, inspected and pressure tested to ensure they are sound. Some are discarded because of damage, which is mostly caused by impact with the ocean. The segments are then dimensionally matched, as the combustion chamber pressure can permanently increase the diameter by several fractions of a millimetre. Once cleared for reuse they are refilled with propellant. The sections are rated for 20 re-uses.

The propellant core shapes are intended to produce a 'sway-backed' thrust curve rather than a constant thrust. This produces a period of lower thrust some 50 seconds into the flight, while the vehicle is passing through the region of maximum dynamic pressure (see Chapter 5). This is the period when the product of velocity and air pressure is at a maximum and when the possibility of damage by aerodynamic forces is greatest, and the risk is minimised if the thrust is reduced for a short time. The Space Shuttle main engines are also throttled back for this period.

At launch the two boosters provide 71% of the total thrust, effectively forming the first stage. Their short nozzles are adapted for sea-level operation, and the huge mass flow rate—almost five tonnes per second—provides the high thrust necessary for lift-off.

4.7.2 The Ariane MPS

The MPS (*Moteur à Propergol Solide*) is similar in many respects to the SRB, and about half the size. It has a 3-metre diameter compared with the 5 metres of the SRB. The thrust, at nearly 6 MN, contributes 90% of the Ariane 5 lift-off thrust. The propellant is very similar, with percent level differences in the composition of ammonium perchlorate oxidant and aluminium powder fuel. The binder—a polybutadiene rubber—may well be somewhat different in detail from the Thiokol rubber used for the SRB. The additives are not specified.

The booster consists of seven sections. The forward section contains the igniter and the aft section the nozzle, and both the forward and aft domes are protected with ablative insulating material. The forward section, which has a rough forged bulkhead to contain the pressure, is charged with 23 tonnes of propellant with a cog-shaped inner void, in Europe, prior to shipment to the launch site at Kourou. The remaining segments are charged, at the launch site, with locally manufactured propellant. The middle segment consists of three casing elements pinned together with factory joints. The MPS joints are overlapping joints with transverse pins, and the inner walls are protected with silica and Kevlar fibre insulator (GSM 55 and EG2) before the propellant is cast into the segment. This thermal protection covers the factory joints. The mandrel for the casting produces a shallow truncated cone shape to the grain void. The lower segment is constructed in the same way, and the mandrel here produces a steeper conical form to the void, opening towards the nozzle. The forward grain burns away in the first 15 seconds, while the two lower grains, each of 107 tonnes, burn for 123 seconds. This produces the 'sway-backed' thrust profile which reduces the thrust around maximum dynamic pressure.

The nozzle is made of composite materials incorporating carbon–carbon and phenolic silica materials. It is supported by a lightweight metallic casing, to which the 35-tonne servoactuators are connected by a strong ring. The nozzle can be traversed by ±6° for thrust vector control.

The boosters are recovered after launch, but presently there is no plan for reuse of casing segments. The main purpose of recovery is post-flight inspection of seals and components to ensure that they are functioning correctly throughout the flight.

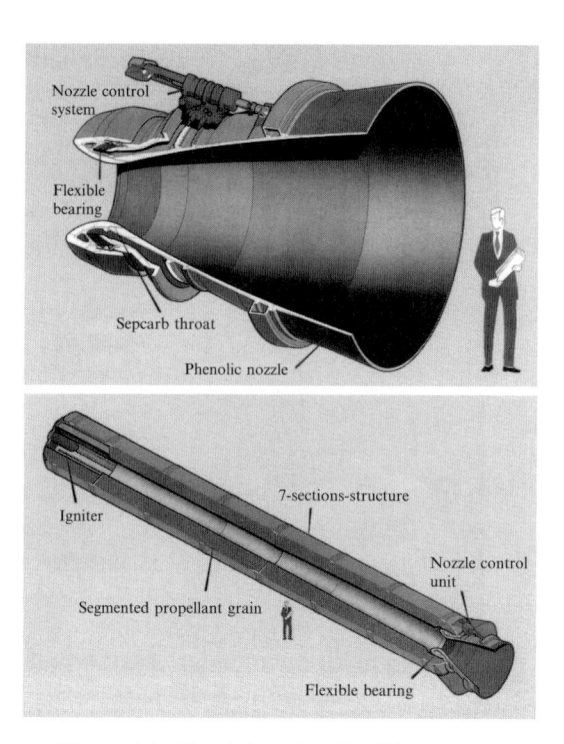

Figure 4.4. The Ariane MPS solid booster.

Post-flight inspection, for example, studies the seals and thermal protection to confirm that it remains intact. Some 10–12 mm of the aft dome protection ablates away during flight, in the region directly adjacent to the nozzle. The throat diameter is 895 mm, and 38 mm of the thermal protection was ablated away during a test firing. This is allowed for in the design, and demonstrates how thermal protection is provided by this technique.

The Space Shuttle and Ariane 5 represent the state-of-the-art heavy launcher capability presently available. The booster technology is mature, as evidenced by the similarity in techniques. Solid propellant boosters of this power represent the best way of increasing the in-orbit payload capability of large expendable launchers. Because of their tough construction and early burn-out, boosters are eminently recoverable, and their reuse is a factor in the economics of launchers. The Space Shuttle has shown the way in reuse, but caution still exercises a strong restraint, as the cost of quality assurance needed to ensure safe reuse is a significant fraction of the cost of new components. The next step in making space more accessible will come from the development of fully reusable launchers and the single stage to orbit. For these, liquid propellants are appropriate, but it is likely that hybrids and intermediate developments will still utilise solid propellant motors. Small launchers make considerable use of all solid propellant propulsion, and this is a market where cost and reliability indicate its continued use.

5

Launch vehicle dynamics

The launch of a spacecraft is fundamental to all space activity, and it is through our development of efficient launch vehicles that the present impact of space on many aspects of science, commerce, and daily life is possible. The launch lasts only a few minutes, and yet during this short period of time, many years of development and investment in the commercial use of the spacecraft can be brought to nothing if just one of the many thousands of components of the launcher fails to perform to specification.

In this chapter we consider the dynamics of launch vehicles, the forces to which they are exposed, and the general nature of the launch process. The multi-stage launch vehicle enables relatively straightforward access to space, and we shall examine some typical launch trajectories and sequences.

Launches take place in the Earth's atmosphere and gravitational field, and much of what we deal with in this chapter is concerned with the effect of gravitational and aerodynamic forces on the launcher. The simple treatment in Chapter 1 avoided these effects by considering only motion in a vacuum perpendicular to the gravitational field. As we shall see, however, the rocket equation included in Chapter 1 is very useful in considering more complicated situations.

5.1 MORE ON THE ROCKET EQUATION

The rocket equation—more properly called Tsiolkovsky's equation—has a relatively simple derivation. It is based on calculating the acceleration of a rocket vehicle with a mass decreasing continuously due to the expenditure of propellant.

The case we have to consider is that of a rocket vehicle of mass M, expelling combustion products at a rate m, with a constant exhaust velocity v_e. The mass of the vehicle is decreasing at the rate m, and, due to the thrust F, developed by the exhaust, the rocket is accelerating. The rocket equation produces the achieved velocity at any time in terms of the initial and current mass of the rocket.

The thrust developed by the exhaust is represented by

$$F = v_e m$$

where

$$m = \frac{dM}{dt}$$

This is a simple application of Newton's third law to the exhaust gases.

The acceleration of the rocket, under the thrust F, is represented by a second application of Newton's law:

$$\frac{dv}{dt} = \frac{F}{M}$$

Substituting for F, from the first equation,

$$\frac{dv}{dt} = v_e \frac{dM}{dt} \frac{1}{M}$$

Cancelling dt, and rearranging, produces

$$dv = v_e \frac{dM}{M}$$

Integrating the velocity between limits of zero and V, for a mass change from M_0 to M, produces

$$\int_0^V dv = v_e \int_{M_0}^M \frac{dM}{M}$$

The solution is

$$V = v_e \log_e \left(\frac{M_0}{M}\right)$$

This is the rocket equation as met with in Chapter 1, where the ratio of initial to current mass defines the current velocity. It is applicable to any velocity increment when the initial and final masses are correctly defined. The assumption of constant exhaust velocity is valid in the vast majority of real cases. Note that the velocity of the rocket vehicle, at any given instant during the burn, is dependent only on the exhaust velocity and the instantaneous mass ratio; the thrust history does not need to be known. In Chapter 1 the equation was used to represent the velocity at burn-out, but here we see it can be used to represent the velocity at any time, while the rocket is still thrusting. We use the term *burn* to indicate a period of operation of a rocket engine, and *burn-out* to indicate the termination of such a period of operation.

It is worth emphasising here that the mass ratio will often be used in this chapter and in Chapter 8 to measure time into the burn of a rocket. This eases the calculations, and because it is independent of the thrust and mass flow rate of the engine, renders the calculations more universally applicable. In all cases where this is applied the mass flow rate is constant, and so there is a direct proportionality to time.

An example of vehicle velocity as a function of mass ratio, for two different exhaust velocities, is shown in Figure 5.1. Note the very strong dependence of

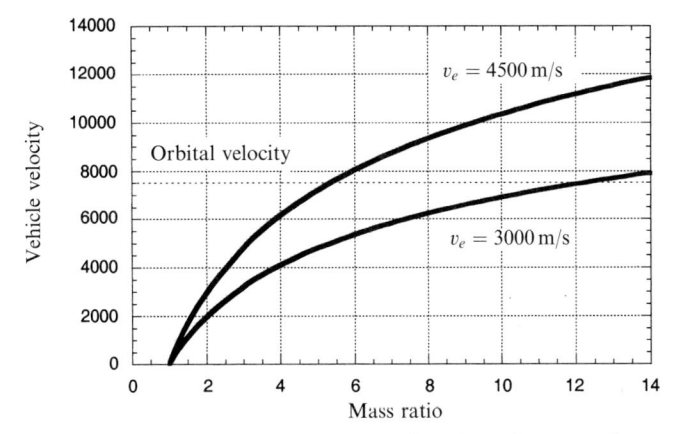

Figure 5.1. Velocity function as a function of mass ratio.

ultimate velocity on the exhaust velocity: for exhaust velocities much below 3,000 m s^{-1} a very high mass ratio is required to reach orbital velocity. Such mass ratios can effectively be achieved by multi-staging. As already noted, the velocity of the rocket vehicle depends *only* on the mass ratio and exhaust velocity. This is true for the situation we have assumed here: the only force acting on the rocket is the thrust developed by the engine; gravity, if it is present, is assumed to act in a direction orthogonal to the direction of the thrust, and therefore to have no effect on the vehicle's acceleration. This situation pertains for orbital manoeuvres, but not in general during space vehicle launches.

5.1.1 Range in the absence of gravity

Range is defined here as the distance travelled along the rocket's trajectory during a burn. It is an important parameter in launch dynamics, particularly when gravity has an effect on the acceleration. It is sometimes useful to know the distance travelled during the burn, even in the absence of gravity—such as during an orbital insertion—or for an interplanetary trajectory injection. In the absence of gravity the range is obtained by integrating the rocket equation over time. The rocket equation expresses the velocity in terms of the mass ratio, which translates into an effective integration over mass. This does not require a knowledge of how the mass flow rate dM/dt varies during the burn.

While the velocity achieved is independent of the mass flow rate, the distance the rocket has to travel in order to reach this velocity is not. It can be seen from the rocket equation that in the limit, if all the propellant were to be expended instantaneously then the final velocity would be the same, but the distance travelled in achieving that velocity would be zero (a shell fired from a gun approximates to this situation). The most common case is where the mass flow rate, and hence the thrust, are constant, and we shall make this assumption here.

The integration has to be over time, so we need an expression for time t in terms of mass loss. For constant thrust this is

$$t = \left(\frac{M_0 - M}{m}\right)$$

where the (constant) mass flow rate m is

$$m = dM/dt$$

Dividing through by M_0 produces

$$t = \frac{M_0}{m}\left(1 - \frac{M}{M_0}\right)$$

This expression for the time is conveniently in terms of the mass ratio, M_0/M.

The distance travelled by the rocket is determined by integrating the velocity, which of course varies with time. This is simply integration of the rocket equation. The distance travelled, s, is expressed by

$$s = \int_0^t V(t)\, dt = \int_0^t v_e \log_e\left(\frac{M_0}{M_0 - mt}\right) dt$$

Evaluation of the integral from time zero to time t leads to

$$s = v_e \frac{M_0}{m}\left[\frac{M_0 - mt}{M_0}\left(\log_e \frac{M_0 - mt}{M_0} - 1\right)\right]$$

Substitution of the expression for t, derived above, produces

$$s = v_e \frac{M_0}{m}\left[1 - \frac{M}{M_0}\left(\log_e \frac{M_0}{M} + 1\right)\right]$$

Therefore, for the distance travelled we have an expression involving the familiar mass ratio M_0/M; but as foreseen, the value of s also depends on the exhaust velocity v_e, and inversely on the mass flow rate m; a high mass flow (a high thrust) leads to a short range. This is intuitive; if all the propellant is exhausted very quickly then the distance travelled will be short. The velocity depends only on v_e and M_0/M, and the range is the distance travelled while reaching this velocity. The above arguments apply only to the range under power; after burn-out the range will continue to increase, but of course the velocity will remain constant in the absence of gravitational effects. Figure 5.2 shows the range as a function of mass ratio for two exhaust velocities. It is important to remember that the functional relationship with mass ratio can apply to any instant during the burn of the rocket, not simply at the moment of burn-out.

The above expression is for zero initial velocity. If the spacecraft already has some velocity—from a previous stage, for instance—then this has to be included in the calculation of the total distance travelled during the burn. The additional distance is

$$\therefore \frac{dS}{dx} = v_e \, \rlap{/}{k} M_o \, \log_e \frac{\bar{x}}{\bar{x}^2} \cdot \frac{d\bar{x}}{\bar{x}^2} \qquad \text{Int of} \qquad \log \frac{x}{x^2}$$

$$\int_{\bar{x}_1}^{\bar{x}_2} \frac{\log x}{x^2} = -\frac{\log(x)}{(2-1)x^{2-1}} - \frac{1}{(2-1)^2 x^{2-1}}$$

$$= -\frac{\log \bar{x}}{\bar{x}} - \frac{1}{\bar{x}}$$

$$\therefore S = v_o \rlap{/}{m} M_o \left[-\log M_o \cdot \frac{\rlap{/}{m}}{M_o} - \frac{\rlap{/}{m} M}{M_o} \right]_{\bar{x}_1}^{\bar{x}_2}$$

$$v_e \, m \, M_o \left\{ 1 - \frac{M}{M_o} - \frac{M}{M_o} \log \frac{M_o}{M} + 1 \right\}$$

$$\therefore v_e M_o \left\{ 1 - \frac{M}{M_o} \left(1 + \log \frac{M_o}{M} \right) \right\}$$

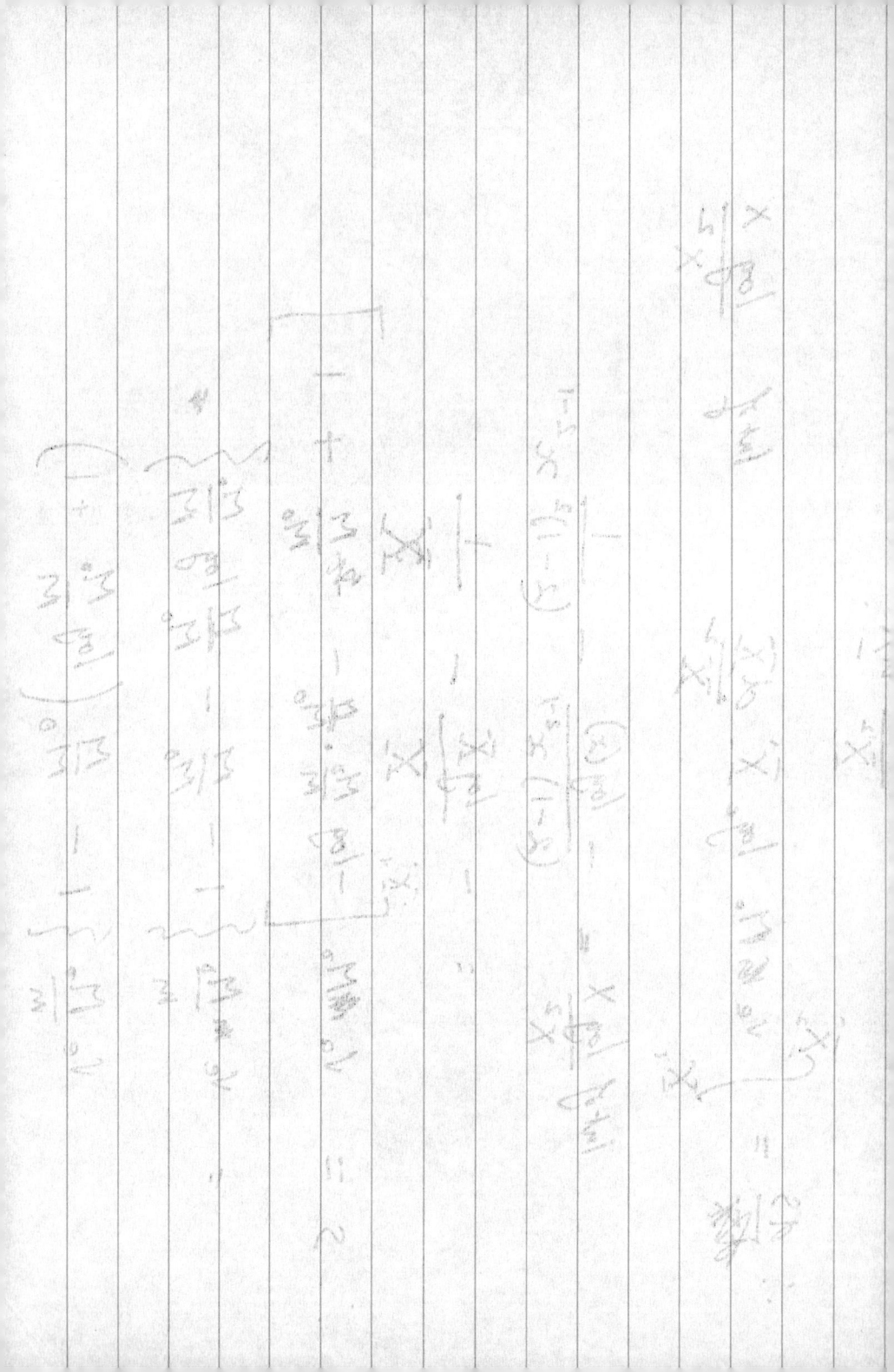

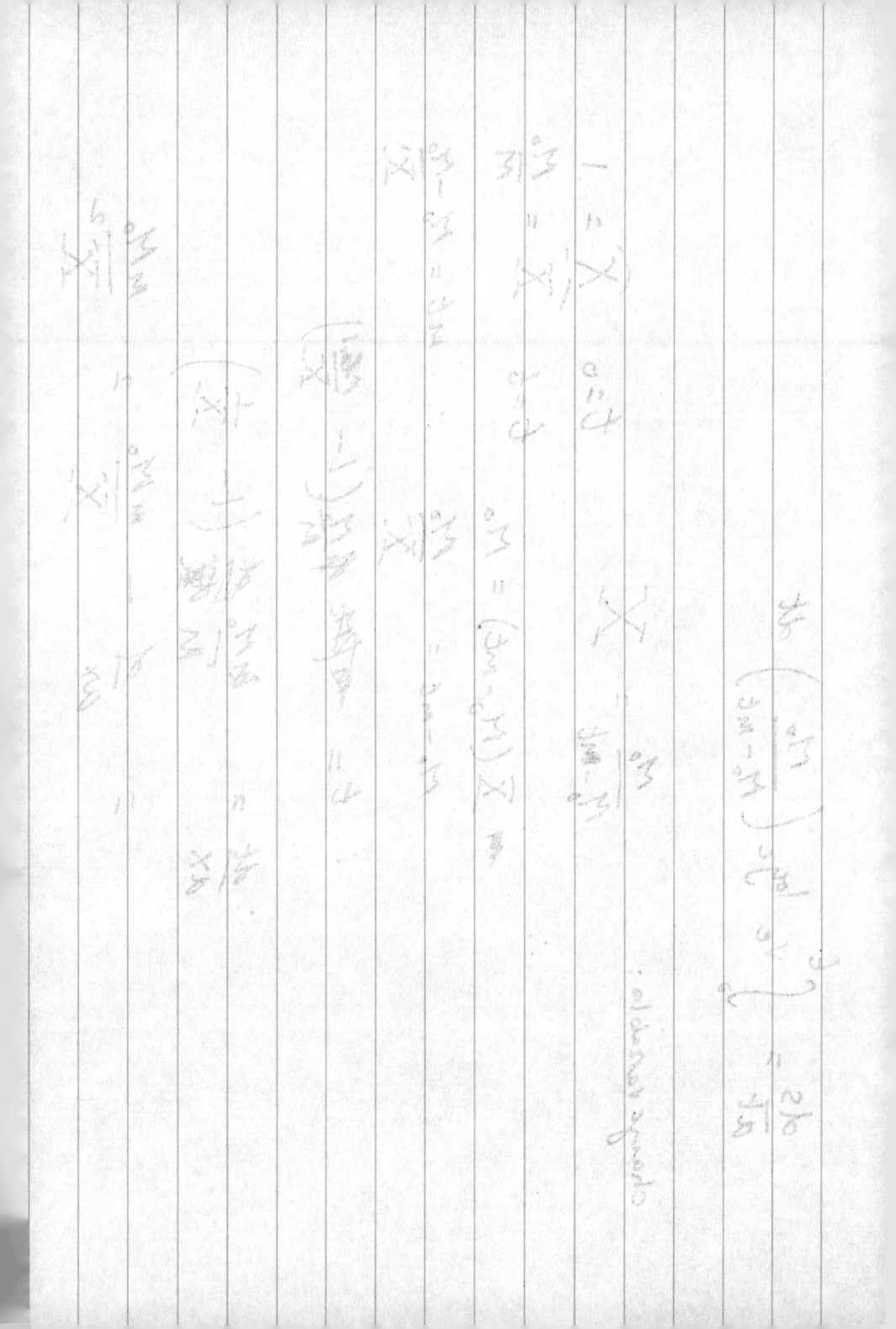

$$\frac{ds}{dt} = \int_0^t v_e \log_e \left(\frac{M_0}{M_0 - mt}\right) dt$$

change variable:

$$\frac{M_0}{M_0 - mt} = X$$

$$\therefore$$

$$\frac{1}{X}(M_0 - mt) = M_0$$

$$M_0 - mt = \frac{M_0}{X}$$

$$\therefore mt = M_0 - \frac{M_0}{X}$$

$t = 0 \quad \frac{X}{1} = 1$
$\frac{1}{X} = 1$

$t = t \quad \frac{1}{X} = \frac{M_0}{M}$

$$\therefore t = \frac{M_0}{m}\left(1 - \frac{1}{X}\right)$$

$$\therefore \frac{dt}{dx} = \frac{M_0}{m} \frac{d}{dx}\left(1 - \frac{1}{X}\right)$$

$$= \frac{M_0}{m} \cdot \frac{1}{X^2} = \frac{mM_0}{X^2}$$

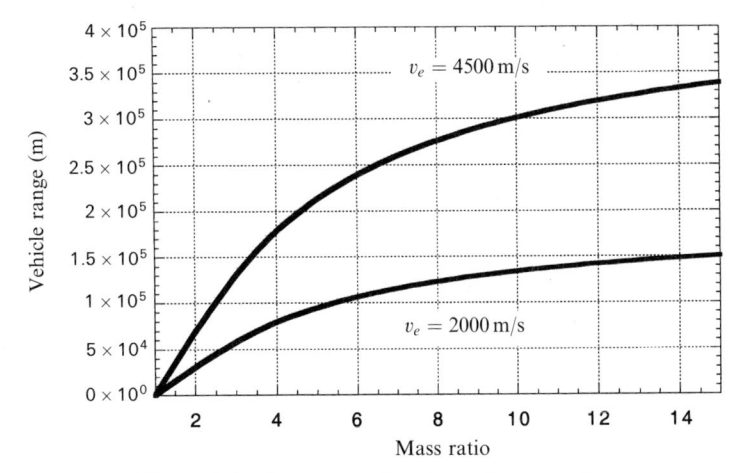

Figure 5.2. Range as a function of mass ratio.

just $V_i t$, or expressing t as above:

$$s_i = V_i \frac{M_0}{m}\left(1 - \frac{M}{M_0}\right)$$

This distance should be added to the distance already calculated.

The above arguments lead to a triad of equations for the velocity, time and range in the absence of gravitational effects:

$$V = v_e \log_e\left(\frac{M_0}{M}\right)$$

$$t = \frac{M_0}{m}\left(1 - \frac{M}{M_0}\right)$$

$$s = v_e \frac{M_0}{m}\left[1 - \frac{M}{M_0}\left(\log_e \frac{M_0}{M} + 1\right)\right] + V_i \frac{M_0}{m}\left(1 - \frac{M}{M_0}\right)$$

These equations can be used for calculations in which gravity plays no part; for example, the injection into orbit of a satellite by the upper stage of a launcher. For orbital injection the thrust and velocity vectors are usually perpendicular to the gravitational field.

As an example of the application of these equations, consider a third stage with an initial horizontal velocity of 2,000 m s^{-1}, containing 4 tonnes of propellant, with a final (empty) mass of 700 kg. This is a typical case for the launch of a small satellite into a 500-km circular orbit. The solid propellant produces a vacuum exhaust velocity of 2,930 m s^{-1}, the mass flow rate is 100 kg s^{-1}, and the mass ratio is 4.7/0.7 = 6.71. Substituting in the above, the velocity increment is 5,560 m s^{-1}, the burn time is 40 seconds, and the range (calculated for zero initial velocity) is 78.5 km. To this must be added the range due to the initial velocity. The total distance travelled during the burn is therefore 158.5 km. Over such a short distance the curvature of the orbit is negligible, and so this calculation is adequate to determine the actual velocity increment. For slower accelerations the curvature of the orbit becomes important, and the third stage would need to be guided to keep the thrust vector perpendicular to the gravity vector, otherwise the orbit would not be circular.

This is a simple case, and for launches from the Earth's surface we need to consider motion at arbitrary angles to the gravitational field and in the presence of the atmosphere. Since most launches begin with vertical flight we shall consider this first, and then examine inclined flight, the effects of the atmosphere, and the gravity turn.

5.2 VERTICAL MOTION IN THE EARTH'S GRAVITATIONAL FIELD

In considering vertical motion, the thrust, velocity and gravitational field vectors are all aligned, and the motion is exclusively one-dimensional. This situation applies in the initial segment of most launches, and is the simplest to treat. As before, we shall derive expressions for the velocity achieved, the distance travelled, and the time, all in terms of the mass ratio.

5.2.1 Vehicle velocity

The above derivation of the rocket equation can easily be adapted to determine the velocity in the presence of gravity. The thrust remains the same, but the acceleration of the rocket is now governed by two forces: the thrust, and the opposing force of gravity.

As before, the thrust developed by the exhaust is represented by

$$F = v_e m \quad \text{where } m = \frac{dM}{dt}$$

The acceleration of the rocket, under the thrust F, and the opposing force of gravity, is represented by

$$\frac{dv}{dt} = \frac{F - Mg}{M}$$

where Mg is the current weight of the rocket. Substituting for F,

$$\frac{dv}{dt} = v_e \frac{dM}{dt} \frac{1}{M} - \frac{Mg}{M}$$

Multiplying through by dt and rearranging produces

$$dv = v_e \frac{dM}{M} - g\,dt$$

Integration between limits of zero and V, the vehicle velocity, for a mass change from

M_0 to M produces

$$\int_0^V dv = v_e \int_{M_0}^M \frac{dM}{M} - g \int_0^t dt$$

The solution is

$$V = v_e \log_e \frac{M_0}{M} - gt$$

Or, using the expression for t in terms of the mass ratio,

$$V = v_e \log_e \frac{M_0}{M} - g \frac{M_0}{m} \left(1 - \frac{M}{M_0}\right)$$

This expression has two terms: the familiar expression from the rocket equation, and a second term involving the acceleration of gravity. The vehicle velocity is equal to that which would have been obtained in the absence of gravity, minus the acceleration of gravity multiplied by the time—a result which could have been arrived at by intuition.

The first part of the expression—the velocity in the absence of gravity—is sometimes referred to as the 'ideal velocity' and the second part as the 'gravity loss'. While the ideal velocity is always independent of the thrust history and the burn time, the gravity loss is not independent. A very short acceleration, with high thrust and high mass-flow rate, leads to a small gravity loss, while a slow acceleration, with low thrust and low mass-flow rate, leads to a high gravity loss. This corresponds with intuition: if the mass flow rate is high, less of the propellant has to be carried to high altitude, or accelerated to high velocity, before it is burned. Since the raising and acceleration of propellant both reduce the amount of energy available to accelerate the payload, exhausting most of the propellant early in the launch is beneficial. Of course high mass-flow rates imply high thrust, and this may be inconvenient, as we shall see later.

The gravity loss is a very important quantity. As discussed in Chapter 1, gaining sufficient velocity to place a spacecraft in Earth orbit is not easy: it requires rockets working at the limits of available performance, and multi-staging. This is because of the deep potential well of the Earth's gravity. The orbital velocity is high—7.6 km s^{-1} for a 500-km orbit—and the work done in transporting the payload to orbital altitude is also high. The work done in transporting the unburned fuel to high altitude is much larger than the work done on the payload, because of the mass ratio. In the example injection given earlier, the mass of the payload was 700 kg, but the mass of the propellant in the upper stage was 4,000 kg. This unburned propellant has to be accelerated and lifted by the other stages. Even when a stage is firing, the propellant not yet burnt has to be lifted. This is wasted energy. Thus, the optimisation of launch trajectories is largely a matter of reducing the gravity loss. A simple estimate of the gravity loss in terms of vehicle velocity can be derived from burn time. For a single stage, typical burn times are around two minutes; the boosters on Ariane 5 and on the Space Shuttle burn for this long. The loss is therefore just the product of time and the acceleration of gravity. The gravity loss while the boosters are firing is 1,200 m s^{-1}, which is a substantial loss in velocity.

122 Launch vehicle dynamics [Ch. 5

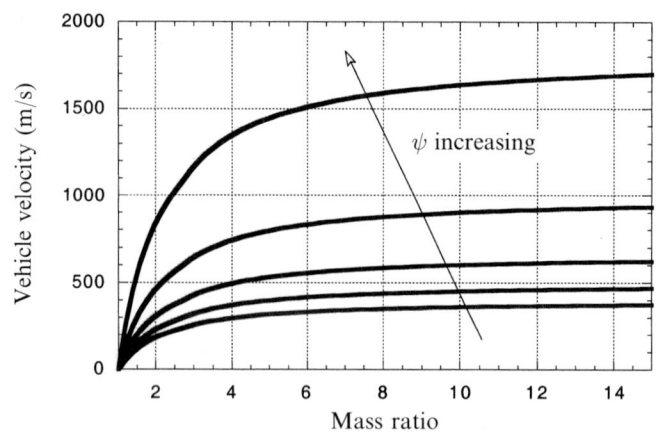

Figure 5.3. Gravity loss: velocity gain and thrust-to-weight ratio.

Referring to the equation for the gravity loss, it is clear that for a fixed mass ratio, and hence fixed ideal velocity, the gravity loss depends on the ratio M_0/m, since m defines the thrust for a fixed v_e. Rocket engineers define a parameter ψ, the 'thrust-to-weight ratio', to refer to the launch or initial conditions of the burn. From the definitions of thrust and weight, ψ is represented by

$$\psi = \frac{F}{gM_0} = \frac{v_e m}{gM_0}$$

Using this substitution the gravity loss is expressed as

$$\frac{v_e}{\psi}\left(1 - \frac{M}{M_0}\right)$$

It can be seen that the gravity loss is governed by the exhaust velocity and the thrust-to-weight ratio. Of course, ψ needs to be greater than unity for the rocket to even leave the launch pad.

The effect of ψ is illustrated in Figure 5.3. There are competing requirements in deciding on an appropriate value for the thrust-to-weight ratio. For maximum velocity, ψ should be large so that the gravity loss is minimised; but this implies high acceleration, which is inappropriate for delicate equipment or manned flight. As we shall see later, large acceleration also implies high velocity, low in the atmosphere, which increases the atmospheric stress on the rocket. A typical value for ψ is about 3.

5.2.2 Range

The range is simply derived, as before, by integration of the velocity expression in the previous section:

$$s = v_e \int_0^t \log_e \frac{M_0}{M} dt - g \int_0^t t\, dt$$

The time t is of course unchanged by the presence of gravity. It depends only on the properties of the rocket, and is represented by the expression derived earlier.

With this information, and by analogy with the previous case, we find

$$s = v_e \frac{M_0}{m}\left(1 - \frac{M}{M_0}\left(\log_e \frac{M_0}{M} + 1\right)\right) - \frac{1}{2}gt^2$$

or substituting $t = M_0/m(1 - M/M_0)$

$$s = v_e \frac{M_0}{m}\left(1 - \frac{M}{M_0}\left(\log_e \frac{M_0}{M} + 1\right)\right) - \frac{1}{2}g\left(\frac{M_0}{m}\right)^2\left(1 - \frac{M}{M_0}\right)^2$$

Comparing this with the non-gravity case, we can again consider the range as the 'ideal range', identical to that in free space, modified by the gravity loss term. As before, the range under power is short when the thrust-to-weight ratio is large.

The equations for velocity, time and range in the presence of gravity are

$$V = v_e \log_e \frac{M_0}{M} - g\frac{M_0}{m}\left(1 - \frac{M}{M_0}\right)$$

$$t = \frac{M_0}{m}\left(1 - \frac{M}{M_0}\right)$$

$$s = v_e \frac{M_0}{m}\left(1 - \frac{M}{M_0}\left(\log_e \frac{M_0}{M} + 1\right)\right) - \frac{1}{2}g\left(\frac{M_0}{m}\right)^2\left(1 - \frac{M}{M_0}\right)^2$$

Note that in all cases, 'range' indicates the distance travelled during acceleration, assuming an initial velocity of zero.

These equations show how the motion of the rocket is altered by gravity when the motion is vertical. This applies to the early stages of most launches, and the effect of gravity can be estimated using these equations. The general effect is that the velocity and the distance travelled are less than would have been predicted by the rocket equation, by the amount of the gravity loss. In launches, the main requirement is to gain horizontal velocity rather than vertical velocity. This is needed to arrive at the necessary orbital velocity. Vertical flight does not contribute to this, and moreover it is very expensive in terms of gravity loss. For this reason, launchers begin to travel horizontally as soon as possible in their flight. In the next section we shall examine the effects of gravity on inclined motion of a rocket vehicle.

5.3 INCLINED MOTION IN A GRAVITATIONAL FIELD

It is obvious that if the whole trajectory of the rocket is vertical, then unless escape velocity is reached the payload will ultimately fall back to Earth. To achieve orbit around a planet requires a high *horizontal* velocity. Thus the majority of the flight path of a launch vehicle is inclined to the gravitational field in order to gain velocity in the horizontal direction. Gravity now affects the direction of flight as well as the magnitude of the velocity. As we shall see, the flight path is curved, even if the thrust

5.3.1 Constant pitch angle

The pitch angle is the angle made by the thrust vector to the horizontal, which in most cases is the same as the angle of inclination of the vehicle axis to the horizontal, since the mean thrust axis coincides with the vehicle axis. This kind of flight is often used during the later stages of the launch because it produces the maximum velocity for a given final injection angle. However, it is not a good flight path for the early parts of the flight, where atmospheric forces are important, and where a pitch angle that varies with time is more desirable.

Since we now have to deal with inclined flight we shall need to consider both vertical and horizontal components of the velocity and distance travelled. The thrust and gravitational force also have to be resolved. The derivation is rather simple because the vertical components are the same as in the previous section, and the horizontal components are unaffected by gravity.

The thrust is independent of the orientation of the rocket:

$$F = v_e m \quad \text{where } m = \frac{dM}{dt}$$

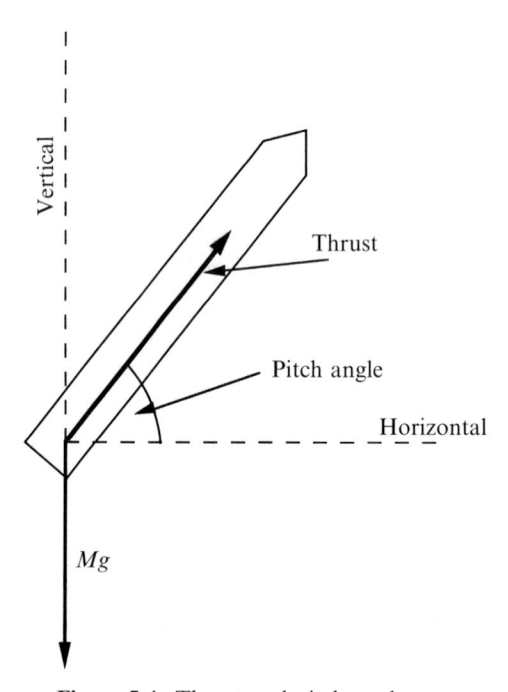

Figure 5.4. Thrust and pitch angle.

The vertical acceleration of the rocket, under the thrust F and the opposing force of gravity, is expressed as

$$\frac{dV_Z}{dt} = \frac{(F \sin\theta - Mg)}{M}$$

where θ is the pitch angle and, as before, Mg is the current weight of the rocket. The thrust is resolved in the vertical direction, and the vertical component of velocity is V_Z. The further steps are identical to those in the previous section, leading to

$$dV_Z = v_e \frac{dM}{M} \sin\theta - g\,dt$$

Integration between limits of zero and V for a mass change from M_0 to M produces

$$\int_0^V dV_Z = v_e \sin\theta \int_M^{M_0} \frac{dM}{M} - g \int_0^t dt$$

and the solution is

$$V_Z = v_e \sin\theta \log_e \frac{M_0}{M} - gt$$

To calculate the horizontal component of velocity V_X, the thrust is resolved on to the horizontal axis, and gravity plays no part, leading to

$$\frac{dV_X}{dt} = \frac{F \cos\theta}{M}$$

Integration as before, over the same limits, leads to

$$V_X = v_e \cos\theta \log_e \frac{M_0}{M}$$

The magnitude of the total velocity V—the speed of the rocket along its direction of motion—can be derived from the triangle of velocities, and is represented by the quadratic sum of the components:

$$V = \sqrt{(V_X^2 + V_Z^2)}$$

Substitution and simplification leads to

$$V = \sqrt{\left(v_e^2 \log_e^2 \frac{M_0}{M} - 2v_e\, gt \sin\theta \log_e \frac{M_0}{M} + g^2 t^2\right)}$$

This is the total velocity along the current velocity vector. The burning time is independent of inclination, and as before

$$t = \frac{M_0}{m}\left(1 - \frac{M}{M_0}\right)$$

Substitution of M = mass at burn-out, would determine the velocity increment from an individual stage if the pitch angle were constant throughout the burn.

At first sight the expression for the velocity appears unfamiliar. The first term under the square root sign is the ideal velocity in the absence of gravity and the second and third terms each contain the gravity and the time, and are clearly related to gravity loss. As the pitch angle approaches 90°, $\sin\theta$ approaches unity. The expression under the square root then approaches the square of the velocity, with gravity loss for vertical flight; which is what would be expected. As the pitch angle approaches zero degrees—horizontal flight—then the expression approaches not the ideal velocity, but the sum of the ideal velocity and the vertical gravity loss term. This is not what one would naively expect. In fact, the term in gt represents the result of the downward force of gravity on the velocity. A rocket projected horizontally from, say, a mountain top, would not reach orbital speed for a long time. Its trajectory would be curved downwards towards the ground, and its speed would *increase* under the acceleration of gravity. In orbital terms the rocket would be in a changing elliptical orbit with apogee at the launch site, and the gt term would represent the conversion of potential energy into kinetic energy which occurs in an elliptical orbit. In fact we have assumed a rectangular set of co-ordinates, with a parallel, not radial, gravity field—this is sufficiently accurate for most (but not all) considerations concerning launch. If the rocket were to be projected horizontally at the correct orbital velocity instantaneously, then it would travel in a circular path, and the gt term would be zero. This is the case considered in Chapter 1.

The gravity loss depends on the pitch angle: the larger the pitch angle, the more thrust is directed to overcoming gravity and is therefore not available to increase the total velocity. This is shown in Figure 5.5, in which the rocket has an exhaust

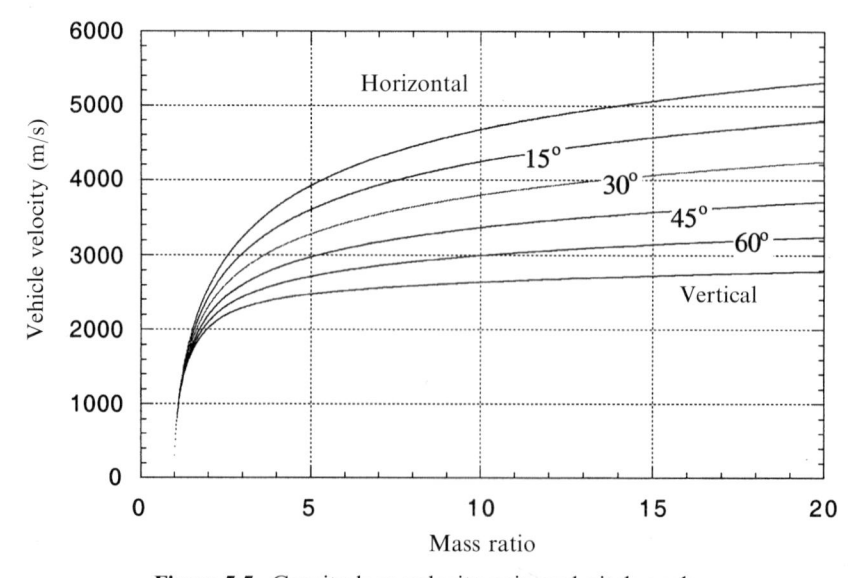

Figure 5.5. Gravity loss: velocity gain and pitch angle.

velocity of 3,000 m/s and travels at a constant pitch angle. There is a very strong pitch angle effect: for a mass ratio of 10—the best that can be expected—the vertical velocity gain is only 57% of the horizontal gain. It is clearly advantageous for a launcher trajectory to move to a small pitch angle as soon as possible, to maximise the velocity gain. There are, however, competing requirements in the early stages of the launch (as we shall see below).

The distance travelled can be calculated by analogy with previous sections. However, it is more tedious and less enlightening than in the one-dimensional case, and is therefore omitted here. It is much more enlightening to look at the flight path.

5.3.2 The flight path at constant pitch angle

It is important to realise that the flight path angle and the pitch angle are not necessarily identical. The pitch angle is the angle of the thrust vector (and the vehicle axis) to the horizontal and the flight path angle is the angle of the velocity vector to the horizontal.

The flight path angle, γ, can be derived using the above expressions for vertical and horizontal velocity, in a triangle of velocities:

$$\tan\gamma = \frac{V_Z}{V_X} = \frac{v_e \sin\theta \log_e(M_0/M) - gt}{v_e \cos\theta \log_e M_0/M}$$

Cancelling produces

$$\tan\gamma = \tan\theta - \frac{gt}{v_e \cos\theta \log_e M_0/M}$$

The second term is always finite, because $M = M_0 - mt$ and is never equal to zero for a practical rocket. So the flight path angle is *always* different from the pitch angle.

This result is of great practical significance, because it indicates that for a constant pitch angle the rocket is forced to travel with its axis inclined to the direction of motion. If we define the angle between the thrust axis and the velocity vector as the *angle of attack*, then this is always non-zero. This difference between the directions of the thrust and velocity vectors is in accord with intuition. Gravity is pulling the rocket down, so some additional vertical thrust is needed to counteract it, which must result in an upward tilt of the vehicle axis from the flight path. Since atmospheric effects on the rocket depend strongly on the angle of attack, constant pitch angle is not the best approach for low in the atmosphere.

The flight path angle for constant pitch angle varies throughout the flight, being at its greatest offset from the vehicle axis immediately after the vehicle axis departs from the vertical. In the limit when $t = 0$, the above equation can be shown to reduce to

$$\tan\gamma = \tan\theta - \frac{gM_0}{v_e m \cos\theta}$$

The initial angle of attack depends inversely on the thrust-to-weight ratio, and is smallest for high thrust-to-weight. Thereafter the angle of attack decreases as the

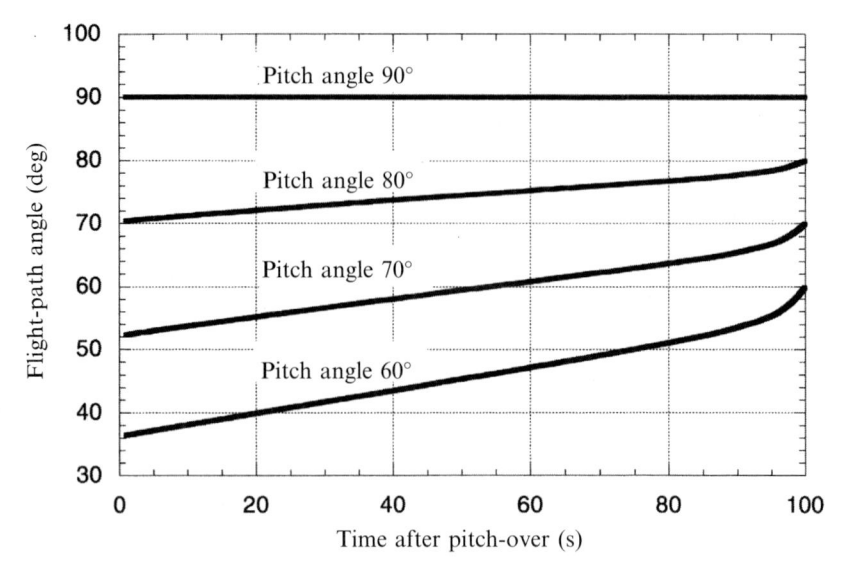

Figure 5.6. Flight path angle as a function of time and pitch angle.

weight decreases, and the thrust, of course, remains constant. Thus, near burn-out the vehicle axis and the flight path are nearly parallel, but they cannot ever be precisely so because of the residual and payload mass. The flight path angle as a function of time for a number of different pitch angles is show in Figure 5.6. The flight path angle changes instantaneously when the thrust axis changes from vertical, and thereafter converges on the pitch angle.

This behaviour is the opposite of that which is desirable: the maximum angle of attack is in the early part of the flight while the rocket is in the denser part of the atmosphere, where aerodynamic forces are greatest. The means for dealing with this problem will be described after aerodynamic effects have been discussed in the next section.

5.4 MOTION IN THE ATMOSPHERE

So far we have described the effects of the gravitational field on the motion of the rocket in some simple circumstances. The atmosphere also has a significant effect on the rocket, which is to be expected, since the velocity quite quickly exceeds the sound speed and the rocket becomes a hypersonic vehicle. Mass ratio arguments require the vehicle to be lightweight, and consequently not well able to withstand the forces so induced. Aerodynamics, especially for hypersonic flight, is a complex subject, but fortunately there are some simple ideas which can be used to estimate the aerodynamic forces on the rocket.

5.4.1 Aerodynamic forces

The motion of the rocket through the atmosphere generates various forces that affect its motion. Discussion of the movements and instabilities which disturb its flight path are beyond the scope of this book, and the effects we shall deal with here are *lift* and *drag*. These both have an impact on the velocity that can be achieved and on the structural integrity of the rocket; they depend strongly on the instantaneous velocity and on the local density of the atmosphere. Figure 5.7 shows how these forces act on the rocket.

The lift is generated by the air flowing over the rocket surface and acts in a direction perpendicular to the flight path of the rocket. The drag is caused by a number of effects, and acts parallel to the flight path and in the opposite direction to the velocity. The transverse force T on the rocket, and the axial retarding force R, are obtained by resolving the lift and drag and adding them, defining the angle of attack as α:

$$T = L\cos\alpha + D\sin\alpha$$

$$R = -L\sin\alpha + D\cos\alpha$$

Here L and D represent the lift and drag respectively, and the minus sign shows that the lift acts as an accelerating force, as shown in the figure. The drag acts as a retarding force, and exists for any angle of attack, including zero. The lift is present only when the angle of attack is non-zero. The magnitude of the lift and drag depend strongly on the velocity, and the form of the dependence is different for velocities below and above the local sound speed. Rockets quickly reach supersonic velocity, in which case the *lift coefficient* is expressed approximately by

$$C_L = 2\alpha$$

For a cylindrical rocket, most of the lift is generated by the nose cone. The drag coefficient is represented by

$$C_D = a + b\mathrm{M}^6$$

for $\mathrm{M} < 1$, and

$$C_D = a + \frac{b}{\mathrm{M}^2}$$

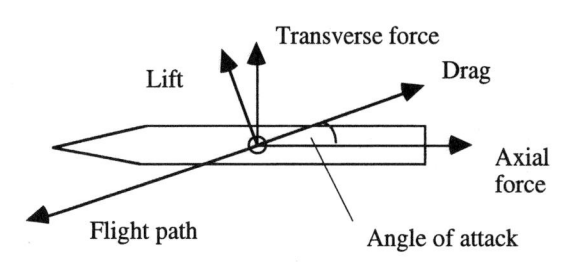

Figure 5.7. The aerodynamic forces acting on a rocket.

130 Launch vehicle dynamics   [Ch. 5

for $M > 1$. Here a and b are constants which depend on α, and M is the *Mach number*—the ratio of the velocity to the local sound speed. The coefficient peaks around the velocity of sound, and typically takes a value of about 0.2.

5.4.2 Dynamic pressure

The lift and drag forces can be expressed in terms of the above coefficients—the velocity V, the atmospheric density ρ, and a *reference area* A. The latter can be regarded as the frontal area of the rocket projected onto the plane perpendicular to the direction of motion. It increases with the angle of attack.

The lift L is expressed as

$$L = C_L A \frac{\rho V^2}{2}$$

and the drag D is represented by

$$D = C_D A \frac{\rho V^2}{2}$$

The quantity $\rho V^2/2$ has the dimensions of pressure and is known as the *dynamic pressure*, often represented as q. Figure 5.8 shows typical profiles of the dynamic pressure, velocity and altitude as a function of instantaneous mass ratio, or time.

Dynamic pressure has two important properties. The V^2 dependence means that the lift and drag, and the disruptive forces on the rocket, increase very rapidly as the rocket accelerates. The effect of drag on first-stage acceleration is quite significant: the acceleration of the vehicle is often almost constant even though the mass is reducing. The dynamic pressure also depends on the atmospheric density, which decreases rapidly as the rocket gains altitude. Thus, with velocity

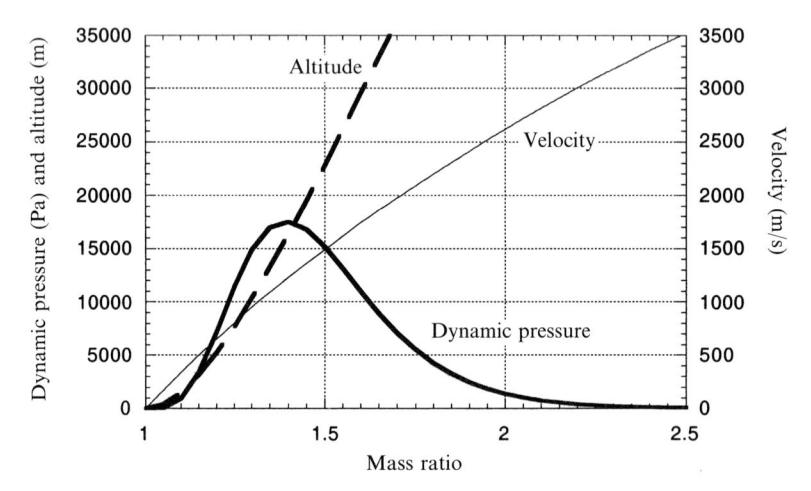

Figure 5.8. Dynamic pressure, velocity and altitude as functions of mass ratio.

increasing, and density decreasing, with time after launch, every rocket passes through a condition known as maximum dynamic pressure, or 'maximum q'. This is the time when the atmospheric forces are at their maximum, and when the risk to the structural integrity of the rocket is greatest. Significant effort is put into reducing this risk as much as possible. For example, the Space Shuttle throttles back the main engines during this period. The booster thrust profile is also tailored to reduce their thrust. A temporary reduction in thrust will reduce the force on the rocket vehicle, but it will continue to accelerate. This approach is possible because the peak in dynamic pressure is so sharp, as a direct result of the exponential reduction in atmospheric density with altitude and the V^2 dependence of the dynamic pressure on velocity.

Reduction in thrust is not always possible, and in any case it does not affect the aerodynamic forces, only the thrust force. It is also advisable to attempt to reduce the aerodynamic forces. A common approach is to hold the angle of attack at zero during the early part of the flight. This not only reduces stress due to lift and drag, but also minimises the drag losses on the rocket when low in the atmosphere, maximising the velocity gain where drag dominates. This means that the rocket velocity vector is axial, so that the area presented to the atmosphere is just the cross section. Because the angle of attack is zero, the lift is also zero. The result is a significant reduction in the aerodynamic forces. This technique results in the rocket following a curved path with the nose gradually dropping, as opposed to the upwardly curved path arising in constant pitch angle flight, and is known as a *gravity turn*.

5.5 THE GRAVITY TURN

In constant pitch angle flight the downward force of gravity is counteracted by an upward thrust component due to the upward tilt of the thrust vector with respect to the velocity vector. If the angle of attack is zero, then the thrust and velocity vectors coincide, and there is no additional upward thrust. The flight path is therefore curved downwards by the influence of gravity—a 'gravity turn'. The gravity turn should be distinguished from the downwardly curved flight path of a ballistic projectile such as a shell, where there is no thrust, and the path depends only on the initial velocity and angle of projection. In the gravity turn a rocket moves under the combined forces of thrust and gravity, and follows a path that is differently curved.

The flight path for a gravity turn has to be computed numerically, but some insight into the motion can be gained through a simplified analysis. The differential equations for the vertical and horizontal motion are the same as for the case of constant pitch:

$$\frac{dV_Z}{dt} = (F - Mg)\frac{\sin\theta}{M}$$

$$\frac{dV_X}{dt} = F\frac{\cos\theta}{M}$$

However, θ is no longer a constant. It is set equal to γ, the flight path angle, which varies and is itself defined from the triangle of velocities by

$$\tan \gamma = \frac{V_Z}{V_X}$$

We must therefore manipulate the differential equations before integrating in order to define expressions for the total velocity V, and for the flight path angle, as a function of time.

Substituting for $\sin \gamma$ and $\cos \gamma$ from the triangle of velocities, the above equations can be written as

$$\frac{dV_Z}{dt} = \frac{F}{M}\frac{V_Z}{V} - g$$

$$\frac{dV_X}{dt} = \frac{F}{M}\frac{V_X}{V}$$

Multiplying the first by V_Z, the second by V_X, and then adding the equations leads, after some manipulation, to an expression for dV/dt, where $V = \sqrt{V_X^2 + V_Z^2}$ is the total velocity along the direction of motion:

$$\frac{dV}{dt} = \frac{F}{M} - g \sin \gamma$$

Multiplying the first by V_X, the second by V_Z, and then subtracting leads in a similar way to an expression for $d\gamma/dt$, where γ is the flight path angle:

$$\frac{d\gamma}{dt} = -\frac{g}{V} \cos \gamma$$

As mentioned above, these equations can only be integrated numerically for the general case. This can be conveniently carried out with a spreadsheet programme. An example calculation of V and γ as functions of time is illustrated in Figure 5.9.

In this particular case the initial flight path angle is 60°, and a gravity turn is followed by setting the flight path and pitch angles equal. Initially the flight path angle changes quickly, but as the velocity increases and the mass decreases the rate of change becomes smaller, and the path stabilises at around 53°.

In the specific case in which $d\gamma/dt$ is a constant (the pitch angle changes uniformly with time) there is an analytical solution. Writing $d\gamma/dt = c$ for the constant pitch rate, the solution is

$$V = V_0 + \frac{g^2}{2} v_e \log_e \frac{M_0}{M}$$

$$\cos \gamma = \cos \gamma_0 + c\frac{g}{2} v_e \log_e \frac{M_0}{M}$$

where V_0 and γ_0 are the initial values of velocity and pitch angle. It can be seen that the pitch angle increases with time, giving the downwardly curved path, and the velocity increases with time, as expected.

The total velocity is increasing with time quite rapidly, but because of the

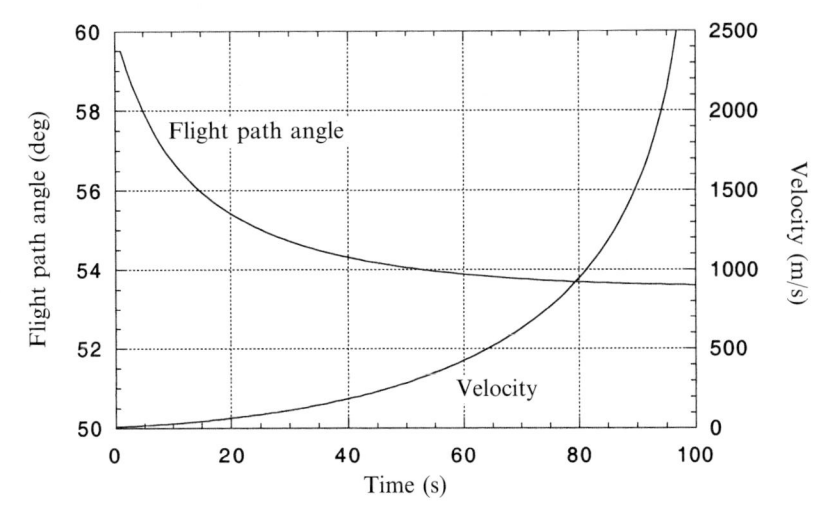

Figure 5.9. Flight path angles and velocity as functions of time for a gravity turn.

downward curve of the flight path, altitude is not gained so quickly. This penalty is justified in the reduced risk to the rocket structure of the zero angle of attack trajectory. Once the maximum dynamic pressure region is passed, a more efficient pitch programme can be followed. Modern launchers use computer controlled pitch programmes to optimise the velocity and altitude achieved during the gravity turn. This reduces the gravity losses while still minimising the effects of dynamic pressure on the rocket.

5.6 BASIC LAUNCH DYNAMICS

The task of a launcher is to deliver a payload to a specific altitude with a specific velocity vector, and the payload must therefore be given both kinetic and potential energy. The potential energy of the payload, is in a sense, a gravity loss, but the actual gravity loss (considered in previous sections) is a real loss of energy from the system, because it results from the potential energy given to propellant which is later burnt and expelled. There is thus a strong argument for burning the propellant early in the flight; that is, for a high thrust-to-weight ratio. The presence of the atmosphere is a counter-argument, since the V^2 dependence of drag on velocity argues for low velocities in the early stages of flight.

5.6.1 Airless bodies

Since there are several factors governing a launch trajectory, it is instructive to distinguish the effects of atmosphere for an initial consideration of complete launch trajectories. Launch from an airless body like the Moon is an instructive problem to examine. In Chapter 1 we saw that the motion of spacecraft is along conic section trajectories, governed by the gravitational field and the angular momentum of the spacecraft about the origin of the field. Neglecting planetary rotation, appropriate for the Moon, a launch is the process of transferring a spacecraft, stationary on the surface, to an orbit above the surface, assumed here to be circular. For minimum energy expenditure the spacecraft should, at the launch site, be given a horizontal velocity sufficient for an elliptical orbit, with perigee at the launch site and apogee at the circular orbit altitude. This is the most efficient way of placing the spacecraft at the altitude of the required circular orbit. The velocity at apogee will be insufficient for a circular orbit, so additional velocity will need to be applied there to circularise the orbit. The reason for the efficiency is that all the acceleration takes place in a horizontal direction, and there is no gravity loss. This idealised approach indicates that for real launches to be efficient, they should as far as possible adopt the same kind of launch trajectory. It is very different from the general perception that rockets are launched vertically.

A purely ballistic launch could take place from the surface, or from a mountain-top, using a modified gun, or, for example, an electromagnetic sled to give the appropriate velocity. In either case the most efficient launch would be horizontal, and sufficient velocity would be given to the vehicle by the sled to put it into an elliptical transfer orbit, with perigee at the launch site and apogee at the desired altitude. If the vehicle had no other motive power, then it would remain in the elliptical orbit, returning to the launch site after one rotation. There must therefore be an onboard rocket engine to circularise the orbit (see Chapter 1).

If a sled or gun is not available, and the launch has to be conventional, then a purely horizontal trajectory is impossible because of the factors mentioned above. The rocket would essentially be in a gravity turn, beginning with its path horizontal, and quickly dropping towards the surface. To counteract this a rocket launch from the surface has to have a vertical component, which could, for example, be a constant pitch trajectory. Immediately this is done, gravity loss comes into play, and optimisation becomes an issue.

The objective is to reach a suitable altitude for injection into orbit, with sufficient horizontal velocity ('horizontal' here meaning orthogonal to the gravitational field vector at the injection point). Energy arguments indicate that the initial orbit should again be elliptical, with a low perigee, and apogee at the altitude of the desired circular orbit. A further burn will be needed to circularise the orbit. For the Moon a perigee at a few tens of kilometres is sufficient, and is just high enough to clear the mountains.

There are two effects of gravity on velocity. The first is the potential energy needed to reach the perigee altitude: of the total velocity increment of the rocket, a fraction will be converted into this potential energy. The second effect is the gravity loss, treated above. The total velocity increment actually achieved will be reduced by the amount of the gravity loss. The first of these is unavoidable, and is part of the dynamical requirement of the orbit; the second can be controlled by appropriate trajectory choice. From our previous arguments, gravity loss is minimised if the pitch angle is small, and the thrust-to-weight ratio is large. The sled launch above is the extreme case in which the pitch angle is zero and the effective thrust-to-weight is infinite. The optimum choice is therefore the lowest perigee, reached by a shallow pitch angle burn, with a high thrust-to-weight ratio. A short burn from the launch pad, with a constant pitch angle, can provide the rocket with sufficient velocity to coast up to the perigee altitude, which is much more efficient from the gravity loss point of view than a slow and continuously powered ascent to the perigee altitude. The pitch angle can be chosen to be the smallest that will allow the perigee altitude to be reached with zero vertical velocity. The horizontal velocity after the burn remains constant, while the vertical velocity decreases to zero as the rocket coasts towards the injection altitude. In this way most of the rocket's energy will be directed into horizontal velocity, making the velocity increment for injection as small as possible. The elliptical orbit injection takes place horizontally with zero gravity loss, as does the final circularisation burn. These burns should use as little propellant as possible, because this propellant has to be carried up to the perigee altitude. The general principle is to burn the propellant as early as possible in the flight with the smallest pitch angle, in order to minimise gravity loss. Once the propellant energy is converted to kinetic energy of the vehicle, it is no longer subject to gravity loss.

The principles developed for the Moon launch apply to a launch from Earth, with the additional complication of the atmosphere and the deep potential well of the Earth.

5.7 TYPICAL EARTH-LAUNCH TRAJECTORIES

Launches from the surface of the Earth are considerably more complicated than those from the surface of the Moon. The gravitational potential well is much deeper, and a multi-stage rocket is required to reach orbital altitude and velocity. The atmosphere causes drag and lift which reduce the acceleration, and are potentially destructive forces. The mass ratio of each stage of the launcher needs to be as high as possible, and this requires the lightest structure. It is considerably easier to design a lightweight structure for a vertical launch than for a horizontal launch. For all these reasons, terrestrial launches begin vertically. It is clear from earlier discussion that ultimately the launcher has to travel horizontally, and so there has to be a transition somewhere in the trajectory.

The lower part of the flight is in the atmosphere. Precautions have to be taken to minimise drag and lift, which define the length of the vertical segment and the nature of the transition from vertical to horizontal. The nature of the trajectory is also different because of the much higher velocity increment needed for Earth orbit. This means a higher mass ratio and therefore a longer burn time for reasonable acceleration. These factors take the trajectory even further from the ideal described above, and greatly increases the gravity loss. Another aspect which may play a part is the capability of stage engines to be restarted. If this is not possible, then the rocket has to be powered continuously into orbit, with further gravity loss. The design of launchers has to take into account these issues, and optimise the trajectory and the launcher itself to maximise the payload capability. This is, of necessity, a very complex process.

5.7.1 The vertical segment of the trajectory

The length of the vertical section varies from vehicle to vehicle. Indeed, the more modern launchers have very short vertical segments; for example, the Space Shuttle just clears the launch tower. A long vertical section reduces stress on the rocket structure by placing the vehicle quickly, and at relatively low velocity, above the densest part of the atmosphere; and, of course, the angle of attack is always zero for vertical flight. The vertical segment, however, does not contribute to the eventual orbital velocity, and the gravity losses associated with transporting most of the propellant, tanks and so on to high altitude are large for vertical flight. The choice is then one of protecting the rocket or losing efficiency. As technology has developed, the length of the vertical segment has tended to decrease, and with modern guidance, engineers can protect the rocket from aerodynamic stress more readily, and take advantage of the increased efficiency of a short vertical segment to launch larger payloads. In the vertical segment, drag rises with the square of the velocity, and limits the acceleration. It does not normally reach its peak value (maximum dynamic pressure) in the vertical segment. This is important in that the densest part of the atmosphere is traversed at a relatively low velocity. The vertical segment can be regarded simply as transportation of the rocket to a region of lower atmospheric density, where the real business of gaining velocity can be carried out under more benign conditions.

5.7.2 The gravity turn or transition trajectory

The gravity turn—or its more modern analogue, the controlled transition trajectory—begins as soon as the rocket departs from its vertical flight, which, with some modern launchers, can be very early. Whatever the precise nature of the trajectory, it minimises the stress on the rocket by keeping the angle of attack close to zero, while the vehicle accelerates through the maximum dynamic pressure zone. The reduction in gravity loss from an inclined trajectory, and the continuing reduction in mass, means that the velocity gain is more rapid. Again, for modern launchers the use of advanced guidance and thrust vector control maximises the efficiency while limiting the stress on the rocket. The use of a controllable thrust engine, as on the Space Shuttle, can be included in this programme to lower the thrust as maximum dynamic pressure is encountered. These advanced guidance programmes are not strictly gravity turns, but the same principles apply. This part of the trajectory is truly a transition region, because the velocity is increasing rapidly, and the atmospheric density is still significant. The objective is to gain as much velocity as possible while at the same time minimising the risk to the structural integrity of the spacecraft by active control of the angle of attack and the thrust. The flight path must not be too flat, because it is important to rise above the region of tangible atmosphere during this portion of the flight.

At this stage, the alternative to a curved path would be either to continue the vertical ascent until the atmospheric density is negligible, or to attempt to gain velocity and altitude quickly by using constant pitch flight. The former would produce a very high gravity loss, while the latter would exert potentially destructive transverse loads on the rocket. Gravity loss should not be underrated: the loss during a typical launch can be as much as 20% of the total velocity.

5.7.3 Constant pitch or the vacuum trajectory

As soon as the atmosphere is sufficiently tenuous to allow the angle of attack to increase, the trajectory can be freely chosen to take full advantage of the rocket performance. Everything is then concentrated on maximising the velocity and altitude. The only restraint is on the acceleration, which should not be too great for the payload or for the astronauts. The nature of the trajectory here can be, at its simplest, constant pitch flight, the objective in this case being to burn the propellant as quickly as possible, with the intention being then to coast to the final orbital altitude. This is somewhat similar to the lunar launch situation. There are, however, other factors to be considered. One of these is the argument of perigee for the orbit: in simple terms, the longitude of the injection point. This may require a more complicated trajectory, which we shall consider after a brief review of orbital injection.

5.7.4 Orbital injection

Once the orbital injection point is reached, the final segment comprises horizontal acceleration to orbital velocity. This is in many cases the simplest segment, since there are no gravitational or aerodynamic disturbing forces: the objective is simply to reach the necessary horizontal velocity. A variety of orbits can be created, depending on the velocity specified (see Chapter 1). If the velocity is equal to $\sqrt{GM_\oplus/r_\oplus + h}$, then the orbit is circular. If the velocity is higher than this, then a variety of elliptical orbits, with the injection point as perigee, can be generated—even a parabolic or escape orbit if the velocity is as high as $\sqrt{2GM_\oplus/r_\oplus + h}$. If the velocity is lower than the circular velocity, then elliptical orbits with the injection point as apogee are possible. Usually the injection point is chosen to be the lowest point of the orbit and to be just high enough to prevent atmospheric drag from causing rapid decay. Orbits with injection point as apogee are therefore usually short-lived and the result of an injection error or motor failure.

In many cases this injection is not final, and the orbit is called a 'parking orbit', which is to allow the final orbit to have different properties. The most common use is to create an orbit with a different argument of perigee; that is, the location of the lowest and highest points of the orbit. Communication satellites, for example, need to be placed in a circular orbit with a 24-hour period, above a particular longitude on the Earth's surface. It is usual to use a parking orbit to allow the spacecraft to reach the correct perigee location before inducing another acceleration to create an

138   **Launch vehicle dynamics**   [Ch. 5

elliptical orbit with its apogee at the final position of the satellite. A further burn at apogee then circularises the orbit with the correct period and phase. Similar considerations apply to interplanetary missions, and to missions which require a change of orbital plane.

The vacuum trajectory segment can be varied from the simple constant pitch burn and coast which is suitable for injection into low circular orbits, or for a low perigee parking orbit. To place a satellite efficiently into a transfer orbit, either the upper stage has to be capable of a restart, or a more complex vacuum trajectory has to be followed. The combination of parking orbit and restart is the most flexible, but it increases the risk and complicates the upper stage. If no restart is used, then the upper stage has to reach the correct injection point for the transfer orbit while still under power. In fact, the trajectory has to be tailored so that it culminates in the transfer orbit injection. A long continuous burn has the potential to generate a large gravity loss, and a slow acceleration to the injection point is therefore not desirable. Alternatively, a rapid acceleration would bring the stage to injection velocity in the wrong place. The solution adopted is to use a rapid and roughly constant pitch acceleration towards an apogee *higher* than the desired perigee of injection. This burns propellant quickly, and stores its energy in the initial form of kinetic energy, and then of potential energy, at the temporary apogee. The subsequent direction of thrust is chosen so that the rocket accelerates downwards, under power, towards the correct injection point, gaining velocity both from the thrust and from its decreasing potential energy. The whole trajectory is designed to bring the vehicle flight path angle and the thrust vector horizontal at the injection point, and to be at the correct altitude. This so-called vertical 'dog-leg' manoeuvre minimises gravity loss, and at the same time allows freedom in the injection point of the spacecraft.

The orbital inclination is defined by the latitude of the launch site and the azimuth of the launch direction. It is easy to see that spacecraft can be launched into orbits with inclination higher than, or equal to, the latitude of the site. For example, a launch site at 30° latitude can provide inclinations of 30° and greater. To inject a spacecraft into an orbit of lower inclination than the launch site requires a large additional velocity increment, because the plane of the natural orbit has to be rotated through the necessary angle, which requires a significant change of orbital angular momentum compared with the original angular momentum. This can be accomplished from a parking orbit by an out-of-plane burn, but is more usually carried out by a horizontal dog-leg manoeuvre. The critical point is the latitude of injection. To achieve, say, an equatorial orbit from a non-equatorial launch site, the injection point has to be vertically over the equator. The vehicle begins by being launched into a simple coplanar trajectory, as if it were intended to enter an orbit with an inclination the same as that of the launch site latitude. At some point in the trajectory it is given lateral acceleration, so that the final injection is over, and parallel to, the equator. Projected on the surface of the Earth, this trajectory shows at least one change in direction. Consideration of the triangle of velocities shows that this change in direction requires considerable transverse velocity increment, and that it is best carried out early in the flight when the vehicle velocity is smaller, but it must be done in a vacuum so that there is no stress on the rocket. This results in its being part of the vacuum trajectory. The most common case is the launch of a geostationary satellite, which must be in an orbit coplanar with the equator.

5.8 ACTUAL LAUNCH VEHICLE TRAJECTORIES

There is a variety of launch vehicle trajectories ranging from simple launches with solid-fuelled rockets to those of highly sophisticated vehicles with advanced guidance programmes. The trajectory can also vary for the same vehicle, depending on the purpose. Here we shall consider only a few examples to illustrate the considerations elaborated earlier.

5.8.1 The Mu-3-S-II launcher

This is a simple solid fuelled Japanese launch vehicle, which has been used to place satellites of about 0.5 tonne into a circular 550-km orbit, from the launch range near Uchinoura, in southern Kyushu. It was developed by the Institute of Space and Astronautical Science, to launch scientific satellites, and is now superseded by the M5 rocket (see Appendix 2). The inclination of the orbit is 33°, given by the latitude of the launch site; the right ascension of the ascending node follows from the time of injection. This launcher, now obsolete, is considered here because it illustrates some of the basic trajectory concepts described above.

The Mu-3-S-II launcher is remarkable in that there is no vertical segment; instead, the launcher is guided by a rail on the launch tower into a flight path with an initial 71° inclination. This breaks all the dynamical rules, but is necessary due to the densely populated nature of the Japanese coast. The fishing village of Uchinoura is only a few kilometres away, and the inclination of the flight path is chosen to ensure that even the most dramatic failure will not result in parts of the rocket falling on the land. The rocket has to be robust in order to survive the aerodynamic forces. The Mu is a very conservative design, and is highly reliable, having launched some 20 satellites successfully, with no failures.

In Figure 5.10, the launch sequence is shown in the form of curves for the velocity, acceleration and altitude, as functions of time. The Mu is a simple three-stage solid propellant rocket with two strap-on boosters. The boosters and the main motors all have cog-shaped charge voids to produce roughly constant thrust.

First stage
The main motor and the solid boosters both ignite at time zero, and the boosters burn for 30 seconds, and separate at zero + 40 seconds. The effect can be seen in Figure 5.10 as a dip in the first-stage acceleration curve at about 35 seconds. The dynamic pressure and pitch angle are shown in Figure 5.11. Maximum dynamic pressure occurs at 30 seconds. Note how the pitch angle is raised while this peak is passed, in order to reduce the angle of attack, while the thrust decreases as the boosters burn out. The rocket then pitches over under thrust vector control until the

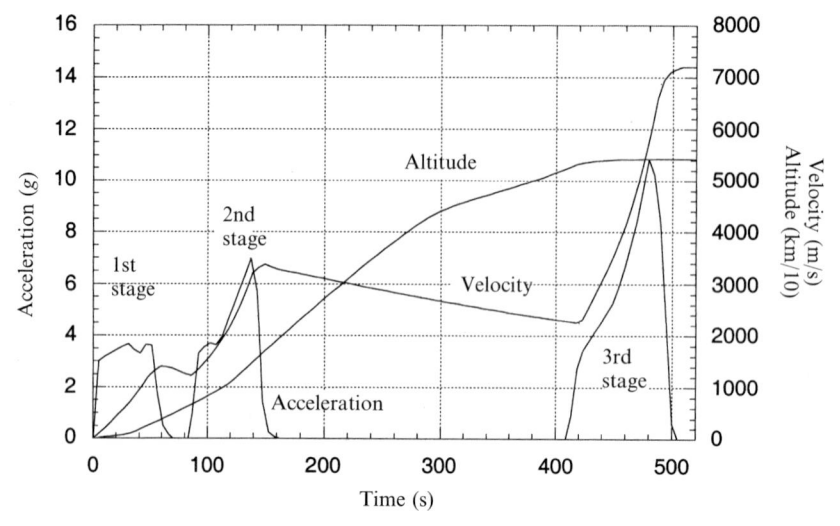

Figure 5.10. Velocity, acceleration and altitude as functions of time.

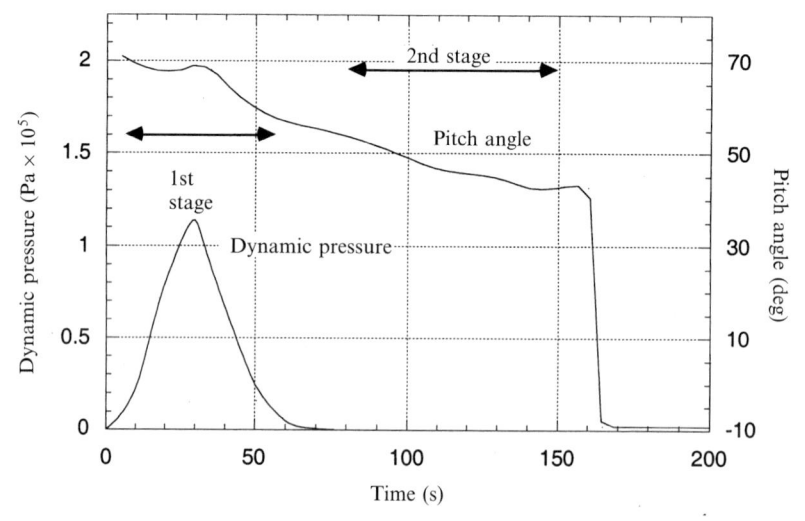

Figure 5.11. Dynamic pressure and pitch angle as functions of time.

burn-out of the main motor of stage 1; the pitch angle is then 58 degrees. Note that the acceleration remains more or less constant during the first stage flight. This is the result of drag (proportional to V^2) in the lower part of the atmosphere, cancelling the effect of the reducing mass of the rocket.

Second stage
After first-stage separation there is a short coast phase, during which the velocity drops. Following second-stage ignition, the acceleration can be seen to increase

Sec. 5.8] **Actual launch vehicle trajectories** 141

rapidly as the mass of the rocket decreases; atmospheric drag is now small. The pitch angle is soon guided to about 45°, and then remains relatively constant. The objective here is to gain maximum velocity and altitude using the second stage. After burn-out the second stage uses small guidance rockets to pitch over until the axis is about 9° *below* the horizontal, which is the correct angle for the final injection into orbit by the third stage, some 1,100 km down-range, allowing for the curvature of the orbit. The second stage then spins up to 60 rpm to stabilise the third stage in this orientation prior to separation.

There is then a long gap in the time sequence before the third stage is ignited. This coast phase is occupied by the rocket in reaching the correct orbital altitude via a ballistic trajectory. Note that the orientation of the rocket is irrelevant during the coast phase. The kinetic energy imparted by the second stage is converted partially into potential energy, as can be seen in the decreasing velocity. This is a good example of the early and efficient burning of the fuel, to avoid the gravity loss associated with carrying unburned fuel to high altitude.

Third stage
At the peak of the ballistic path, where the residual velocity is purely horizontal, the third stage ignites and produces the increase in horizontal velocity necessary to secure the orbit. In another example of efficiency, this stage is unguided and has no thrust vector control; all the guidance is carried out by the second stage before separation. This maximises the mass ratio of the third stage by reducing its dry weight, and enables it to generate the necessary velocity efficiently, maximising the payload mass. The third stage ignites after separation. Its orientation has been set by the second stage, and is maintained by the stabilising effect of the 65 rpm spin. Note the very rapid acceleration; there is no gravity loss in this orientation, and the thrust-to-weight ratio is also high.

Following burn-out, the spacecraft is separated from the third stage, despins, and assumes autonomous existence for the life of the mission. The simple nature of the final injection means that there is some spread in the possible final velocity and direction, mostly from variability in the total velocity increment of the third-stage solid motor. This has to be allowed for in the specification of the orbit.

5.8.2 Ariane 4

The Mu is a very simple and robust design, intended to obtain the best performance from small solid-fuel rocket motors which have low exhaust velocity. This places a corresponding burden on the designer, to maximise efficiency, for small payloads into low Earth orbit (LEO). It serves well to illustrate the basic launcher dynamics discussed above. The Ariane rocket uses powerful liquid-fuelled motors to place large spacecraft in geostationary orbit. The guidance is more sophisticated, and the parallels with the simple dynamics we have discussed are not exact. Figure 5.12 shows the Ariane 4 dynamic parameters. While the Mu weighs about 92 tonnes and launches a payload of 0.5 tonnes into LEO, the Ariane weighs about 470 tonnes and launches a payload of 4.5 tonnes into geostationary transfer orbit (GTO).

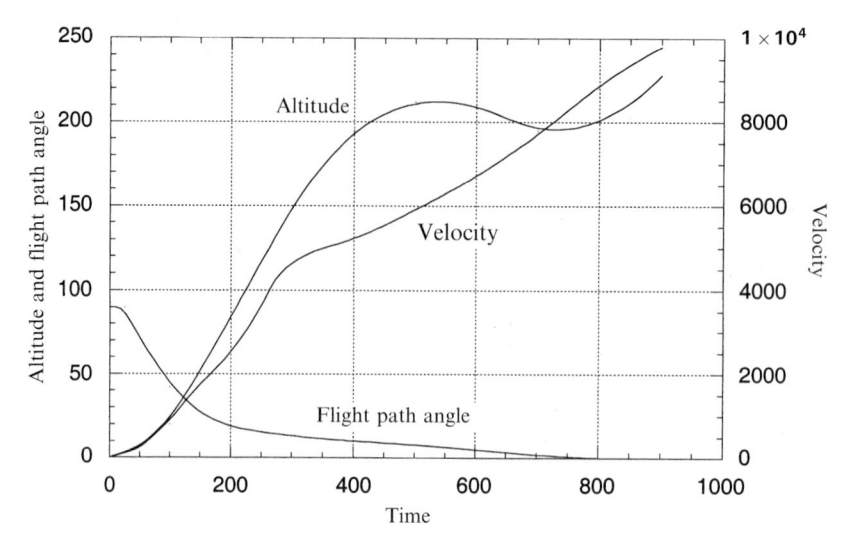

Figure 5.12. Ariane 4 dynamic parameters.

The Ariane 44L—the largest in the Ariane 4 series—has four strap-on liquid-fuelled boosters, a first and second stage using storable liquid propellants, and a third stage with liquid hydrogen and liquid oxygen. The exhaust velocities are much higher than those of the Mu. The Mu launch lasts about 500 seconds and includes a coast phase, while for Ariane the time to injection is about 900 seconds. The Mu has burned both first and second stages during the time required for first-stage burn-out on Ariane, and the acceleration of Ariane is much less that of the Mu. 60 seconds after launch, when the Mu first stage has burned out and the velocity is 1.4 km/s, the Ariane velocity is only 200 m/s. By the time the first stage of Ariane has burnt out, the Mu velocity is 3.4 km/s, while Ariane has reached around 1.2 km/s.

First stage
The first stage burns for 204 seconds, and the boosters for 135 seconds. There is a short vertical segment lasting about 60 seconds, followed by a rapid guided gravity turn during which the pitch angle drops to 20° – much flatter than the Mu. The altitude gain is small because of this, and indeed peak altitude during injection is only 210 km. This results in a considerable saving in gravity loss.

Second stage
The second stage burns for 124 seconds, the pitch angle decreases further while the velocity rises rapidly to 4.7 km/s, and the rocket climbs to 150 km. The fairing is jettisoned at a height of about 90 km.

Third stage
The cryogenic third stage has the task of increasing the vehicle velocity to more than 10 km/s—in order to place the payload in geostationary transfer orbit—an ellipse

Sec. 5.8]   Actual launch vehicle trajectories   143

with apogee at 36,000 km. The acceleration is very slow, and never exceeds 1.75 g. The objective is to build up sufficient velocity to enter GTO. The rocket passes its maximum altitude at about 500 seconds and enters a 'dog-leg' manoeuvre during which some of the potential energy gained by early burning of fuel is converted to kinetic energy. The injection point is at a lower altitude of around 190 km. During the final stages the vehicle is accelerated both by the high exhaust velocity cryogenic engine and by gravity. This whole trajectory is designed to minimise gravity loss and to give maximum kinetic energy during the long burn of the third stage. Since the final orbit will be at geostationary altitude, there is no need for the initial injection point to be high. This approach is closer to the lunar launch strategy outlined above, with a low injection point to minimise gravity loss and potential energy requirements, and with no requirement of restart capability.

5.8.3 Pegasus

A contrasting flight path is adopted by the Pegasus small launcher, which is carried to significant altitude by an aircraft and launched horizontally. This uses the lift of the aircraft wings to gain the initial altitude and so reduce the expenditure of rocket propellant; being above the densest part of the atmosphere reduces drag and dynamic pressure. A further advantage is the ability to launch from anywhere, provided there is a suitable airfield nearby. The Pegasus parameters are illustrated in Figure 5.13.

First stage
The three-stage Pegasus is slung below a Lockheed L1011 aircraft and carried to an altitude of 11.6 km. It is then dropped from the aircraft, and glides for 5 seconds

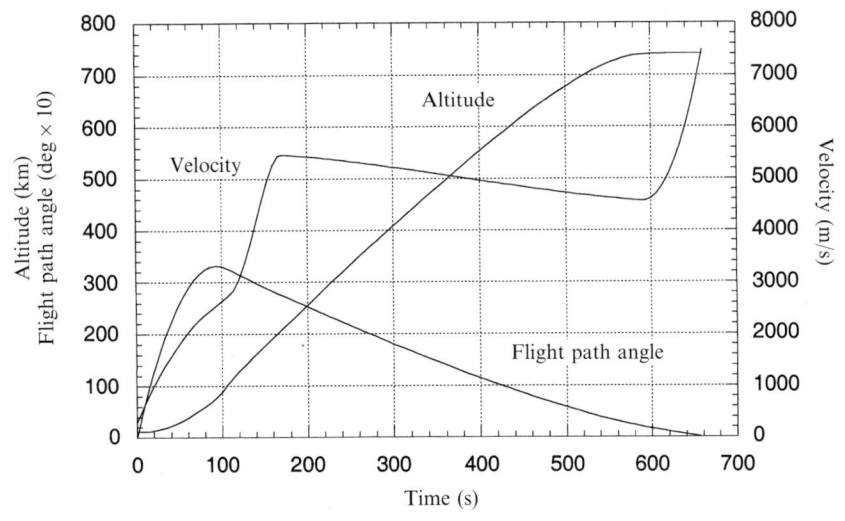

Figure 5.13. Pegasus dynamic parameters.

before first-stage ignition. For the whole of the first-stage burn it gains altitude and velocity as a hypersonic aircraft, passing through the maximum dynamic pressure after about 40 seconds. At this altitude the atmospheric pressure is of course much lower than at sea level, and the dynamic pressure at peak is therefore only 48.8 kPa. The flight path angle is initially a few degrees, and rises, after maximum dynamic pressure, to reach 33° by the end of the first-stage burn. The Pegasus gains velocity rapidly, reaching Mach 7.9 (2,300 m/s) by the time the first stage burns out after 76 seconds.

Second and third stages
Second-stage ignition takes place after 95 seconds, and both velocity and altitude increase until burn-out at 166 seconds. The vehicle then coasts until 594 seconds. During this coast phase, kinetic energy obtained by early burning of fuel is converted into potential energy. As the altitude increases to the required orbital insertion value (in this case, 740 km), the velocity drops from 5,469 m/s to 4,564 m/s. The third stage then ignites to generate the required insertion velocity of 7,487 m/s.

The Pegasus is an example of new launchers exploiting modern technology and filling a niche in the market, in this case for small satellites. The guidance throughout uses sophisticated control algorithms which constantly monitor the velocity and flight path and readjust the pitch angle to ensure optimum insertion. The high-altitude launch reduces dynamic pressure, and the low pitch angle throughout maximises horizontal velocity. Burning the fuel early and coasting to orbital altitude is a further efficiency gain.

Reviewing the other launchers, it is obvious that the first stages are very heavy, and in general burn out quite low in the atmosphere; boosters typically separate at about 30 km. The velocity is not very high because the flight is near vertical and because of atmospheric drag, and for these reasons it is inefficient to use a rocket to reach this altitude. The use of an aircraft is better, because lift is a very much more fuel-efficient way of gaining altitude than is thrust, and to launch Pegasus in this way is therefore a very sensible approach. Neither the velocity nor the altitude at launch are as large as can be achieved with boosters, but the saving is still considerable. Aircraft are, of course, not large enough to launch very big rockets, and can therefore only be used with small launchers.

6

Electric propulsion

Chemical rockets use the energy stored in the propellants to create a hot gas, which then becomes the working fluid in a heat engine, and is expelled, generating thrust. There is an elegant simplicity in this triple function of the propellant and its combustion products, which is reflected in the simple nature of the rocket engine. There is, however, a fundamental limitation which results from combining the functions of working fluid and energy source: no more energy can be put into the rocket than is contained in the propellant flowing into the engine. This means that the power output of the rocket is rigidly defined by the chemical energy and flow rate of the propellant. The exhaust velocity and thrust are defined by the thermodynamic relationships in Chapter 2, and there is no possibility of exceeding these values. As has previously been pointed out, the arrival of the space age was dependent on stretching the ability of chemical rockets to the limit, through multi-staging, and on engines that perform very close to their theoretical best. More ambitious space programmes—manned missions to Mars, for example—could be achieved with the same technology, but would require a very large effective mass ratio because of the velocity increment involved. Moreover, all the necessary propellant would need to be raised to Earth orbit. It would be preferable if somehow more propulsive power could be extracted from the propellant, and the exhaust velocity could be increased beyond the 4.5 km/s that is the best available from chemical rockets.

6.1 THE IMPORTANCE OF EXHAUST VELOCITY

To determine how important a higher exhaust velocity is for future space missions, it is enough to invert the rocket equation. Here we use $R = \dfrac{M_0}{M}$ to represent the mass ratio of the rocket vehicle:

$$V = v_e \log_e R$$
$$R = e^{V/v_e}$$

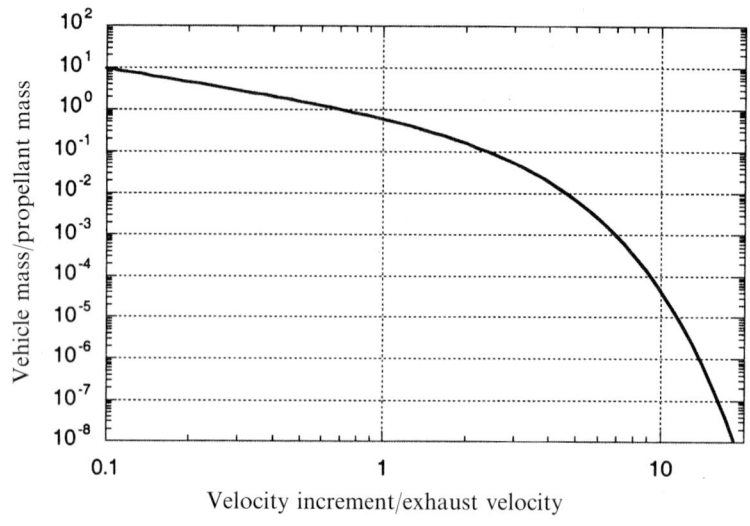

Figure 6.1. Vehicle velocity and payload fraction as a function of exhaust velocity.

Since we are concerned with the available payload, R should be expressed in terms of the ratio of the 'dry' vehicle mass to propellant mass:

$$\frac{1}{R-1} = \frac{1}{e^{V/v_e} - 1}$$

This function is plotted in Figure 6.1.

In Chapter 1 we observed that the final velocity of a rocket vehicle can greatly exceed the exhaust velocity, provided that the mass ratio is high enough. It is immediately obvious from Figure 6.1 that if the ratio of velocity increment to exhaust velocity departs very much from unity, the ratio of propellant mass to vehicle mass quickly becomes unreasonably high. For a velocity increment ten times the exhaust velocity, the ratio is five orders of magnitude. This means that for high velocity increment missions the exhaust velocity 'barrier' of about 4.5 km/s must be broken, and higher exhaust velocities achieved.

The solution to this problem—which has been known for a long time—is to separate the energy input to the engine from the propellant flow. More energy can then be given to a kilogramme of propellant than is available from its chemical reaction. There is then no limit to the achievable exhaust velocity, provided that sufficient power is available and that the engine itself can survive the energy flow. The necessary power can be supplied electrically, or by direct heating from a nuclear fission reactor.

6.2 REVIVED INTEREST IN ELECTRIC PROPULSION

The concept of electric propulsion has been known for a considerable time, and different types of electric thruster have been developed and tested in space. However, they have remained a curiosity until comparatively recent times, when it was realised that the requirement for high velocity increment did not apply only to ambitious space exploration missions, but to station keeping for communications satellites. Over the satellite's lifetime, drift from the correct orbit, induced by solar radiation pressure and gravity gradients, has to be constantly corrected. This requires a significant amount of propellant, the mass of which could be used for more communications equipment, leading to higher profitability. Increased exhaust velocity from the thrusters translates directly into decreased propellant mass.

Once electric thrusters were in commercial use, interest in their use for exploration missions revived, and there are now vehicles—Deep Space 1, having completed a comet rendezvous, and SMART-1 on its way to the Moon—both powered by electric propulsion. The advantages, even for unmanned planetary missions, are significant. The alternative is to carry the extra propellant required for a chemical thruster into Earth orbit with the probe, which has a dramatic effect on the mass ratio at launch, and results in a serious reduction in payload. Use of electric propulsion enables unmanned interplanetary missions, requiring large velocity increments, which would otherwise be difficult using present-day launchers.

6.3 PRINCIPLES OF ELECTRIC PROPULSION

As mentioned above, the basic principle of electric propulsion is to apply electrical energy to the propellant from an external power source. This can be done in several ways. The simplest is to heat the propellant with a hot wire coil, through which an electric current passes. This elementary approach, used in some commercial thrusters, is very successful. More energy can be delivered from the electric current if an arc is struck through the propellant, which generates higher temperatures than the resistive approach and produces a higher exhaust velocity. Finally, electric or magnetic fields can be used directly to accelerate propellant ions to very high velocities, producing the highest exhaust velocity of all. These *ion* thrusters, and *Hall effect* thrusters are seen as the most promising for deep space applications, and they are already coming into commercial use for station keeping and interplanetary propulsion.

While for a chemical rocket the link between energy supply and propellant simplifies analysis, for electrical propulsion the power supply introduces free parameters for which we have to make estimates when deriving expected vehicle performance. Electric power can come from a battery, solar panels or an onboard nuclear or solar generator, each of which has its own advantages and disadvantages. What is important, from the vehicle performance point of view, is the power-to-mass ratio—W/kg. In most cases the power does not diminish with progress through the flight, while the mass of propellant decreases in the familiar way as the vehicle accelerates. This is in direct contrast to the chemical rocket, in which both the propellant and the available energy decrease together. Using these ideas, simple estimates of vehicle performance can be produced.

6.3.1 Electric vehicle performance

The propulsive force developed by an electric thruster has the same physical origin as that developed by a chemical thruster: it is the momentum transferred to the propellant stream, and the rocket equation still applies. However, the mass ratio now has to include the mass of the power supply, and the exhaust velocity depends on the power delivered, the nature of the propellant, and the way that the thruster transfers momentum to the propellant. The simplest approach is to consider the vehicle as having three components: the structure, including payload, propellant tanks and thrusters; the propellant; and the power supply. The thrusters—whether electrothermal or electromagnetic—have a certain efficiency in converting electric power to thrust, and the power supply has a certain power-to-mass ratio. Expressing these efficiencies as η for the thruster efficiency, and ξ (W/kg) for the power-to-mass ratio, the following relationships apply:

$$V = v_e \log_e R$$

$$R = \frac{M_S + M_P + M_E}{M_S + M_E}$$

$$\xi = \frac{P_E}{M_E} \text{ (w/kg)}; \qquad \eta = \frac{mv_e^2}{2P_E}; \qquad m = \frac{M_P}{t}$$

$$v_e = \sqrt{\frac{2\eta P_E}{m}} = \sqrt{\frac{2\eta \xi M_E}{m}} = \sqrt{\frac{2\eta \xi t M_E}{M_P}}$$

$$F = mv_e = \sqrt{2m\eta \xi M_E} = \sqrt{\frac{2\eta \xi M_E M_P}{t}}$$

where the subscripts S, P and E refer to structure, propellant, and electric power supply respectively, P_E is the electric power, and F is the thrust. The power is assumed to be proportional to the mass of the power supply, and the mass flow rate, m, is assumed to be constant. The burn time in seconds is represented by t.

It will be apparent from the above that the exhaust velocity is no longer a free parameter. It is fixed by the power and the mass flow rate, which is in turn related to the burn time and the mass of the propellant. This is easy to understand if we think of the energy carried away per second by the exhaust. This is just $\frac{1}{2}mv_e^2$, where m is the mass flow rate in kg/s. This is equal to the energy per second given to the propellant by the thruster, or the power converted in the thruster. Increasing the exhaust velocity or the mass flow rate requires an increase in the power supplied to the thruster, and a higher mass flow rate leads to a shorter burn time.

The dry mass of the rocket depends on the mass of the power supply, and hence on the power. For the chemical rocket we can choose these parameters independently: the exhaust velocity is defined by the choice of propellant and engine design, and the vehicle velocity then depends on the mass ratio alone. For the electric rocket there is a complex interrelation amongst the design parameters, and so the rocket equation is no longer simple to use. It can be expressed in several forms:

$$V = \sqrt{\frac{2\eta\xi M_E}{m}} \log\left(1 + \frac{M_P}{M_S + M_E}\right)$$

where the power is shown in terms of the power supply mass; or

$$V = v_e \log\left(1 + \frac{2\eta\xi t}{2\eta\xi t \frac{M_S}{M_P} + v_e^2}\right)$$

where the exhaust velocity is the independent variable, using the substitution:

$$\frac{M_E}{M_P} = \frac{v_e^2}{2\eta\xi t}$$

where t is the burn time. The burn time is important because it defines the rate at which propellant is used, and hence the power that has to be applied. For the same onboard mass of propellant, a short burn time requires higher power and a heavier power supply.

6.3.2 Vehicle velocity as a function of exhaust velocity

The last equation for vehicle velocity (above) shows that the mass ratio for a given propellant mass decreases as the exhaust velocity increases, due to the increased power supply mass. Again this is not true for a chemical rocket, in which the exhaust velocity and mass ratio are, in principle, independent. This means that for an electrically propelled vehicle, an increase in exhaust velocity requiring an increase in power, and associated mass of the power supply, could result in no improvement in vehicle velocity, due to the increased mass ratio. Figure 6.2 shows the velocity of the vehicle as a function of the exhaust velocity, assuming a fixed relationship between exhaust velocity and power supply mass, with the burn time as a parameter. The ratio of structural mass to propellant mass is fixed at 0.15, equivalent to a mass ratio of 6.6. In effect, the 'classical' mass ratio is fixed.

Figure 6.2 shows that the vehicle velocity does not increase monotonically with exhaust velocity, and peaks for a certain value. It can also be seen that increasing the burn time increases the peak value, both of the vehicle velocity and the optimal exhaust velocity. The decrease in vehicle velocity beyond a certain point is due to the increasing mass of the power supply and hence a reduction in mass ratio, as mentioned above.

The effect of burn time (recalling that it has no effect on a chemical rocket) is interesting. Given that the 'classical' mass ratio is fixed for this rocket, and that only the mass of the power supply changes, changes in burn time indicate changes in mass flow rate. The exhaust velocity for a given power depends inversely on the mass flow rate, and so low mass flow rates or long burn times are beneficial. Thrust is inversely

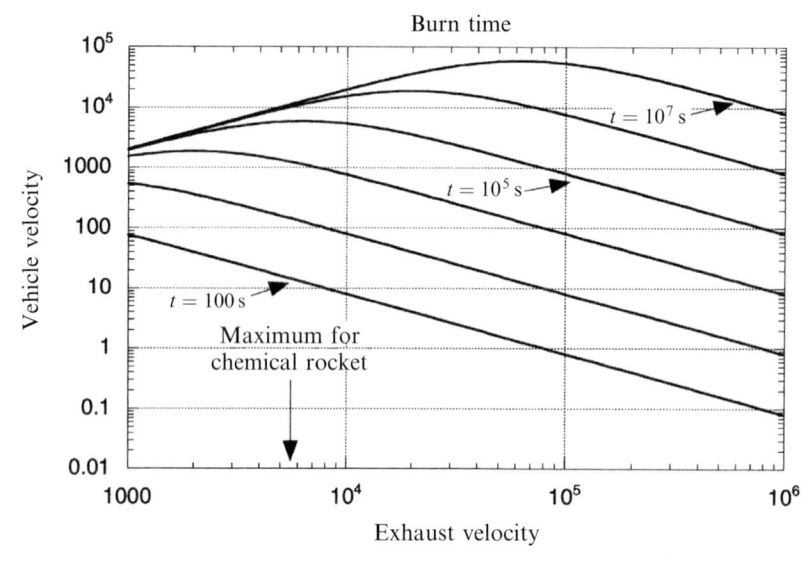

Figure 6.2. Vehicle velocity as a function of exhaust velocity and burn time.

proportional to burn time, and so long burn times and high exhaust velocities imply low thrust. As we consider the different kinds of electric propulsion systems we shall see that all types have very low thrust, but that this is offset by their high exhaust velocity.

6.3.3 Vehicle velocity and structural/propellant mass

Since it is the saving in propellant mass which is the object of using electric thrusters, it is useful to examine vehicle velocity as a function of the ratio of structural (or payload) mass to propellant mass. This is displayed in Figure 6.3, with the burn time fixed at 1 million seconds and the power-to-mass ratio of the supply fixed at a highly optimistic 500 W/kg. The ratio M_S/M_P is shown as a parameter. As might be expected, the vehicle velocity increases as the fraction of propellant increases. It should also be noted that the peak vehicle velocity moves to higher exhaust velocities as the payload mass increases, confirming that high exhaust velocity is advantageous for planetary missions. As an example, a velocity increment of around 6 km/s, which would be needed for a nine-month one-way journey to Mars from LEO, can be achieved with a vehicle-to-propellant mass ratio of 5 if the exhaust velocity is 60 km/s. For a liquid hydrogen–liquid oxygen engine, with an exhaust velocity of 4.5 km/s, the ratio is 0.36.

Given the important part played by the power supply, it is also useful to examine the role of the power-to-mass ratio of the supply. Solar cells can provide up to 200 W kg^{-1}, and a very optimistic value of 500 W/kg has been assumed in the foregoing. Figure 6.3 shows the effect of this ratio on the vehicle velocity, with the burn time again fixed at 1 million seconds, and the payload-to-propellant mass ratio

Colour plates

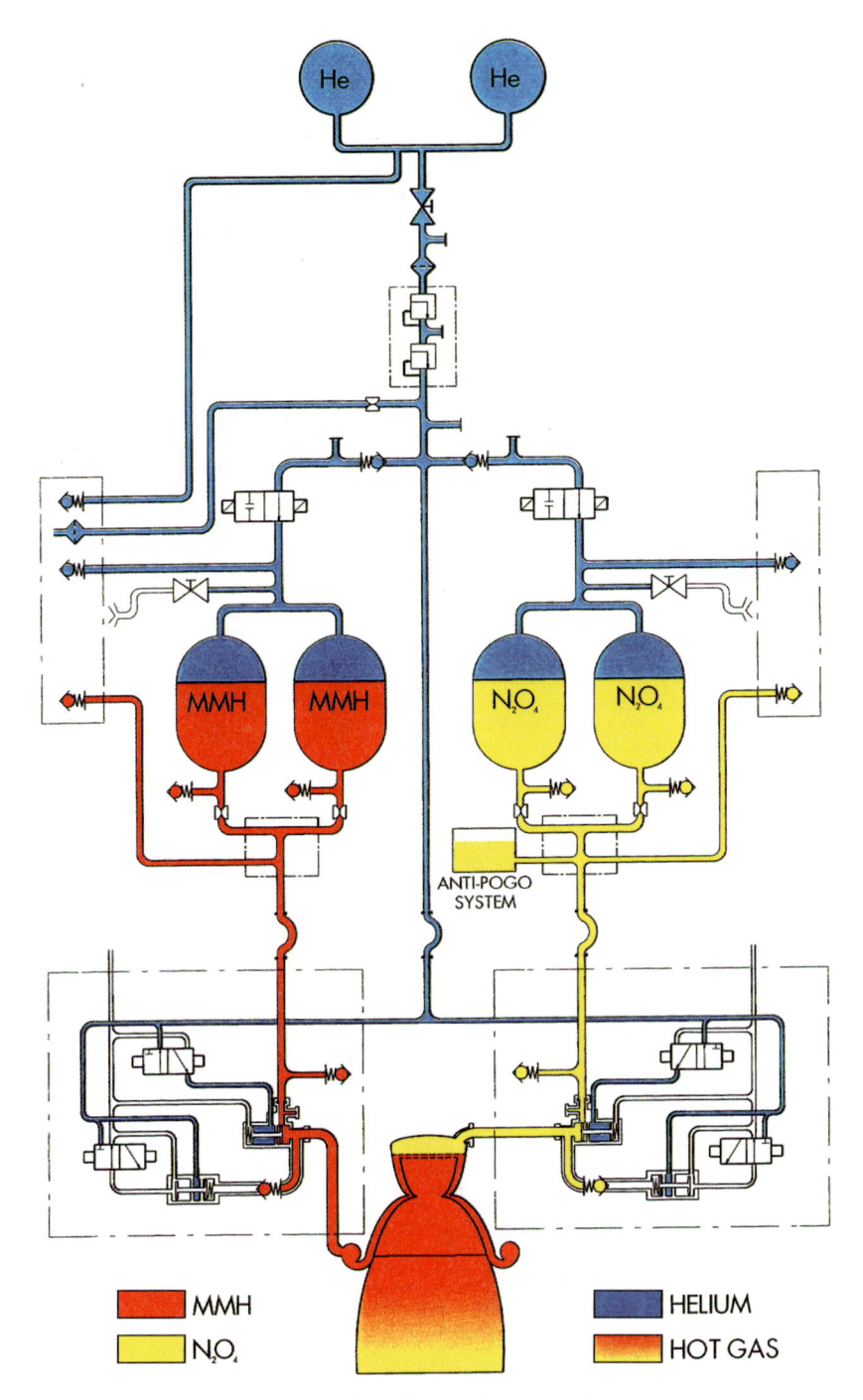

Plate 1 The Ariane 5 Aestus engine.
Courtesy SNECMA.

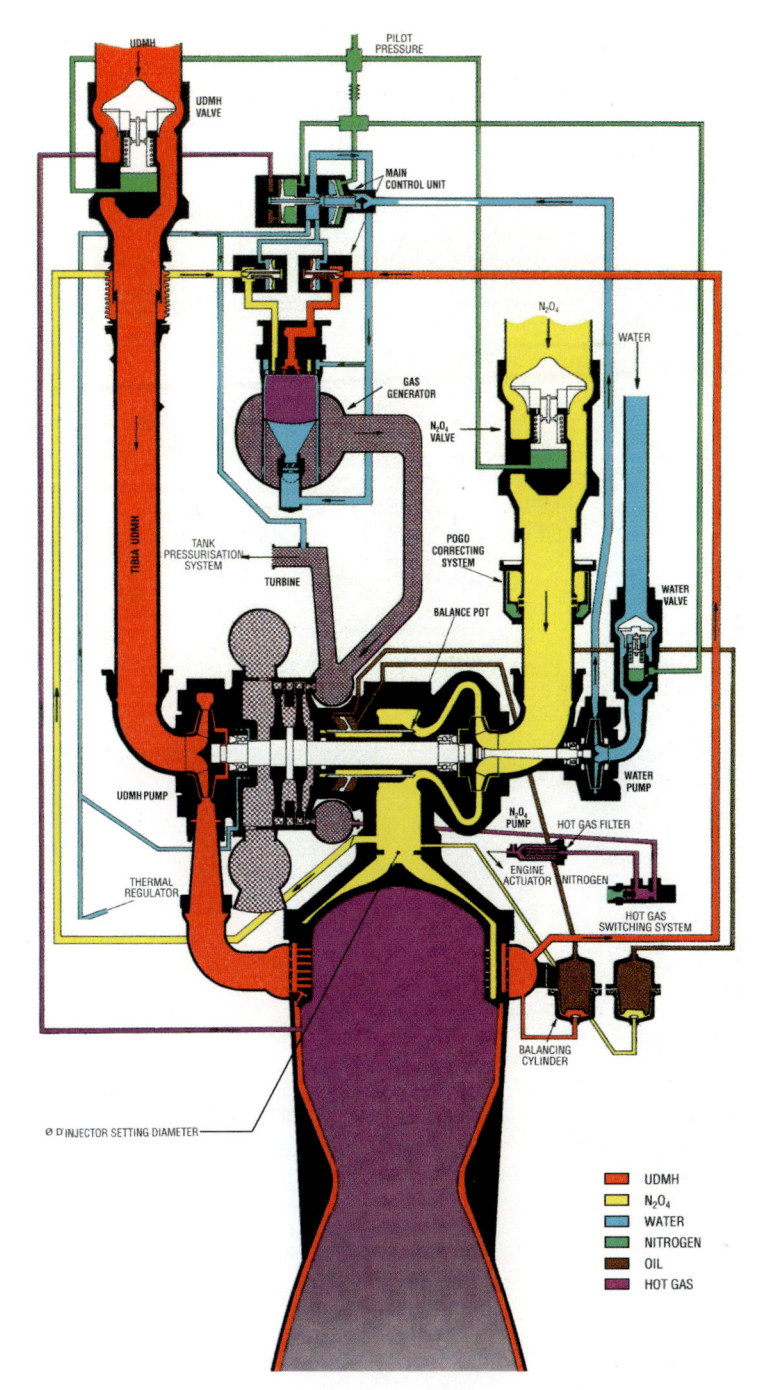

Plate 2 The Ariane Viking engine.
Courtesy SNECMA.

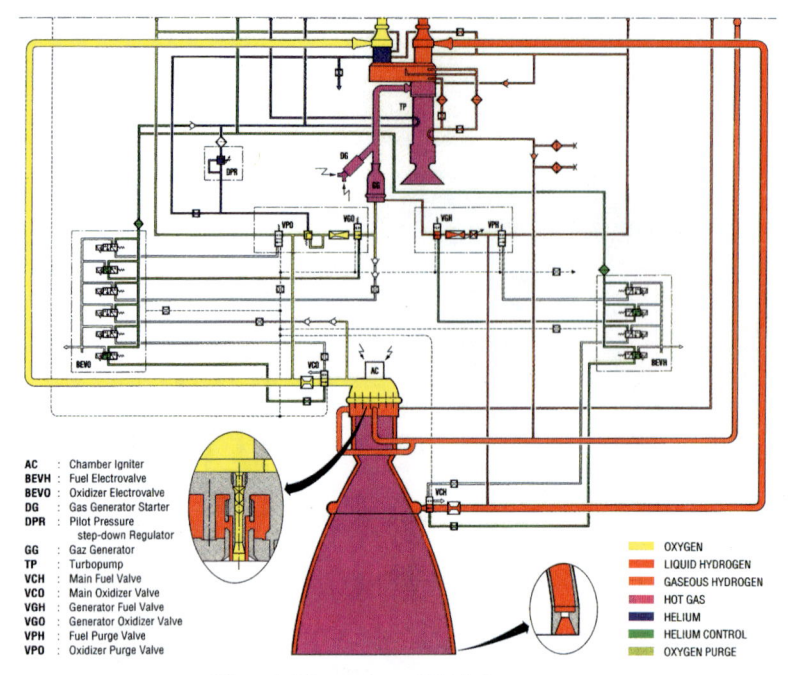

Plate 3 The Ariane HM7 B engine.
Courtesy SNECMA.

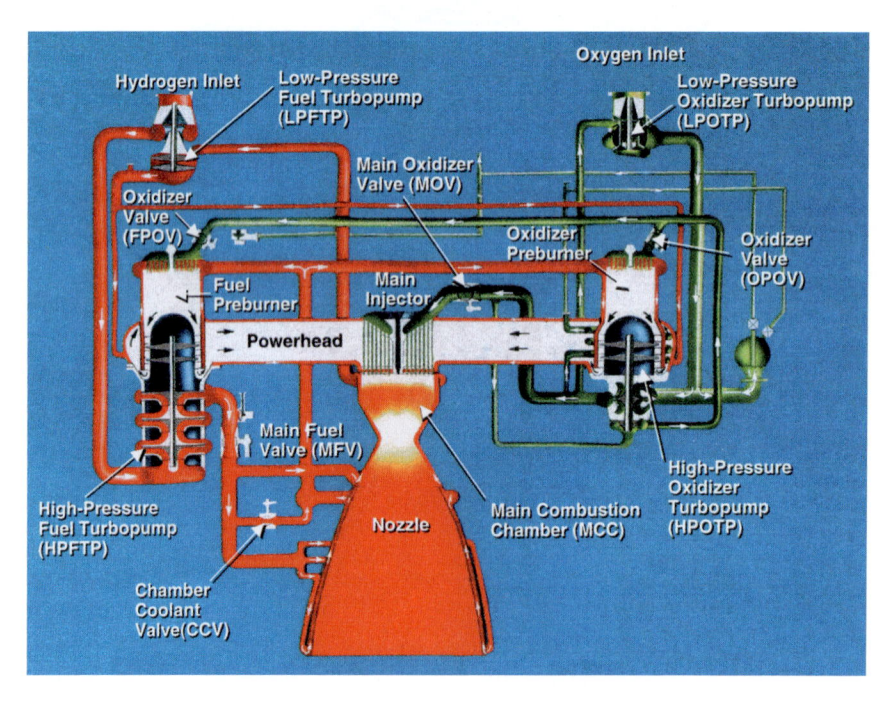

AC : Chamber Igniter
BEVH : Fuel Electrovalve
BEVO : Oxidizer Electrovalve
DG : Gas Generator Starter
DPR : Pilot Pressure step-down Regulator
GG : Gaz Generator
TP : Turbopump
VCH : Main Fuel Valve
VCO : Main Oxidizer Valve
VGH : Generator Fuel Valve
VGO : Generator Oxidizer Valve
VPH : Fuel Purge Valve
VPO : Oxidizer Purge Valve

OXYGEN
LIQUID HYDROGEN
GASEOUS HYDROGEN
HOT GAS
HELIUM
HELIUM CONTROL
OXYGEN PURGE

Plate 4 The Space Shuttle main engine.
Courtesy NASA.

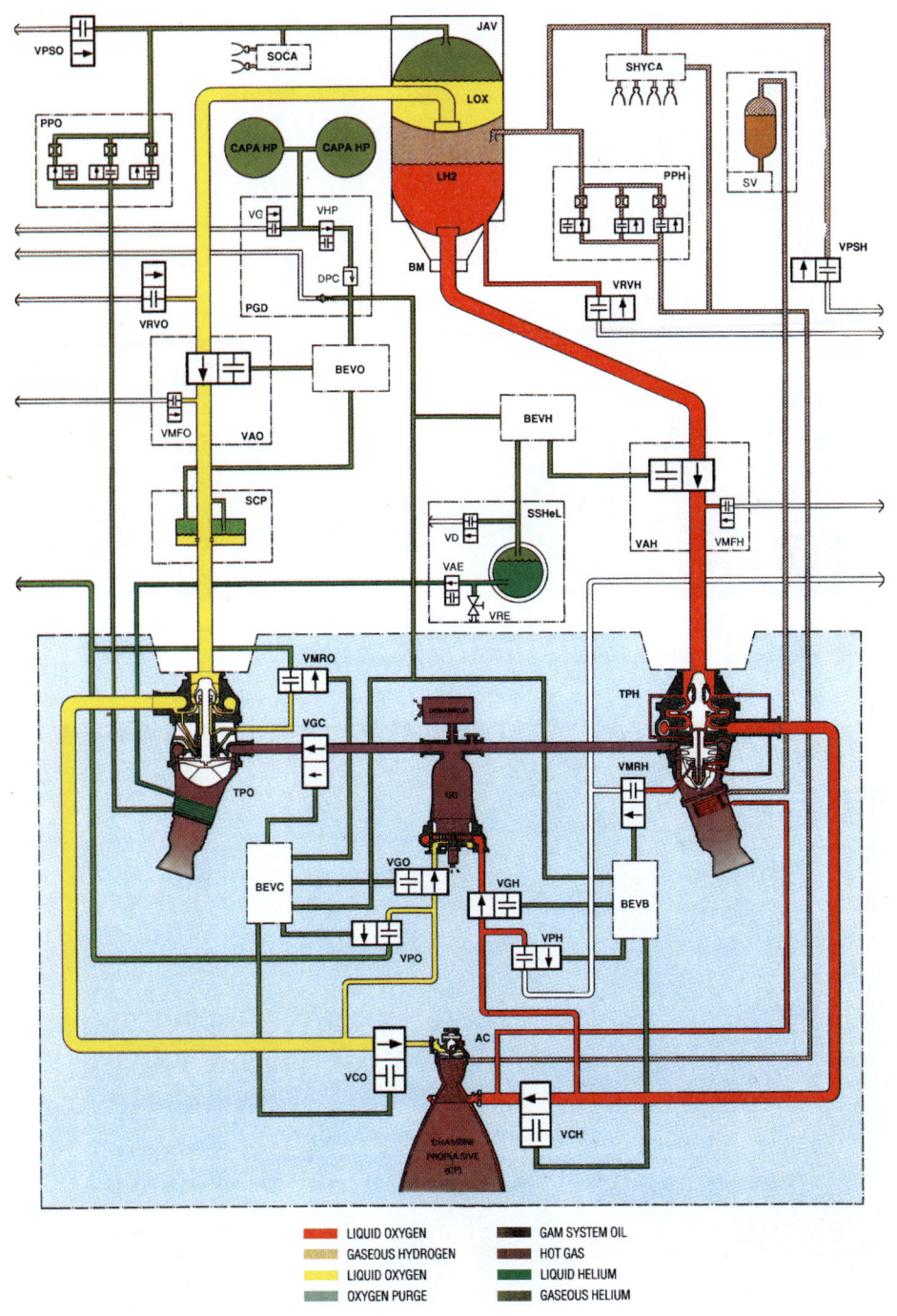

Plate 5 The Ariane 5 Vulcain cryogenic engine.
Courtesy SNECMA.

Plate 6 The launch of Apollo 16 on the Saturn V rocket. The five F1 liquid oxygen–kerosine engines produce a combined thrust of 34 MN to lift the Saturn V off the launch pad.
Courtesy NASA.

Plate 7 Test firing of RL 10 engine.
Courtesy NASA.

Plate 8 Firing test of the Space Shuttle main engine. Foggy cold air drifts downwards cooled by contact with the delivery system for the cryogenic propellants. The inverted cone shape of the exhaust is typical of an engine, designed for vacuum, firing at sea level—it is over-expanded.

Courtesy NASA.

Plate 9 Testing the thrust vector control system on a Space Shuttle main engine while firing.
Courtesy NASA.

Plate 10 Titan IV launcher.
Courtesy NASA.

Plate 11 The NSTAR ion engine mounted on Deep Space 1 prepared for testing in a vacuum.
Courtesy NASA/JPL.

Plate 12 The NSTAR ion engine firing in a vacuum tank.
Courtesy NASA/JPL.

Plate 13 Artists impression of Deep Space 1.
Courtesy NASA.

Plate 14 The PPS 1350 Hall effect engine used for SMART-1, under test.
Courtesy ESA.

Plate 15 Artists impression of the SMART-1 spacecraft on its way to the Moon.
Courtesy ESA.

Plate 16 Experimental ion propulsion system under test.
Courtesy NASA.

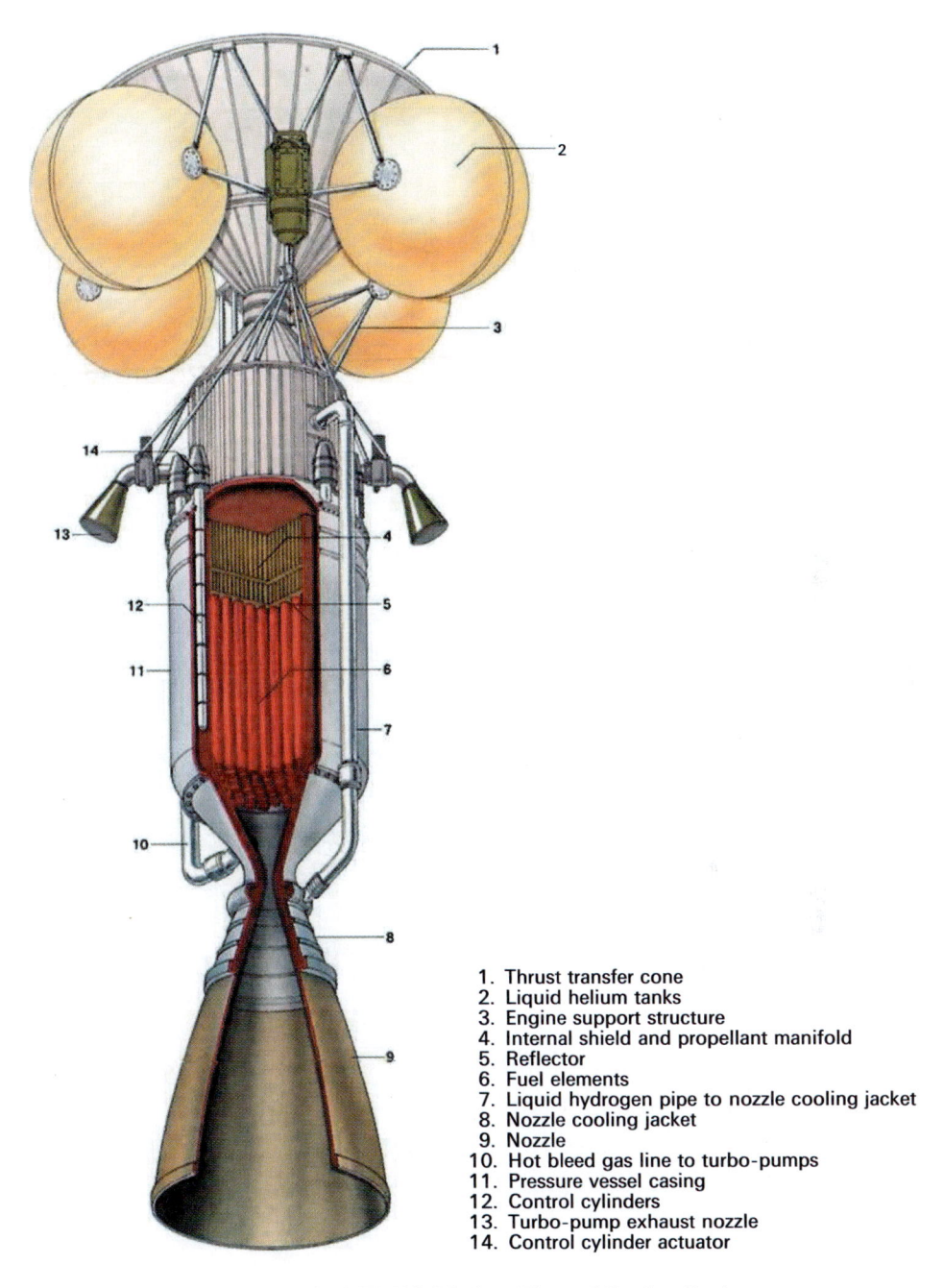

1. Thrust transfer cone
2. Liquid helium tanks
3. Engine support structure
4. Internal shield and propellant manifold
5. Reflector
6. Fuel elements
7. Liquid hydrogen pipe to nozzle cooling jacket
8. Nozzle cooling jacket
9. Nozzle
10. Hot bleed gas line to turbo-pumps
11. Pressure vessel casing
12. Control cylinders
13. Turbo-pump exhaust nozzle
14. Control cylinder actuator

Plate 17 The NERVA Nuclear Thermal Rocket Engine.
Courtesy NASA.

Plate 18 A NERVA programme engine on the test stand.
Courtesy NASA.

Plate 19 A possible Mars expedition vehicle powered by three nuclear thermal rocket engines (artists impression).
Courtesy NASA.

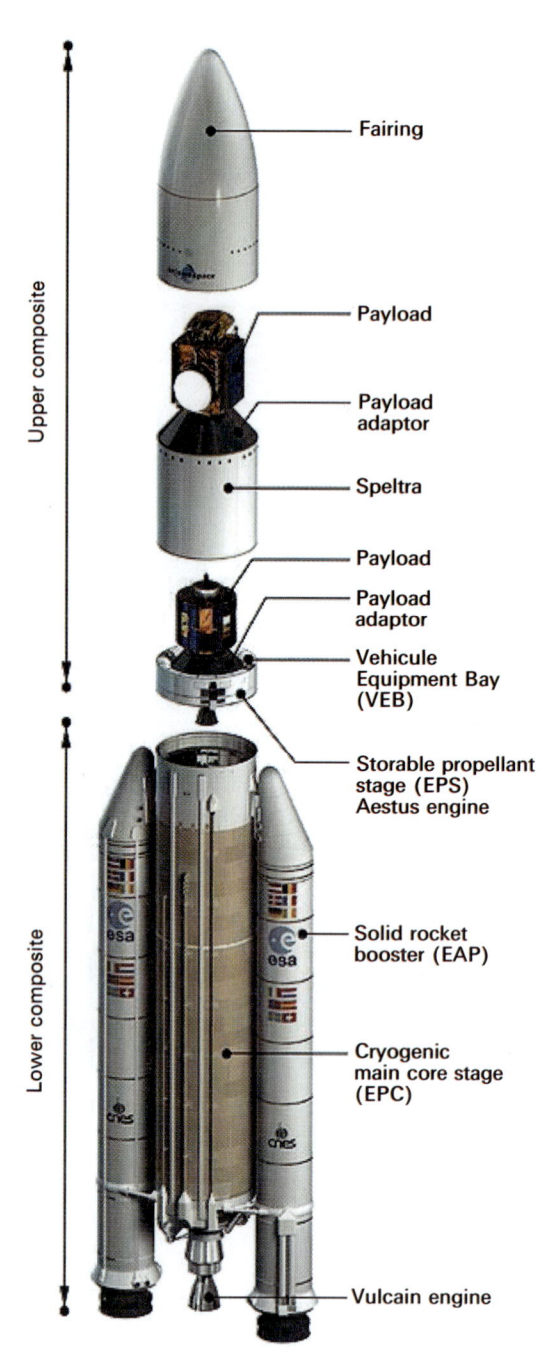

Plate 20 An exploded view of the Ariane 5 launcher—see Appendix 3. The principal components are the main cryogenic stage, the two solid boosters, and the upper stage. Ariane 5 is normally adapted to carry two spacecraft, mounted in tandem, using the *SPELTRA* adaptor within the shroud.

Courtesy Arianspace.

Sec. 6.3]   **Principles of electric propulsion**   151

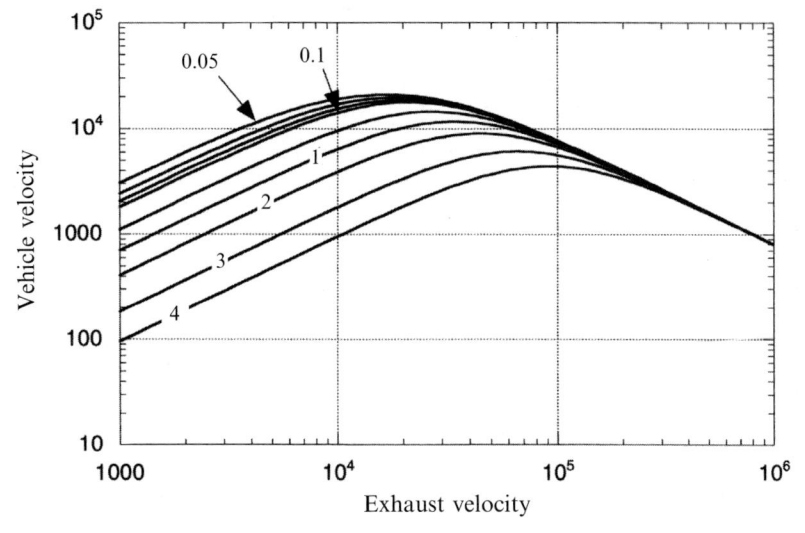

Figure 6.3. Vehicle velocity as a function of payload/propellant mass and exhaust velocity.

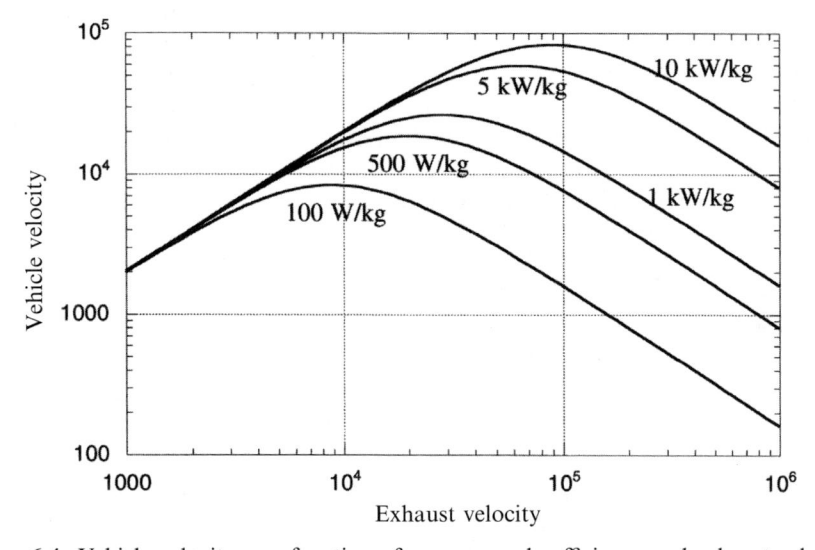

Figure 6.4. Vehicle velocity as a function of power supply efficiency and exhaust velocity.

set to 0.15. The power-to-mass ratio is shown as a parameter ranging from 0.1 to 10 kW/kg.

Figure 6.4 shows that a higher power-to-mass ratio increases the vehicle velocity, as expected, and that the peak velocity moves towards higher exhaust velocities as the power-to-mass ratio increases. This shows that for really large velocity increments, a high power-to-mass ratio must be matched by high exhaust velocity.

The importance of exhaust velocity is therefore obvious; high exhaust velocity allows much higher payload-to-propellant mass ratios, and the power-to-mass ratio of the power supply is crucial in obtaining the best performance. We shall now examine the different kinds of electric thruster, returning to the question of vehicle performance with specific examples and missions. It is as well to remember throughout this chapter that while high exhaust velocity is a characteristic of electric thrusters they uniformly have very low thrust and therefore long burn times, as shown in Figures 6.2, 6.3, and 6.4.

6.4 ELECTRIC THRUSTERS

Electric thrusters can be divided into two broad categories: those that use electricity to heat the propellant, which emerges as a neutral gas, and those which use electric or magnetic fields to accelerate ions. The functional form and analysis of these two classes differ.

6.4.1 Electrothermal thrusters

The basic electrothermal thruster, or 'resisto-jet', consists of a nozzle with a high expansion ratio, connected to a chamber in which the propellant is heated by a hot wire through which passes an electric current. This type of electric thruster uses the same thermodynamic effects to generate a high-velocity exhaust stream as does a chemical rocket. For high exhaust velocity, the pressure and temperature of the gas entering the nozzle need to be high, which implies efficient heating of the gas. Since gases are bad conductors of heat, only a thin layer, in contact with the heater, becomes hot; moreover, the wire radiates heat to the chamber walls, and so some power is lost. The heat lost to the walls is essentially a loss of power in the thruster: η is reduced. Poor transfer of heat to the gas results in lower temperature and pressure when the gas enters the nozzle, and from the results in Chapter 2 it is clear that the exhaust velocity and thrust are then reduced. Heat losses to the chamber walls can be reduced by a low-mass radiation shield made of concentric metal foils. To maximise heat transfer to the gas, a multichannel heat exchanger is used to bring as much of the gas volume as possible into contact with the heater. Figure 6.5 illustrates a section of such a thruster.

The performance of electrothermal thrusters and other thermal thrusters can be calculated using the equations in Chapter 2. In particular, the exhaust velocity can be calculated using the thrust coefficient and the characteristic velocity, as for a chemical rocket:

$$v_e = C_F c^*$$

The characteristic velocity is expressed as

$$c^* = \left\{ \gamma \left(\frac{2}{\gamma+1} \right)^{\frac{\gamma+1}{\gamma-1}} \frac{\mathfrak{M}}{RT_c} \right\}^{-1/2}$$

and is independent of the way the gas is heated.

Sec. 6.4] Electric thrusters   153

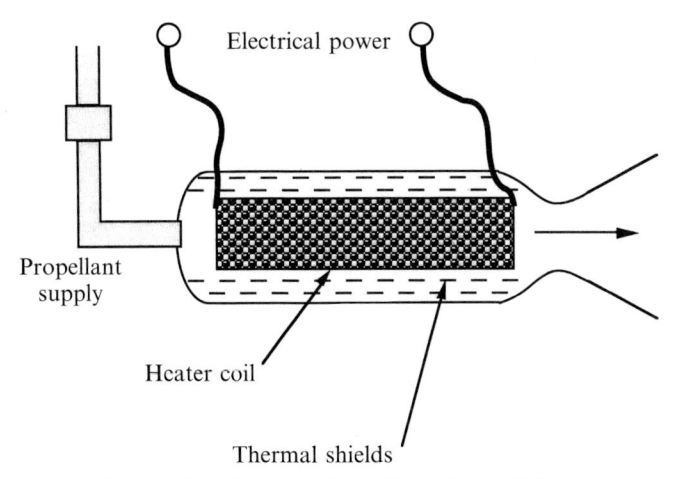

Figure 6.5. Schematic of an electrothermal thruster.

The thrust is a function of the nozzle and the combustion chamber pressure, as before. Since these thrusters are exclusively used in a vacuum, with a high-expansion ratio nozzle, a value close to the ideal (2.25 for $\gamma = 1.2$) may be assumed. As with a chemical rocket, the characteristic velocity depends on the temperature and molecular weight. Unlike the chemical rocket, however, since there is no combustion this is just the molecular weight of the propellant gas. The temperature reached by the gas depends on the power input and the mass flow rate. Note that for a chemical rocket the temperature is defined by the nature of the propellants, and to first order does not depend on the flow rate. In contrast, for an electrothermal thruster the temperature depends inversely on the flow rate. For a very low flow rate it is clear that the temperature could reach an arbitrarily high value, which would destroy the thruster, and in general a maximum temperature for the thruster is specified, based on its construction. This fixes the ratio of flow rate to power input. There are two design issues: the heating filament itself has to remain intact, and the chamber wall must not fail under high pressure and temperature. Filament temperatures beyond the melting point of refractory metals such as platinum or tungsten cannot be used. The chamber walls can be protected, as in the case of a chemical rocket, by active cooling with the propellant, possibly regenerative. The limit on temperature, imposed by the melting point of the filament or heater is serious. Chemical rockets can have combustion chamber temperatures well above the melting point of the chamber walls, because the walls can be cooled. A heating filament, however, cannot be cooled—by definition—and so the gas cannot be hotter than the service temperature of the filament.

As an example, consider a 1 kW thruster working with hydrogen gas having $\gamma = 1.2$ and a high expansion ratio nozzle, so that the thrust coefficient is 2.25. The maximum temperature is 2,200 K—about the limit for a resistive heater. The molecular weight is 2 for hydrogen. Substituting in the equation, the characteristic velocity is 4,659 m/s. This is about three times that obtainable with a chemical thruster, because of the low molecular weight of hydrogen. In a chemical rocket, combustion takes place and produces heavy oxides; for hydrogen and oxygen, water is produced, with a molecular weight of 18. With ideal expansion this thruster has an exhaust velocity of

$$C_F c^* = 2.25 \times 4659 = 10483 \, \text{m/s}$$

which is much higher than can be attained by chemical means. This simple device—a hot wire in a chamber, connected to a nozzle—generates an exhaust velocity more than twice as high as the most efficient chemical rocket engine, simply by separating the functions of energy input and propellant flow. This is a dramatic indication of the potential of electric propulsion.

We can now calculate the mass flow rate from the power:

$$\tfrac{1}{2} m v_e^2 = \eta P_E$$

$$m = \frac{2\eta P_E}{v_e^2} = \frac{2 \times \eta \times 1000}{10483^2} = 1.8 \times 10^{-5} \eta \, \text{kg/s}$$

This is very small compared with the flow rates of chemical rockets, and the thrust is also small—only 0.2 N.

This type of thruster can generate an effective exhaust velocity in excess of that achievable by chemical means, but the mass flow rate and thrust are low. This means that a vehicle using such a thruster can ultimately reach a very high velocity, but the time taken to accelerate to that high velocity is much longer than with a chemical thruster. This is the fundamental nature of electric propulsion.

If the thruster is used for station keeping or attitude control, then it is the low propellant usage that is important; and again, the high exhaust velocity minimises propellant usage, but the thrust is small, and so manoeuvring times may be long.

A particular advantage of the electrothermal thruster is that the propellant can be chosen free of other constraints. Anything that is compatible with the materials of the chamber and heater can be used. Hydrogen has the lowest molecular weight, and this advantage is clear from the above calculation. It does, however, dissociate to monatomic hydrogen at high temperatures, extracting heat from the exhaust stream and decreasing the exhaust velocity, which would, in practice, reduce the performance from that we have calculated. Helium is perhaps better as a low molecular weight gas, because it is already monatomic. Water is a possible propellant, and although the molecular weight is high and there is the possibility of dissociation, it is 'free' on manned space vehicles. It has been proposed that electrothermal thrusters could be used for attitude control on the International Space Station, using waste water as the propellant.

Another conveniently storable liquid is hydrazine, which dissociates into a mixture of hydrogen and nitrogen, together with a little ammonia. In a particular form of electrothermal thruster, standard catalytic decomposition is used, together with electrical heating, to inject more energy into the reaction. Care must be taken that the filament service temperature is not exceeded, and so this approach does not lead to a higher exhaust velocity but to lower power requirement. While electrical thrusters are establishing themselves as reliable units, the presence of hydrazine as a back-up chemical mono-propellant provides extra security. Typical hydrazine electrothermal thrusters produce a thrust between 200 and 800 mN and lower exhaust velocities around 3,000 m/s, because of the higher molecular weight of the nitrogen. Power ranges from 0.2 to 0.8 kW. The low exhaust velocity is still much greater than the 1,000 m/s obtainable with a catalytic hydrazine thruster, and so less propellant is used for a given velocity increment.

Electrothermal thrusters are relatively uncomplicated, and have the advantages adumbrated above. Amongst electric thrusters they produce a moderately high thrust, and a modest increase in exhaust velocity, over that obtainable from chemical rockets. The electrical efficiency is high, with η close to 90%. Higher exhaust velocities and powers are difficult to achieve because of the difficulty of transferring heat from the filament to the gas. The upper limit to gas temperature is, in fact, about half what can be accommodated in a chemical rocket combustion chamber. This is because in the latter the gas is heated in the volume of the chamber by the chemical reaction. The walls, protected by an insulating layer of cooler gas, and actively cooled, can remain below the service temperature of their material. For the filament-heated electrothermal thruster, the hottest part is the filament itself, otherwise heat could not be transferred to the gas. The service temperature of the filament is therefore the limiting factor. Electrothermal thrusters can achieve high exhaust velocity because a low molecular weight propellant can be chosen. The transfer of heat to the gas from the filament places a fundamental limit on the performance of resistive thrusters. More efficient heat transfer is needed, and direct heating of the gas by the electric current is the obvious approach.

6.4.2 Arc-jet thrusters

In the arc-jet thruster, the propellant gas is heated by passing an electric arc through the flow. For a neutral gas exposed to an electric field, the resistance is initially very high until, as the potential across it is raised, ionisation occurs, and the gas begins to conduct. The resistance drops rapidly, and the current increases until all the gas is ionised or until the supply resistance dominates. Electrons and positive ions move in opposite directions, and transfer their charge to the anode and cathode respectively. Neutral gas atoms are heated by collision with the ions and electrons. The thermodynamic behaviour of a plasma is complicated: neutral atoms and positive ions can take part in expansion, but the electrons serve only to render the plasma neutral. Recombination of electrons and ions will release the electron energy in the form of additional hot neutral molecules. For these reasons, the analysis of an arc-jet is complicated.

Figure 6.6 illustrates a typical arc-jet thruster. The anode and cathode are made of tungsten, which has a high melting point. The pointed cathode rod is supported in a boron nitride insulator, which also holds the anode. The anode is shaped to create the gap across which the arc is struck and through which the propellant flows. Downstream of the arc, the anode is shaped to form the nozzle, for the expansion of

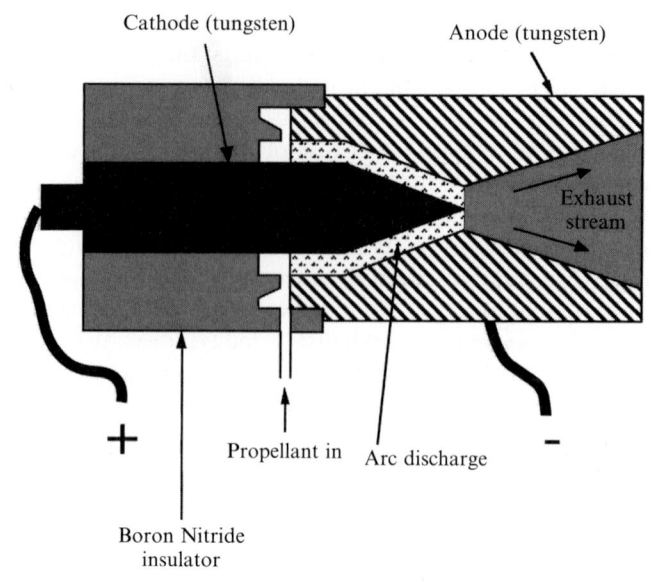

Figure 6.6. Schematic of an arc-jet thruster.

the exhaust. A steady DC potential of 200–300 V is applied across the gap by a power supply, which can regulate the current to match the flow rate. The propellant gas is introduced to an annular plenum chamber, and enters the arc region through a narrow annular slit (the width of which is exaggerated in the figure).

The power which can be applied to an arc-jet can be up to 100 times greater than can be applied to the filament of an electrothermal thruster. The current passes through the gas itself, and so temperature limits can be much higher. The difficulty is the effect of the ions in the arc itself; these carry half the current, and strike the surface of the cathode at high speed, causing vaporisation of the material. The electrons affect the anode to a lesser extent; in fact, some of the electrode material becomes ionised, and takes part in the current transport. This limits the life of the anode and cathode, and places a limit on the current that can be passed through the arc. The temperature limit imposed by the filament in the electrothermal thruster has its analogue in the limit imposed by the material of the electrodes in the arc-jet. It is, however, higher because the main heating occurs in the bulk of the propellant gas and not in the metal parts.

For high efficiency the rapidly moving ions in the arc should transfer their energy to neutral gas atoms, or recombine before they reach the expansion part of the nozzle. This means that the current passed should be limited so that not all the propellant is ionised. High-current arcs are also naturally unstable, because the ions respond not only to the electric field but also to the magnetic field caused by their own motion. This 'pinch' effect may cause the arc to break up into a number of columns, concentrating the energy flow into small spots on the electrodes and reducing the efficiency of energy transfer to the neutral atoms. The hot-spots so produced increase the erosion of the electrodes and shorten their life.

Heat loss is similar to that in other rocket engines. The hot gas is confined to the arc and the central regions of the nozzle, and a layer of cooler gas forms along the inner faces of the nozzle which insulates it. Because of the high temperature, radiation loss of heat from the hot gas is significant. Typical efficiencies—taking into account conduction, radiation and losses due to dissociation and ionisation—are 30–40%. This is the ratio of electrical power to exhaust-stream power, and is lower than that of an electrothermal thruster.

Despite these difficulties, the arc-jet allows a much higher power dissipation than the electrothermal thruster, and a higher propellant temperature (4,000–5,000 K), due to the direct interaction of the electric current and the propellant.

Performance can be determined in the same way as for an electrothermal thruster. Again, propellants with low molecular weight are best. Maximum exhaust velocities can be as high as 20 km/s (with hydrogen), reflecting the higher temperature. Arc-jets are used with hydrogen, ammonia and hydrazine. The latter two dissociate to nitrogen and hydrogen, producing a higher average molecular weight, but they have the advantage of being non-cryogenic liquids. If hydrazine is used then the exhaust velocity is only 4,500–6,000 m/s, because of the molecular weight effect. This is greater than can be achieved with any chemical rocket engine.

Power levels used in station-keeping thrusters are 2–3 kW, and high-power arc-jets have operated at 200 kW for short periods. The electrical efficiency is lower than that of the electrothermal thruster, and a heavier power supply is therefore required, although this is usually offset by the higher exhaust velocity and hence the smaller quantity of propellant required. The arc-jet seems to be establishing itself as an effective station-keeping thruster, and there are several models available commercially. It is relatively simple, having only two electrodes and a low operational voltage, which makes for reliability in the arc-jet itself and also in the power converter. Established life-times are greater than 800 hours of operation, which is long enough to take full advantage of the low thrust (see Figure 6.3).

6.5 ELECTROMAGNETIC THRUSTERS

If we wish to exceed the exhaust velocities achievable using electrical heating of the propellant, it is necessary to abandon thermodynamic effects and act directly upon the atoms of the propellant by using the electromagnetic field. This implies that the propellant has to be ionised—which is already happening in the arc-jet, where it is a nuisance, reducing the efficiency. If the propellant is fully ionised, then direct acceleration of the ions by electric and magnetic fields can produce a very high bulk velocity indeed. In contrast to the thermal rocket—whether chemical or electrical—we are no longer concerned with the conversion of random high molecular speeds (heat) into the bulk motion of the gas in the exhaust. Instead, every ion is constrained by the field to move in the same direction, creating the bulk flow. The difficulty is that the density of the exhaust has to be low enough for the ions not to be slowed by collisions or to recombine with electrons. This means that the mass flow will be small—even smaller than in electrothermal thrusters.

The highest exhaust velocity is achieved by accelerating positive ions in an electric field created between two grids having a large potential difference. This is also the simplest conceptually. Calculated velocities can be in excess of 50,000 m/s, and practical thrusters achieve 25,000–32,000 m/s. Other techniques make use of the magnetic field created when a current passes through the plasma to accelerate it, which can provide a steady flow of plasma or a pulse. The steady-state version—a Hall thruster—derives from the Russian space programme, where it has flown on many spacecraft. The pulsed system is still under development in the laboratory. Both effectively use magnetohydrodynamic effects to produce exhaust velocities and mass flow rates intermediate between arc-jet thrusters and ion thrusters.

6.5.1 Ion propulsion

This is the simplest concept: propellant is ionised, and then enters a region of strong electric field, where the positive ions are accelerated. Passing through a grid, they leave the engine as a high-velocity exhaust stream. The electrons do not leave, and so the exhaust is positively charged. Ultimately this would result in a retarding field developing between the spacecraft and the exhaust, and so an electron current is therefore discharged into the exhaust to neutralise the spacecraft. The electrons carry little momentum, and so this does not affect the thrust.

The schematic (Figure 6.7) shows the thruster is divided into two chambers. The propellant enters the ionisation chamber in the form of neutral gas molecules. There is a radial electric field across the chamber, and electrons are released from the cathode (which can be a thermionic emitter). The electrons are accelerated by the radial field, and reach energies of several tens of electron volts, which is enough to ionise the neutral propellant atoms by collision. To extend the path length of the electrons and ensure that they encounter as many neutral atoms as possible, an axial magnetic field is provided, which makes them move in a spiral path. The ionisation therefore becomes efficient; that is, the number of ions produced, as a function of the electron current, is maximised. In theory, all the electrical energy in an electrothermal thruster enters the exhaust stream, but in an electromagnetic thruster each ion in the exhaust has to be created with an energy of about 20–30 eV per ion. This energy does not go into propulsion, and is lost. Thus, it is important to maximise the ionisation efficiency.

The ionised propellant atoms drift under a small negative field through the first grid into the accelerating chamber. The grids have a high potential across them, and are separated by 1–2 mm. The ions gain energy in the strong electric field and, passing through the outer grid, form the ion beam.

There is no need for a nozzle to generate the thrust, because the motion of the ion beam is ordered and not chaotic. The theory developed in Chapter 2 is not valid for field-accelerated exhaust jets. The thrust itself is exerted on the accelerating grids (by the departing ions), and is transferred through the body of the thruster to the spacecraft (Figure 6.8).

Sec. 6.5] **Electromagnetic thrusters** 159

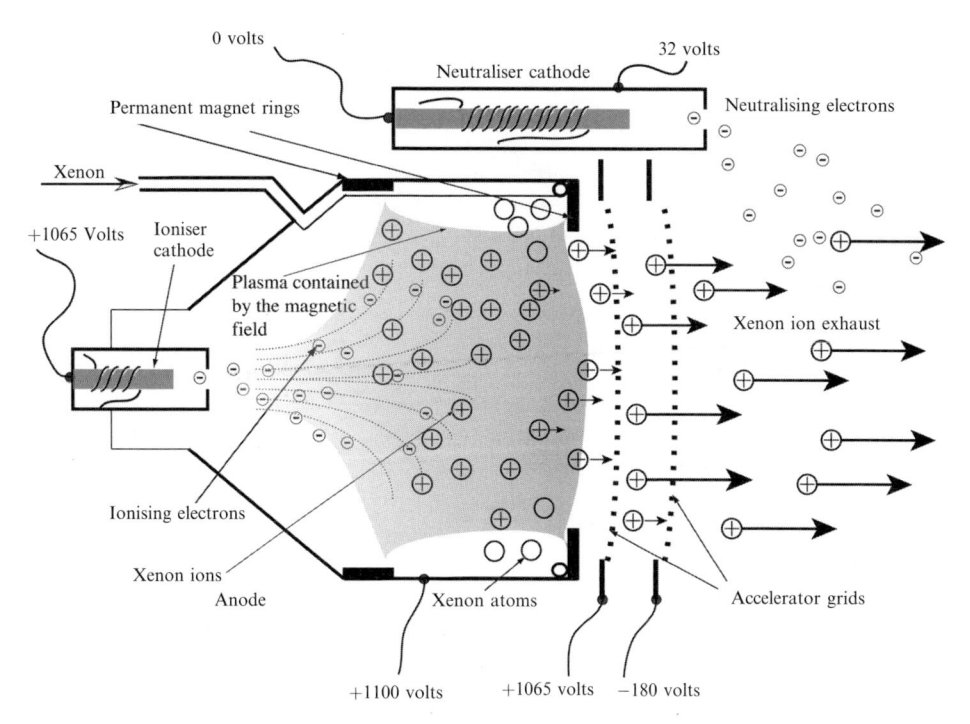

Figure 6.7. A schematic diagram of the NSTAR ion thruster, as used on Deep Space 1. The xenon gas enters the chamber from the left, and is ionised in the shaded region, confined by a magnetic field, generated by strong permanent magnets. The xenon ions drift into the gap between the two grids on the right and are accelerated by the electric field. They pass out, to form the exhaust stream, and electrons are released from the neutraliser cathode to keep the spacecraft neutral.

Ion thruster theory

The ion thruster is simple in concept, as described above. The theory of operation is also relatively simple, and because it is so different from that of a thermal rocket it is useful to include a brief description here, so that the strengths and limitations can be appreciated.

As in all reaction propulsion systems, the thrust depends ultimately on the transfer of momentum from an exhaust stream to the vehicle. The exhaust velocity is straightforwardly given by the potential difference between the grids. Ions dropping through this potential difference each gain a fixed amount of energy, and this converts directly into a velocity. The other parameter in the thrust is the mass flow rate. For an ion thruster this is directly related to the current flowing between the grids, and the ion current itself becomes the exhaust stream. To increase the thrust of a given ion thruster, the current has to be increased; but it cannot be increased indefinitely, as there is a natural limit. It is this limit which we can examine theoretically.

Figure 6.8. The NSTAR engine mounted on Deep Space 1 for testing. The curved acceleration grid can be seen as well as the neutralising electron gun.
Courtesy JPL/NASA.

6.5.2 The space charge limit

As depicted in Figure 6.7, the ion thruster is in some ways similar to a thermionic valve, and some of the same considerations can be used to estimate performance. The accelerating grids have an electric field between them, which, in the absence of the ions, is constant, and depends only on the potential difference and separation between the grids. When the ions are introduced into the space between the grids, they alter the field; effectively, they partially shield the first grid. Thus, the profile of the accelerating field depends on the number of ions in the beam, and therefore on the mass flow rate. As the exhaust stream density increases, there will be a point when the accelerating field at the first grid drops to zero, because the positive charge of the downstream ions cancels the field. This is the *space charge* limit, representing the maximum ion current that can flow. Note that ion acceleration can still occur further into the cell, but any further increase in the *current* of ions would generate a retarding field at the cell entrance, preventing the ingress of further ions.

Because the velocity of the ions at the second grid depends only on the potential drop between the grids, there is no effect of space charge on the exhaust velocity, only on the mass flow rate.

The modification of fields by charges is expressed by Poisson's equation:

$$E = \frac{d^2V}{dx^2} = -\frac{Nq}{\varepsilon_0}$$

where E is the electric field and N is the ion density at a given point, and q is the individual charge on the ions. The ion velocity derives from equating their kinetic energy with the energy drop associated with the change in potential:

$$\tfrac{1}{2}Mv^2 = q(V_1 - V)$$

$$v = \sqrt{\frac{2q(V_1 - V)}{M}}$$

Representing the flow of ions as a current density j—ion flux per unit area—the differential equation becomes

$$\frac{d^2V}{dx^2} = -\frac{j}{\varepsilon_0 v} = -\frac{j}{\varepsilon_0}\sqrt{\frac{M}{2q(V_1 - V)}}$$

Multiplying both sides by $2(dV/dx)$, the equation can be integrated to yield

$$\left(\frac{dV}{dx}\right)^2 - \left(\frac{dV}{dx}\right)_1^2 = \frac{4j}{\varepsilon_0}\left(\frac{M(V_1 - V)}{2q}\right)^{1/2}$$

Here $\left(\dfrac{dV}{dx}\right)^2$ is the square of the electric field, and $\left(\dfrac{dV}{dx}\right)_1^2$ is the value at the first grid.

For the space charge limit—where the ions shield the downstream electrode—the field at the first grid is exactly zero. This can be substituted, and the equation integrated again by taking the square root. This produces an expression for the electric field at any point where the potential is known:

$$E = \frac{dV}{dx} = 2\left(\frac{j}{\varepsilon_0}\right)^{1/2}\left(\frac{M}{2q}\right)^{1/4}(V_1 - V)^{1/4}$$

The potential itself can be determined by separating the variables, and carrying out a further integration, to yield

$$V = V_1 - \left[\frac{3}{2}\left(\frac{j}{\varepsilon_0}\right)^{1/2}\left(\frac{M}{2q}\right)^{1/4}x\right]^{4/3}$$

This determines the potential at any point between the grids as a function of the current density. If we fix the potential drop across the electrodes by setting the potential of the second grid to zero, and the first to V_1, and substitute in the above, an expression for the current density at the space charge limit is easily derived:

$$j = \frac{4\varepsilon_0}{9} \left(\frac{2q}{M}\right)^{1/2} \frac{V_1^{3/2}}{x_2^2}$$

Here x_2 is the separation of the grids. If we define $E_0 = \frac{V_1}{x_2}$ as the quiescent field (in the absence of ions), then the current can be expressed as

$$j = \frac{4\varepsilon_0}{9} \left(\frac{2q}{M}\right)^{1/2} \left(\frac{E_0^3}{x_2}\right)^{1/2}$$

This shows that the current density, and therefore the thrust, depends strongly on the electric field, and inversely on the electrode separation; that is, for the same field a smaller gap produces a higher current and hence a higher thrust. For a given potential drop, decreasing the gap length also increases the field, and so small gaps are very advantageous.

6.5.3 Electric field and potential

The maximum field strength is likely to be determined by the breakdown characteristics of the electrodes and insulators. Higher potentials will in general be reflected in more massive power supplies, and so a small gap is important. The electric field in the gap is modified by the ion current; it is lower than the quiescent field near the first electrode, and higher near the second electrode. The maximum value can be derived by substituting for the current, in the field equation, to produce

$$E_2 = \frac{4}{3}\frac{V_1}{x_2} = \frac{4}{3}E_0$$

Figure 6.9 shows an example calculation for xenon ions accelerated between two grids separated by 1 mm, with a field of 10^6 V m^{-1}. The charge-to-mass ratio of singly-charged xenon ions is $7.34 \; 10^5$ Coulombs kg^{-1}, and ε_0, the permittivity of free space, is $8.85 \; 10^{-12}$ farads m^{-1}. Substituting these values into the equation for current density, a value of $j = 47.65$ amps m^2 is obtained. This can then be substituted in the equation for potential to determine its value as a function of distance in the gap, and a similar substitution in the field equation determines the field. The quiescent field is constant at 10^6 V m^{-1}, and the quiescent potential falls linearly across the gap from 10^4 V to zero. The modified field is zero at the first electrode, as predicted for the space charge limit. It rises rapidly away from the first electrode, and is equal to the quiescent field at about 0.4 mm. Thereafter, it rises to be 30% greater than the quiescent field at the second grid. The accelerating force on the ions thus increases through the gap.

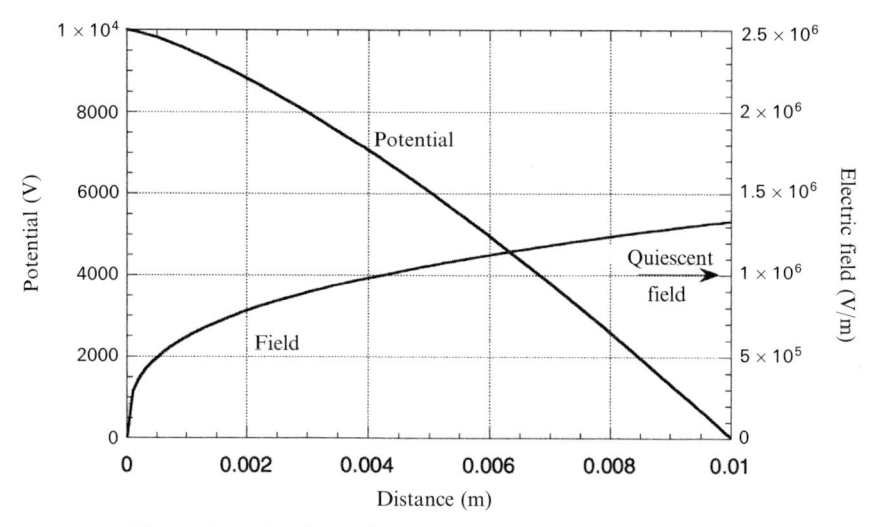

Figure 6.9. Electric field and potential in space charge limit.

6.5.4 Ion thrust

This theory shows that for a given potential drop and gap size, the current density—effectively the number of ions per unit area of the grids—reaches a limiting value. The current density is the equivalent of the mass flow rate in a chemical rocket, and here we see that an increase in the mass flow rate requires an increase in the field or a decrease in the gap. These two parameters cannot be varied indefinitely. As the gap decreases, the electric field increases, and non-uniformities in the grid construction will eventually cause a breakdown between the grids, due to field concentrations. This means that a given ion thruster will have a fixed ion current. To increase the mass flow rate, the diameter of the thruster needs to be increased. The combustion chamber pressure and temperature in the chemical rocket find their analogues here in the potential drop and gap size, and the analogue of the throat diameter is the diameter of the electrodes.

The thrust of an ion thruster of a given dimension is therefore fixed by the electrode configuration and potential drop. Using the above equations it is possible to calculate the thrust and the exhaust velocity as a function of the power and potential difference, which will allow the performance of ion thrusters to be assessed in the terms used earlier when discussing the mission parameters.

The mass flow rate per unit area is related to the current by

$$m = j \frac{M}{q}$$

The thrust per unit area is therefore represented by

$$\frac{F}{A} = m v_e = \frac{8}{9} \varepsilon_0 \left(\frac{V_1}{x_2} \right)^2 = \frac{8}{9} \varepsilon_0 E_0^2$$

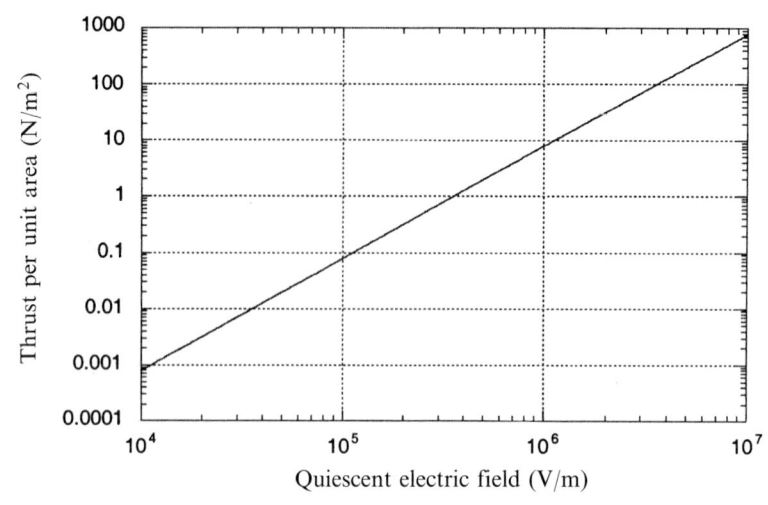

Figure 6.10. Thrust per unit area as a function of quiescent field for an ion thruster.

This is obtained from the familiar Newtonian equation by substituting the expression for the ion velocity, derived earlier, for the exhaust velocity, and the current density for the mass flow rate per unit area.

From this it can be seen that the thrust is proportional to the area of the thruster (the open area of the second grid) and to the square of the quiescent field. Figure 6.10 shows the thrust per unit area as a function of the quiescent field: the latter is simply the ratio of potential drop to gap size.

6.5.5 Propellant choice

The thrust per unit area is independent of the charge-to-mass ratio of the ions; that is, it is independent of the nature of the propellant. On the other hand, the exhaust velocity itself depends on the charge-to-mass ratio and the potential drop:

$$v_e = \sqrt{\frac{2qV_1}{M}}$$

For high exhaust velocity a high charge-to-mass ratio is therefore required, together with a large potential drop. As we shall see, ion engines naturally produce a very high exhaust velocity—often too high for many types of mission—and so the need for, say, hydrogen as a propellant rarely arises. In fact, the most successful engines use quite heavy ions such as xenon, caesium or mercury.

Figure 6.11 shows the exhaust velocity as a function of potential drop and ion species for several propellants. Comparing this figure with Figure 6.3, it can be seen that the typical optimum exhaust velocities range between 10^4 and 10^5 m/s. This is very much at the lower end of the exhaust velocity scale in Figure 6.11. Moreover, since thrust depends on the quiescent field, and there has to be a lower limit to the

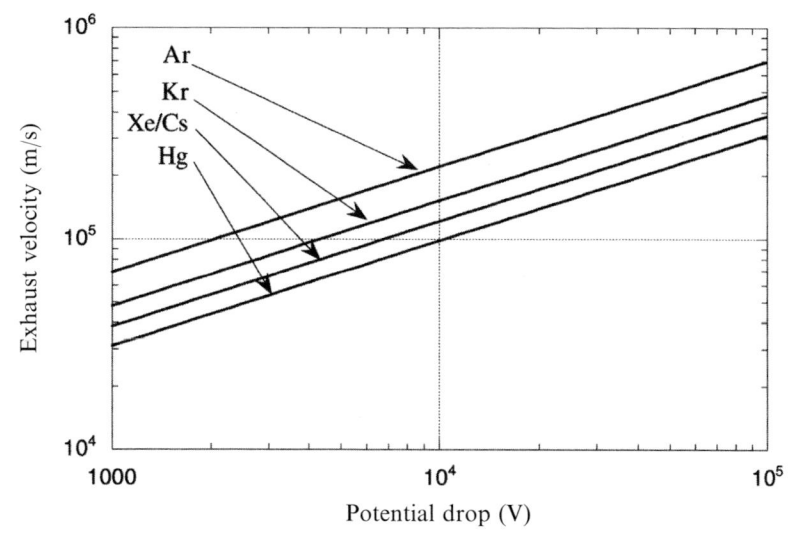

Figure 6.11. Exhaust velocity and ion species for an ion thruster.

gap size, a high potential drop is needed for reasonable thrust. This potential drop tends to produce even higher exhaust velocity, which may be inefficient for many missions. This is the fundamental dilemma of ion propulsion: the engine itself is naturally a high exhaust velocity and low-thrust device. Attempts at optimisation therefore focus on lowering the exhaust velocity rather than raising it.

The thrust itself is developed on the electrode grids, and this force, together with the electric field, will tend to cause the grid to distort. If the gap is too small this distortion will be reflected in a distorted field, and possibly electrical breakdown. Thus the gap can be no smaller than about 0.5 mm. This means, for example, that even a thrust of $1\,N\,m^2$ requires a field of $4\,10^5\,V\,m^{-1}$, or a potential drop of 2,000 V. For these conditions, argon ions already produce an exhaust velocity of $10^5\,m/s$, which for most missions is well above the optimum.

The obvious solution is to use ions of low charge-to-mass ratio. Early engines used mercury and caesium, and Figure 6.11 shows that these give more reasonable performance. Mercury and caesium are toxic, and tend to cause contamination of the spacecraft. The other disadvantage of these metallic propellants is the energy needed to evaporate them; as conductors they are difficult to ionise in the liquid or solid phase. Typical mercury or caesium thrusters have a boiler attached to the ionisation chamber, where the metals are evaporated to form a gas prior to entering the chamber. This is an additional drain on the power supply, and leads to reduced performance.

Modern large ion thrusters use xenon as a propellant. The charge-to-mass ratio is reasonable, and the exhaust is non-toxic and cannot contaminate the spacecraft. It has to be stored as a liquified gas, and with reasonably thick walls it can be kept as a liquid in a sealed tank, making it suitable for long-duration missions.

6.5.6 Deceleration grid

Even with the heaviest ions the velocity is still too large to be optimal for present-day missions. Rather surprisingly, there is a considerable advantage to be gained by slowing the ions down to bring their velocity closer to the optimum for moderate velocity increments. To appreciate this, consider that the thrust—which is essentially the space charge limited current—depends on the square of the potential drop, while the velocity depends on the square root. The technique adopted is to place a third electrode downstream of the second grid, at a somewhat higher potential. The ions are accelerated, and the current is the same, inside the region defined by grids one and two. On leaving the second grid, the ions are decelerated by the third grid to a lower exhaust velocity, but the current remains the same. The net thrust is intermediate between the thrust from the two-grid engine and the (lower) thrust that would have been developed if the second grid were at the third grid potential in a two-grid configuration. The advantage of lower exhaust velocity, with unchanged mass flow rate, outweighs the loss of thrust. For a given beam power the three-grid system produces a higher thrust than the two-grid system.

The third grid also deals with another problem which so far has not been mentioned: the upstream migration of electrons. Any electrons which enter the region between the first two grids will be accelerated backwards toward the ion source. This constitutes a current drain, with no propulsive effect, and can also damage the ion source. Electrons have to be injected into the beam, after the last grid, to neutralise the exhaust. In the absence of the decelerating field they can leak backwards into the gap. The decelerating field forces these electrons back into the exhaust stream, where they can do no harm.

6.5.7 Electrical efficiency

The electrical power consumed by the beam is simply the product of the current and the potential drop. Power per unit area is

$$\frac{P_E}{A} = \frac{4\varepsilon_0}{9} \left(\frac{2q}{M}\right)^{1/2} \frac{V_1^{5/2}}{x_2^2}$$

The ratio of power to thrust is then

$$\frac{P_E}{F} = \frac{1}{2} \left(\frac{2q}{M} V_1\right)^{1/2} = \frac{1}{2} v_e$$

This can be used to compare the performance of different electrical propulsion systems. However, we must first consider the losses in the system. There is power lost due to radiation from the ions and neutral gas molecules. The ions are often created in an excited state, and relax, emitting a blue or ultraviolet photon. Similarly during ionisation, neutral gas molecules are often only excited by an electron collision and not ionised, and this energy is radiated away and is wasted.

The electrons injected into the beam for neutralisation are an additional power drain. There is not much recombination in the exhaust; and besides, this occurs after the propellant has left the spacecraft. The characteristic blue colour of the exhaust plume is due to recombination and de-excitation of ions (Plate 16).

The main quantifiable energy loss is that required to ionise the neutral gas molecules. While the ionisation potential of propellant atoms is quite small— 10–20 eV—the actual energy expended per ion in all types of ion engine is closer to 500 eV, because of the energy lost in the above processes. The electron bombardment ioniser illustrated in Figure 10.7 can ionise 80% of the neutral gas atoms for an expenditure of 450–600 eV per ion. The remaining neutral atoms are a nuisance in that they absorb energy from the ions and defocus the beam, but to increase the ionisation efficiency to 90% would require about 800 eV per ion. This energy loss per ion must be kept as low as possible, because it is of the same order as the energy of an individual ion in the beam. For example, consider a xenon ion in a beam of velocity 10^4 m/s. Its kinetic energy is 1.1 10^{-17} J, or 68 eV; for such a low velocity the ion engine is very inefficient, with given energy losses per ion of around 400 eV. For higher velocities the efficiency increases rapidly, so that at 10^5 m/s it is around 90% efficient.

To appreciate the effect of the ionisation power we can compare the energy in the ions making up the beam to the total energy input:

$$\eta = \frac{\frac{1}{2}Mv_e^2}{\frac{1}{2}Mv_e^2 + E_{Ion}}$$

This shows that for a constant ionisation energy the efficiency depends on the ion mass: heavier ions have a greater power efficiency. Replacing P_E with ηP_E in the thrust-to-power ratio equation produces

$$\frac{F}{P_E} = \frac{2Mv_e}{Mv_e^2 + 2E_{Ion}}$$

which is shown for several ion species in Figure 6.12. The figure shows that for exhaust velocities below 10^5 m/s the nature of the propellant is important, and that significantly higher thrust-to-power ratios are produced by heavy ions. Beyond 10^5 m/s the ion species is irrelevant, because the ionisation energy loss becomes negligible compared with the kinetic energy of the ions, and the efficiency approaches 100%.

Figure 6.12 also shows that the thrust of an ion engine is really small. Ion engines are best used for very high-velocity increment missions in which the time penalty associated with small thrust is not important, and they are therefore suitable for interplanetary missions and for station keeping, where the low thrust is not a disadvantage (Figure 6.13). For attitude control the slow response consequent upon the small thrust may be a disadvantage, and other types of electric propulsion are more suitable. As mentioned above, for reasonably high exhaust velocity the electrical efficiency of ion engines can be as much as 90%, and so the power supply mass can be relatively small. It has to be more complex than the supply for an electrothermal engine, because of the high voltages required and the number of

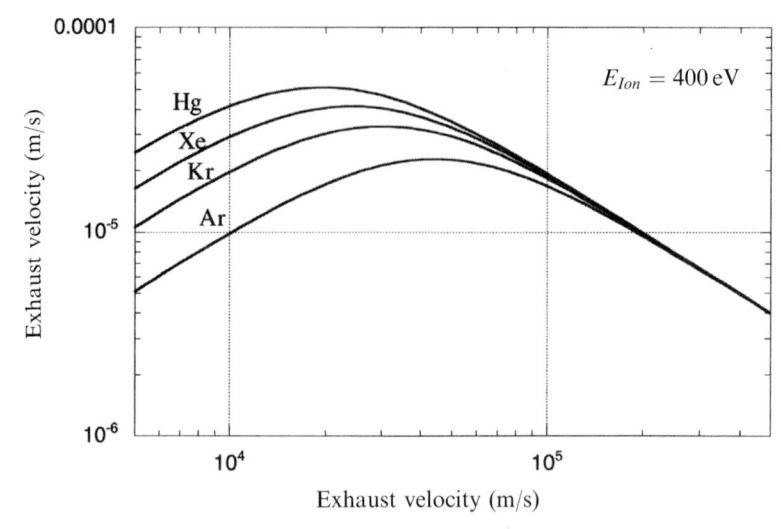

Figure 6.12. Thrust-to-power ratio for various ions as a function of exhaust velocity.

Figure 6.13. Two ion engines that were used on the ESA Artemis spacecraft to raise the perigee.
Courtesy ESA.

electrodes to be fed. In terms of mass, this is normally not as important as the high efficiency of the engine.

6.6 PLASMA THRUSTERS

The low thrust and very high exhaust velocity of ion engines are a disadvantage for many applications where the efficiency of electric propulsion would be beneficial.

Sec. 6.6] Plasma thrusters 169

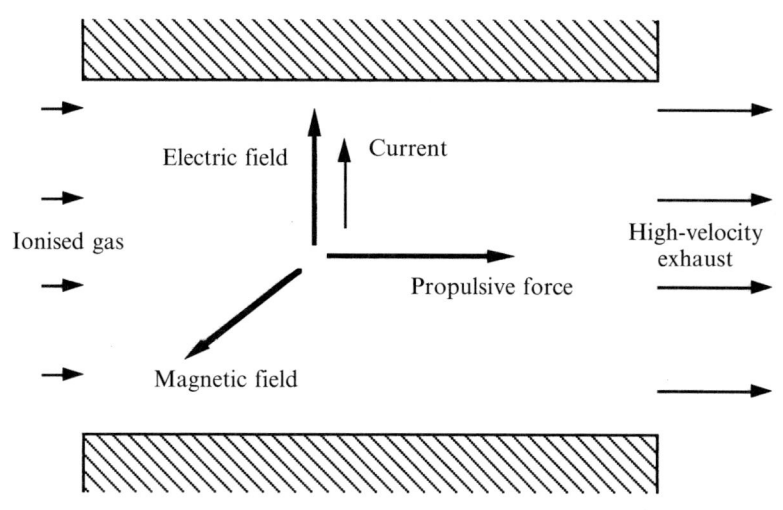

Figure 6.14. Principle of the plasma thruster.

These disadvantages are attributable to the fact that only the positive ions contribute to thrust, and the ion current is limited to a low value by the space charge effect. If the ion flow could be increased beyond the space charge limit, then a much more versatile engine could be developed. It would have higher thrust and a somewhat lower exhaust velocity more in keeping with the requirements of a wide range of missions.

For many years it has been the dream of engine designers to produce a practical plasma thruster. The principle is simple. An ionised gas passes through a channel across which are maintained orthogonal electric and magnetic fields (see Figure 6.14). The current, carried by electrons and ions, which develops along the electric field vector, interacts with the magnetic field to generate a propulsive force along the channel. The force acts in the same direction for both electrons and ions, and so the whole plasma is accelerated; and to a first order, the accelerating force is not limited by the density of the plasma, so there is no limit analogous to the space charge limit. The gas does not, in fact, need to be completely ionised; even a few percent of ions is sufficient, because they transfer their energy to the neutral gas molecules by collision. Energy lost by the ions in this way is immediately restored by the electric and magnetic fields.

Such a device would have exactly the characteristics required: a higher mass flow giving higher thrust, and exhaust velocities in the 10–20 km/s range, ideal for many missions. The principle is analogous to that of a linear electric motor, where the current flowing through the plasma is represented by the current in the armature, and the accelerating force acts in the same way.

The process is, however, nowhere near as simple as this: the transverse current generates its own magnetic field, the gas is heated, and the ions are acted upon by electric and magnetic fields which they themselves generate. A combination of thermodynamics and electromagnetic theory is required to predict the outcome.

This science—magnetoplasmadynamics—was gleefully seized upon by theoreticians, with no fewer than 16 interrelations between parameters. At the same time many experimental devices were made and tested, but nothing emerged which could be used in space—until one of those occurrences arose which restore faith in experimental science.

Workers experimenting with arc-jets at very low gas pressures (millibars) observed that currents of the order of 3,000 amps could be made to flow with no erosion of the electrodes, and with exhaust velocities of up to 100 km/s and an efficiency of 50%. This was obviously a magnetoplasmadynamic effect, but the mechanism was obscure. The experimental results were undeniable, and this remained the best hope for the development of a practical plasma thruster until the end of the Cold War.

Nevertheless, there have been many designs of plasma thrusters based on the ideas presented here, but few have got beyond the design stage. One device that has performed satisfactorily in the laboratory is the pulsed magnetoplasmadynamic thruster. The anode forms the external cylindrical electrode, and a solid rod along the axis forms the cathode. Thousands of amps are passed between the electrodes, generating both an E-field, and an azimuthal B-field. The ions and electrons are ejected from the cavity by the Lorentz force between the ion current and the crossed fields. The reaction to this acceleration is felt on the anode structure and transferred to the spacecraft. While steady state versions have been operated in the laboratory, only a pulsed version has been operated in space. In general the thrust of these devices is much higher than other kinds of electric thruster, being in the range 20–200 Newtons, compared with milliNewtons for most electric thrusters; the exhaust velocity is also high—in the range 10 to 110 km/s. This is the archetypal high-power thruster, with some laboratory versions operating at powers as high as 10 MW; efficiency and exhaust velocity increase with the power dissipated. A variety of propellants has been used, including the noble gases and lithium. This latter figured in a major programme in Russia. In the 1970s this thruster operated for 500 hours at 500 KW and several thousand seconds at over 10 MW. Recent tests have also been conducted at lower power (120 kW) and lower exhaust velocity (35 km/s); the efficiency here was 45%.

Particularly for interplanetary missions, perhaps even for manned ones, this type of thruster is ideal. Nevertheless, so far it has proved impossible to solve operational problems associated mainly with the very high currents. Erosion of the cathode is a particularly intractable problem. The only device to have been flown is the pulsed type, where the current is switched on and off every few milliseconds. The Japanese Institute of Space and Astronautical Science flew a device like this, which survived 40,000 pulses in space, and 3 million in ground testing. The power was much lower (1 kW) and the exhaust velocity was 11 km/s, with an integrated thrust of 20 mN.

6.6.1 Hall effect thrusters

Over the years, Russia developed a plasma thruster based on the Hall effect, and implemented it on more than 100 satellites. This device has now become available worldwide, and is simple and practical. The Hall thruster belongs to the family of magnetoplasmadynamic devices, and has been shown to share the practical properties of the low-pressure arc-jet described above—and it works. However, it was only in the Russian space programme that it had been brought to a practical and space-qualified form.

In an ionised gas the Hall effect, as it applies to thrusters, can be understood in the following simple way, developing from the situation shown in Figure 6.14 but taking into account the collisions between the electrons, ions, and neutral gas molecules. In free space with the crossed fields (shown in Figure 6.14) the electrons and ions follow spiral tracks with diameters dependent on their charge-to-mass ratio; electrons move in tight spirals, and ions move in wider spirals. The net current in the *axial* direction is zero, because the electrons and ions move in the same direction. It is, in fact, this motion which generates the plasma flow shown in the figure. If we now consider that this spiral motion will be interrupted every time an ion or electron collides with a gas molecule, we can see that if these collisions are very frequent then very little motion along the channel can occur, and the predominant drift will be along the E vector. If the collisions are infrequent, then the spiral motion continues uninterrupted, and the predominant drift is along the axis of the channel.

The parameter which determines whether collisions are frequent or infrequent is the Hall parameter—the ratio of the gyro frequency of the particle in the magnetic field to the particle collision frequency. For large values of this ratio, axial drift occurs because collisions are infrequent; for small values, collisions dominate and the flow is along the electric field.

If electrons and ions are affected equally, there will be no net current in the axial direction, and a neutral plasma flow will arise. In fact, because of the large difference in electron and ion charge-to-mass ratio, electrons and ions behave very differently. With their small gyro radius, electrons can, under certain density conditions, drift freely in the axial direction, while the ions, undergoing many more collisions per cycle, are constrained to drift along the electric field. This generates a net axial electron current called the Hall current. This current, and the fields it sets up, can be used to accelerate the plasma in a Hall thruster.

There are two possible configurations, the simplest of which is shown in Figure 6.15. Note that in this figure the electric field is axial, unlike the situation depicted in Figure 6.14, and so the Hall current is transverse. Here there is an axial electric field and a transverse magnetic field. If the gyro-frequency of the electrons is sufficiently large, their Hall parameter is significantly greater than unity and they drift in a direction orthogonal to both fields (upwards in the diagram) as the Hall current j. This current again interacts with the magnetic field to produce an axial force, which accelerates the plasma. These effects are described as acting sequentially, although in reality all act together as an internally self-consistent system of forces and fields; the result is a high-velocity exhaust stream.

This configuration easily adapts to a coaxial geometry, and it is in this form that the Russian Hall thruster has been successful. The coaxial form has a major advantage, because the Hall current can form a closed loop and the electrons never interact with the walls.

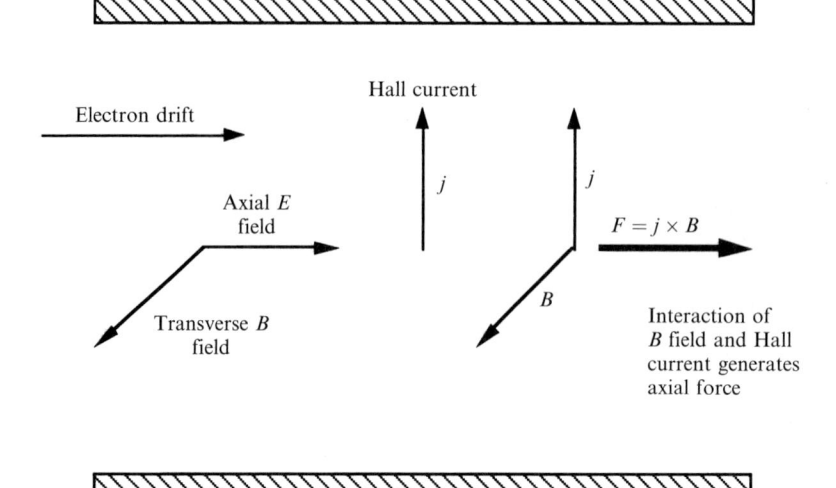

Figure 6.15. Principle of the Hall effect thruster.

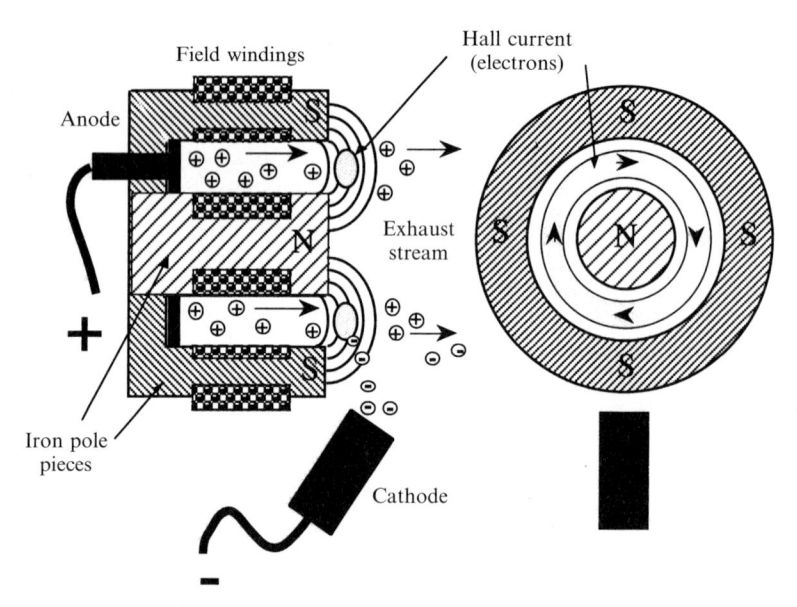

Figure 6.16. Schematic of the Hall thruster.

The general scheme is illustrated in Figure 6.16. The principle is exactly the same as illustrated in Figure 6.15, except for the coaxial geometry. The iron poles, energised by field windings, generate a cylindrically symmetric magnetic field of a few hundred Gauss, which is radial across the annular discharge cell. This cell is fully lined with insulating material—usually alumina or boron nitride. The propellant is introduced through fine holes distributed uniformly around the base of the cell, and is partially ionised by the discharge, which is developed between the annular anode and the cathode electron gun. The Hall effect ensures that the electrons acted on by the crossed electric and magnetic fields set up a circular Hall current at the exit of the cell. The interaction of the Hall current and the radial magnetic field generates an outward force which is transmitted to the ions and neutral gas atoms by collision, and this generates the exhaust stream.

The magnetoplasmadynamics is complicated, and it is in some ways simpler to think of this device as an electrostatic accelerator in which the outer grid is replaced by the electrons circulating in the Hall current. Ions can be thought of as being accelerated by the electric field developed between the Hall current ring (negatively charged) and the anode annulus at the base of the cell. It has the inestimable advantage over the ion drive of there being no space charge limitation, and partial ionisation is acceptable; in fact the exhaust stream is more or less neutral. Much higher thrust is thus possible.

The Hall thruster is a practical device, and the theory is complex and by no means secure, so an example will be more illuminating than an attempt to calculate the performance. By analogy with the electrostatic engine, the overall efficiency will increase with exhaust velocity, and the exhaust velocity itself will increase with the applied potential difference between anode and cathode.

The Russian SPT 140 5 kW thruster—a large Hall thruster—develops an exhaust velocity of 22.5 km/s for a discharge potential of 450 V, and a thrust of 250 mN. The discharge current is 10 A, and the efficiency is 57%. The thrust-to-power ratio is close to the theoretical value for electrostatic acceleration given in Figure 6.12. However, the thrust per unit area is much larger than for the electrostatic engine—about a factor of 10 (Figure 6.10). This is a direct consequence of the space charge limitation of the electrostatic engine, which limits the current flow. The exhaust velocity is typical for this kind of thruster: 15–25 km/s is the normal range, compared with the electrostatic thruster, which is really optimum in the 50–100 km/s range.

It is clear that the development of Hall thrusters and other types of plasma device is in a very active phase, stimulated by the input from Russia. Improved systems are continuing to emerge, and will find ready application in station keeping and inter-orbit transfer.

6.6.1.1 Hall thruster variants

There are two kinds of practical Hall thruster, both emerging from the Russian programmes. The first kind, as described above, has the annular cavity lined with insulator, usually boron nitride, and the cavity itself is rather deep. This is often described in the Russian literature as the Stationary Plasma Thruster, or SPT (Figure 6.17). The other type, which came from a different laboratory in Russia, has a much shallower annular cavity, lined with metal rather than insulator. This is

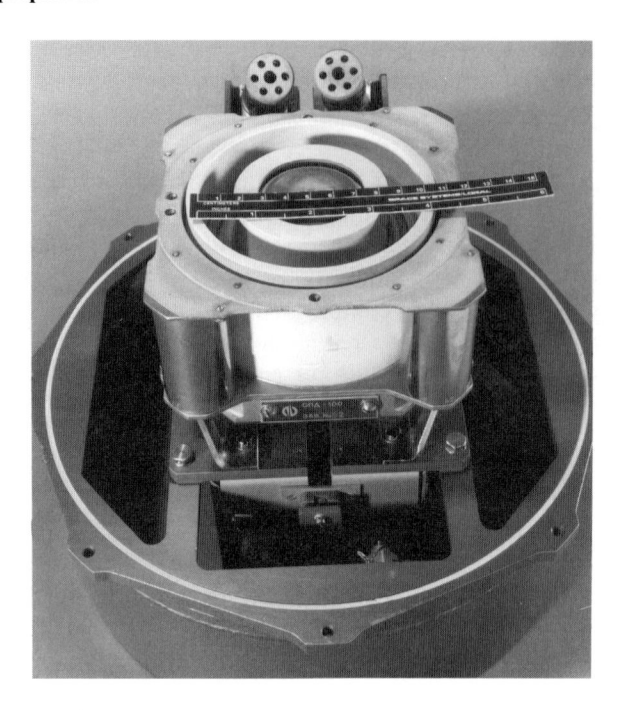

Figure 6.17. The Russian SP-100 Hall effect thruster. 100 mm diameter of the insulated cavity is shown by a ruler. The twin electron guns can be seen as well as two of the four coils that power the outer ring magnet. This is an SPT rather than a TAL device, it has a deep discharge cavity, insulated with boron nitride.
Courtesy NASA.

known as the Thruster with Anode Layer, or TAL (see Figure 6.18). Both kinds work well, and produce similar performance, but by examining the differences, some insight into the detailed performance of the Hall thruster can be gained.

So far, we have only considered that a discharge occurs down the annular cavity between the anode at the bottom and the electron generator placed outside the thruster. The electrons of course cannot travel down the cavity to the anode easily, because of the magnetic field; they form the ring of electron current at the mouth of the cavity. The question then arises as to how the xenon atoms are ionised in the cavity. This becomes clear when we realise that although the electrons are constrained by the magnetic field to drift azimuthally to form the Hall current, they also have an axial component to their velocity, caused by the electric field, and some penetrate to the anode. In fact, through most of the cavity, the plasma is neutral, with a current of high-velocity ions moving out of the cavity compensated by a slow moving—because of the magnetic field—higher density electron current, moving towards the anode. Ionisation is accomplished by these electrons, and most of the acceleration of ions takes place in this neutral region—note that, although the charge state is neutral, there is a strong field here. There are also secondary electrons,

Figure 6.18. A Russian D-100 TAL Hall thruster with a metallic anode layer.
Courtesy NASA.

produced by impact of the primary electrons and ions on the walls of the cavity; these contribute to the electron current flowing towards the anode. In the SPT, the role of the insulator lining the cavity, is twofold. It enables an axial field to be maintained down the cavity, and it provides a surface with a high secondary electron production coefficient, so that many electrons are produced from this surface, which help to maintain the discharge and to ionise the xenon atoms. The large secondary electron cross section of the insulator produces many low-energy electrons, and the neutral acceleration region extends deep into the channel.

At first sight, the idea of including a metal liner in the cavity, as in the TAL device, seems counterproductive; metals have low secondary electron coefficients, and, of course, a conductor in the discharge region forces the potential at its surface to be constant. In fact, this device works well, but has significantly different properties. It does so because the presence of the metal liner reduces the secondary electron flux and forces the neutral acceleration region out of the channel, and into the region just above it. Thus, the Hall current ring and the neutral acceleration region are close together, and the metal-lined channel can be very shallow indeed. The advantage of this is that collisions between the electrons and the channel walls are reduced and so is the erosion of the channel caused by this process. Because the acceleration takes place very close to the anode, which has been raised almost to the top of the channel, this is called the Thruster with Anode Layer. Both kinds of device are efficient and stable in operation, with it being more a question of different flavours of device, than a fundamental advantage with one or the other. Both have a substantial heritage in the Russian space programme, as shown in

Table 6.1. Development status and heritage of some Hall effect thrusters.
Courtesy of NASA/JPL and SNECMA.

Thruster	Power (kW)	I_{sp} (sec)	Efficiency	Thrust (mN)	Development status	Flight heritage
SPT-50*	0.3	2,000	0.4	17	Flight tested	Meteor
D-20** (TAL)	0.3	2,000	0.4	17	Laboratory model	None
SPT-70*	0.7	2,000	0.45	40	Flight qualified	Kosmos, Luch
D-38** (TAL)	0.7	2,000	0.45		Laboratory model	Meteor
SPT-100*	1.4	1,600	0.5	100	Flight qualified	Gals, Express
D-55** (TAL)	1.4	1,600	0.5		Flight qualified	Flew 1997‡
T-100†	1.4	1,600	0.5		Laboratory model	None
SPT-160*	4.5	2,500	0.6	400	Under development	None
D-100** (TAL)	4.5	2,500	0.6		Under development	None
T-40^A	0.1–4	1,000–1,600	0.6	5–20	Development tested	In preparation for system qualification
T-140†^A	1.8–4.5	1,800–2,200	0.6	160–300	Under development	Preparation for system qualification
T-220^A	7–20	1,500–2,500	0.6	500–1,000	1,000 hours operation in tests	Preparation for system qualification
PPS 1350^B	1.5	1,800	0.55	92	Flight qualified	SMART-1

* Design Bureau Fakel, Kaliningrad (Baltic region), Russia
** TsNIIMASH, Kalingrad (Moscow region), Russia
† NIITP, Moscow, Russia
‡ United States classified military flight application
A Tested at Air Force lab or NASA-Glen
B SNECMA France

Table 6.1. The typical performance of these devices is as follows: input power 1,400 W, efficiency 50%, exhaust velocity 16 km/s, and thrust 83 mN. As such, they fit very well to applications where higher thrust is needed, coupled with a moderately high exhaust velocity. Reference to Figures 6.2 to 6.4 shows that the major gain in efficiency occurs when the exhaust velocity is greater than 10 km/s.

The heritage of these devices in the Russian programme is extensive. From 1971 through to 1974, four SPT-60 thrusters flew on Meteor satellites for station keeping; four SPT-50 thrusters flew on a further Meteor satellite in 1976; and Cosmos and Luch satellites carried a total of sixty SPT-70 devices, between 1982 and 1994. SPT-100 thrusters were introduced for the Gals and Express telecommunications satellites for north–south station keeping; a total of 32 being used between 1994 and 1996.

6.6.2 Radiofrequency thrusters

Most high power—and hence high thrust—systems use electrodes, of one kind or another, to generate the current in the gas that provides the ions and hence the thrust. These always erode in the discharge, being worse for high power systems. Several attempts have been made to increase the power input by using microwaves to provide the internal energy source. The simplest device is analogous to the electrothermal thruster: microwaves are used to heat the gas in a 'combustion' chamber connected to a de-Laval nozzle, which converts the hot gas into an exhaust stream. The 'combustion' chamber is in fact a microwave cavity designed to set up standing electromagnetic waves that heat the gas by accelerating electrons, which in turn ionise the propellant, allowing higher microwave induced currents to flow in the gas and heat it to propellant temperatures. Some laboratory thrusters of this type have been made.

A much more complex scheme is the variable specific impulse device, called **VASIMIR**; it uses radio frequency electromagnetic fields to ionise and accelerate the plasma. The process here begins with the gas, hydrogen and helium mixed, being exposed to the electromagnetic radiation from an RF antenna that ionises most of the atoms. It then passes into a cavity with strong magnetic fields, where cyclotron resonance is excited by a high-power RF generator; resonance here gives a very high efficiency of power transfer from the RF field to the gas. The electrons and ions oscillate within the plasma and raise its overall temperature to a very high value. Temperature here is somewhat different from temperature in an un-ionised gas, because the ions and electrons behave somewhat differently from neutral molecules at high temperatures. Nevertheless, the assemblage of ions and electrons behaves somewhat like a gas at high temperature and pressure. It then enters a *magnetic nozzle*, which behaves like a conventional nozzle, in that it allows the hot, high-pressure plasma to expand, and so to generate a high-velocity exhaust stream, and thrust, in the conventional way. The difference is that the nozzle has no mechanical presence at all; it is made up from a carefully shaped, strong, static, magnetic field. The electrons and ions in the plasma are forced to travel along the diverging magnetic field lines, so that the plasma expands as it emerges from the engine. The thrust is generated by the reaction of the ions on the magnetic nozzle (Figure 6.18).

This is the basic scheme, but for the VASIMIR engine, unionised gas is injected into an outer, mechanical nozzle, which surrounds the magnetic field nozzle. This gas is heated by contact with the ionised gas in the magnetic nozzle, and adds to the thrust. Variability in the thrust is achieved by changing the size of the magnetic nozzle throat, simply by changing the field. At the same time, the amount of heating and the propellant flow rates can be separately adjusted to change the exhaust velocity. In many ways this device is analogous to a chemical rocket, but it has the advantages of electric propulsion. It is claimed that this device can operate at very high powers, and can be optimised for high-thrust missions. So far, only laboratory demonstrations of the processes involved have been made.

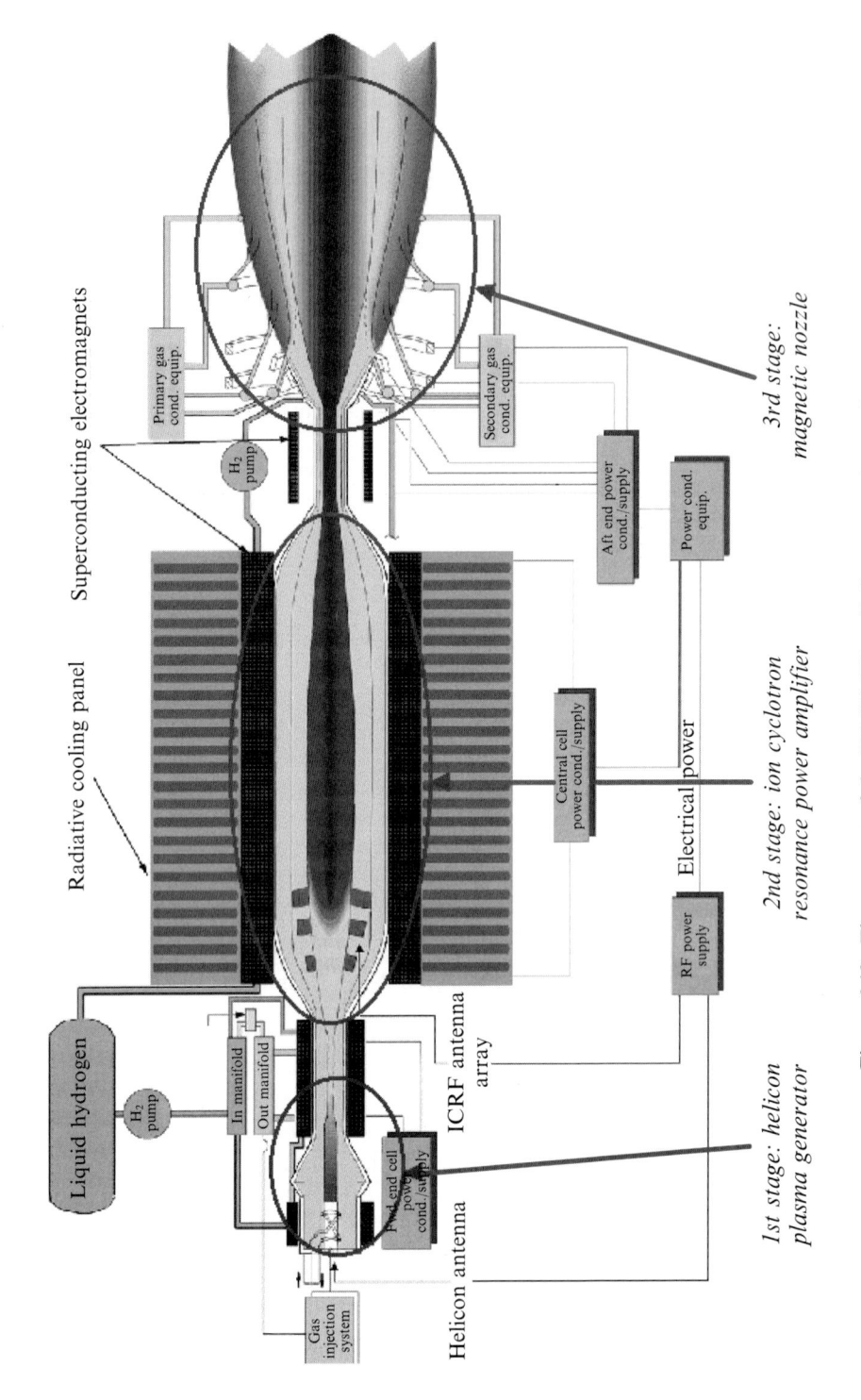

Figure 6.19. The concept of the VASIMIR radiofrequency plasma thruster.
Courtesy NASA.

6.7 LOW-POWER ELECTRIC THRUSTERS

A disadvantage of most of the thrusters described so far is that they cannot be switched on and off quickly, but rather the discharge takes some seconds to become established. This is no disadvantage for large delta-V manoeuvres, where, if anything, very long thrusting times are required. For high-precision station keeping, however, this is a major problem. What is needed for high precision manoeuvring is the ability to make many, small, metered, changes to the momentum of the spacecraft, in a short time. Delay in building up thrust could be fatal to such a scheme; or the uncontrolled thrust developed during a few seconds of build-up, might be wasted, and require a compensating thrust in the opposite direction, leading to low fuel efficiency. Also, many kinds of electric thruster cannot be operated below a certain beam current; they become unstable. In all these cases, the high propellant efficiency of electric propulsion, vital for high-precision station keeping, is lost.

A device that accomplishes the necessary vernier thrusting, for precision station keeping, or formation flying, is the Field Effect Emission Thruster or FEEP. Here a very low and continuously variable thrust is possible, using the field effect principle to provide ions. A liquid metal, usually caesium—because of its low melting point (29°C)—coats one of a pair of electrodes, across which a very high field is maintained. The liquid metal is drawn up into a number of conical protrusions, by the electric field. The field between the tip of the protrusion increases as it grows and the gap between it and the other electrode decreases. At the limit, the cross section of the point becomes of the order of atomic dimensions, and atoms become ionised by the very strong local field, and are picked off the tip of the cone; they are accelerated across the gap to form an ion current. This device is analogous to the field-effect microscope, used to image atomic arrangement in materials deposited on a very fine needle-point. A field strength of 107 V/m, at the tip, is necessary to ionise caesium by this method. An important principle here is that a field strength, sufficient to produce ions from a single liquid metal 'needle', will occur for any potential difference applied across the electrodes—above a certain threshold, because the 'needle' will simply grow until the gap between its tip and the accelerating cathode is sufficiently small for an ionising field to be created. For a low potential difference between the electrodes, only a few 'needles' will be produced, before the field is reduced by their presence. For a high potential difference, many 'needles' can be produced. This gives the fundamental property of the FEEP thruster: the ion current can be precisely controlled by changing the potential difference between the electrodes. Since the ion current produces the thrust, the thrust can be precisely controlled. A separate electron gun is needed, as in the case of an ion thruster, to keep the spacecraft neutral.

Current concepts of FEEP thrusters use a slit cathode as the accelerating electrode, and the caesium is allowed to flow over a flat anode parallel to the cathode, and separated by about one millimetre. The caesium is fed from a reservoir through a capillary and flows over the anode in a thin layer, controlled by surface tension. Caesium is advantageous here because it melts easily, has a low ionisation potential, and wets metal surfaces. The ions are accelerated through the slit to produce the thrust. Typical performance characteristics are: thrust from 250 mN upwards, depending on the power; continuously thottleable power levels, up to several hundred watts; and an efficiency of 60%. The most important characteristic, after controllability of thrust, is the very high exhaust velocity, typically 60 to 100 km/s. This makes high-precision station keeping and formation flying very fuel-efficient. The disadvantages are the potential for contamination of the spacecraft by the emitted caesium, and the requirement for a high operating voltage, which the FEEP shares with the ion thruster.

6.8 ELECTRICAL POWER GENERATION

Figure 6.4 shows that, as the exhaust velocity increases, the optimal performance of an electric propulsion system moves towards higher power-to-weight ratios for the electrical supply. At the same time we know that the power required depends on the product of the thrust and the exhaust velocity—see Chapter 7. Thus a high-thrust, high exhaust velocity, engine, has a high power requirement. The available sources of electrical energy, in space, are few. Up to the present-day, the only sources used to power electric thrusters have been batteries or solar cells. Since the mass of the power supply adds directly into the payload mass for any manoeuvre, it is important to find power sources that are capable of high power delivery, and have a high power-to-mass ratio.

6.8.1 Solar cells

The maximum efficiency of solar cells, in converting solar energy to electricity, ranges from 15 to 20% depending on the type. Typically, for a 30-kW array, the mass per kilowatt is about 13 kg. The areal extent would be about $210 \, m^2$, achieved by deploying a folded structure, once in space. For lower power, 5–6 kW, a mass per kilowatt of 7 kg can be achieved. This reflects the mass needed for the structure of the larger deployable array. With improved solar cells, especially gallium arsenide, and the use of solar concentrators, which focus the sunlight collected by lightweight reflectors on to a smaller area of solar cells, a mass per kilowatt of about 3 kg is thought to be achievable. The immediately obvious disadvantage of solar power is the limit to the total power available imposed by our inability to build very large deployable arrays. Powers much above 100 kW are unlikely to be achievable with the current technology. A less obvious disadvantage is the fact that sunlight diminishes in intensity with the square of the distance from the sun; a spacecraft travelling away from Earth orbit towards the outer solar system faces a constantly decreasing power availability. It is difficult to imagine that enough power can be extracted from solar panels to drive an electrically propelled spacecraft much beyond the orbit of Mars. On the other hand the power output drops significantly with increased temperature, so that travelling much inside the orbit of Mercury is not possible with solar cells as the main power source.

There is a further problem with solar cells, and this is their sensitivity to radiation damage. Energetic protons, in the Earth's radiation belts, or emitted by the sun during solar storms, displace atoms in the silicon and change its properties so that the power available decreases significantly with time, depending on the radiation exposure. Normally a 130% over-size requirement ensures 5 years of adequate power for satellites in Earth orbit. For electric propulsion, using low thrust, a spacecraft could spend several months in the radiation belts of the Earth, and the cells would degrade significantly during that time. This again indicates the need for over-sizing of the panels, and correspondingly lower power-to-mass ratios.

Despite these disadvantages, all current electrically propelled space missions use solar cells; the system is called Solar Electric Propulsion, or SEP.

6.8.2 Solar generators

Given the conversion efficiency of 15–20% for solar cells, it is clear that a conventional mechanical generator set, with an efficiency of 30–40% would provide double the power-to-area ratio. These systems have yet to be deployed, but are under active consideration for power generation. The basic arrangement is to concentrate solar energy on to a 'boiler' containing a working fluid, which then drives an engine connected to a generator, just as in a conventional terrestrial power system. The efficiency would be 30–40% as already mentioned, and the system would be immune to radiation damage, and could possibly work with smaller solar intensity for deep space missions. The difficulties are the usual ones that arise with the use of mechanical systems in space: seals, glands, bearings, all perform badly in space, because of the vacuum, and zero-gravity environment; radiation damage to organic materials is another factor. However, a considerable amount of development work has gone into closed-cycle heat engines, mainly for refrigeration, but equally well applicable to power generation. In Europe, and especially the UK, the Stirling engine has been developed. This has an oscillating piston which manages, without glands or organic seals, to generate cooling from electrical power, or, if reversed, to generate electrical power from heat. In the United States, while the Stirling engine is used, a similar device, the Brayton turbine, has been developed as well. There is no reason why these devices, alongside lightweight solar concentrators, could not improve the overall mass-to-power ratio, and allow greater electrical power to be extracted, from the same area of solar illumination.

6.8.3 Radioactive thermal generators

The problem of low solar illumination at the outer planets, even on Mars, prompted the development of electrical power generators employing nuclear systems. Introduced in 1961, these RTGs have been used for many missions: the Viking landers on Mars, the Mariner missions, the Voyager missions to the outer planets, and the present-day Galileo and Cassini missions. RTGs (Figure 6.20) are the only current solution to power supply in the Saturnian and Jovian systems,

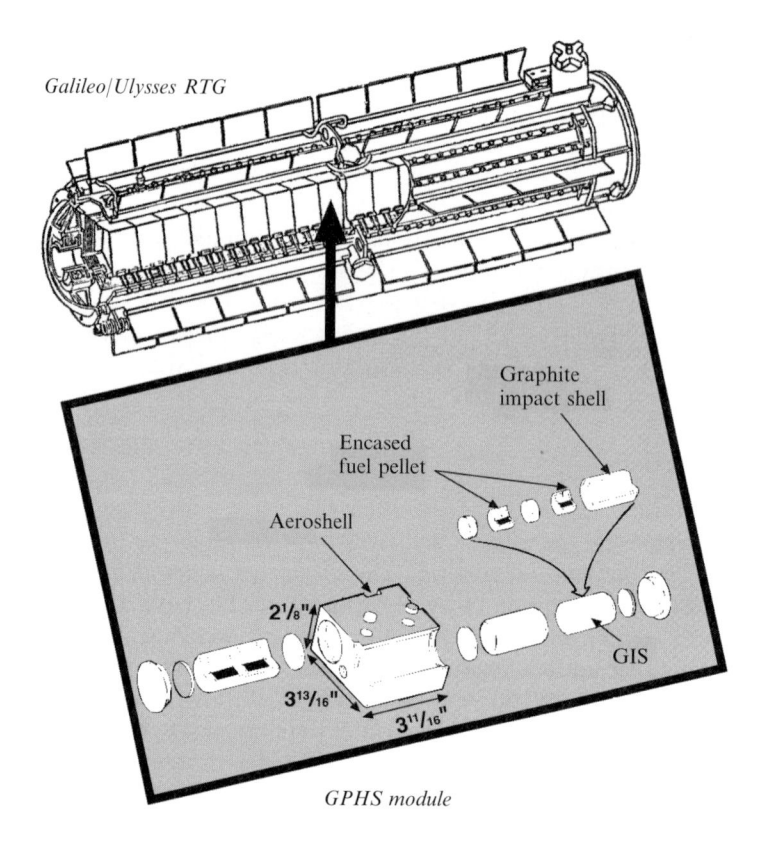

Figure 6.20. A complete RTG, cutaway to show how the fuel pellets are inserted. Note how the thermoelectric converters are connected between the individual sections of heat generator and the small radiators associated with each section.
Courtesy NASA.

and beyond. The concept is to convert the heat, generated by radioactive decay of a suitable element, into electrical power. Plutonium (Pu^{238}) is generally used, because it has a suitable half-life of 80 years: too long, and it would not generate enough power per kilogram; too short, and it would not generate power for long enough. There are two components, the heat source, and the power converter. The heat source comprises small discs of plutonium oxide—a ceramic material designed to break-up on impact rather than to release dust, contained within iridium capsules. Iridium is remarkable in that, while very strong, it can be distorted and stretched without rupture. Two iridium-encased pellets are enclosed within a graphite impact shell, to protect them from damage on impact, or from flying fragments following an explosion. Two of these assemblies are encased in a further graphite *aeroshell* to protect the assembly from heating during re-entry (Figure 6.21). Apart from the plutonium oxide pellets, the rest of this assembly is there for safety reasons, in case of an accident during launch, or an unforeseen re-entry of the generator. The rule,

Sec. 6.8]   Electrical power generation   183

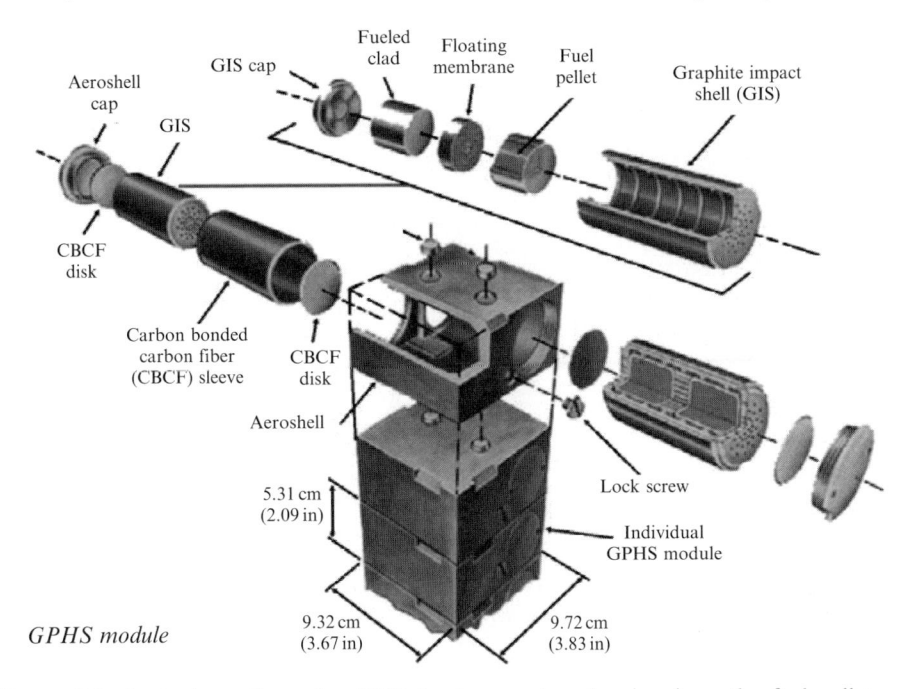

Figure 6.21. A single section of a RTG heat generator showing how the fuel pellets are protected with iridium capsules, graphite impact shields, and a graphite re-entry shield.
Courtesy NASA.

demonstrated by tests, is that no radioactive material is to be released in a worst-case scenario involving either or both of these events.

In all current RTGs, the heat generated by the plutonium is converted into electricity, using thermoelectric generators, with their hot junctions connected to the heater units, and their cold junctions connected to radiators that dump the heat away to space. A pair of such hot junctions is connected to each of the aeroshell units—containing four plutonium oxide pellets—the corresponding cold junctions are connected to a single radiator for each unit, as shown in Figure 6.20. Early devices used lead-tellurium junctions and produced a few watts of electrical power. More recent devices use silicon-germanium junctions, and produce up to 285 W of electrical power, with the plutonium running at 1,235° C. Even the latest versions have an efficiency of conversion of thermal energy into electrical energy of only 6%. With a mass of more than 50 kg, the mass per kilowatt is 175 kg, far in excess of the solar panel ratio. Nevertheless, these are currently the only way to produce electrical power in the outer solar system.

The low efficiency of thermoelectric generation—only a few percent—has led to designs being developed for the use of mechanical generators to convert the heat into electrical power; as mentioned above, such generators can have between 30 and 40% efficiency. This would improve the mass per kilowatt by a factor of 5 to 7, reaching 25 kg/kW. A scheme using a Stirling engine and a linear alternator is shown in

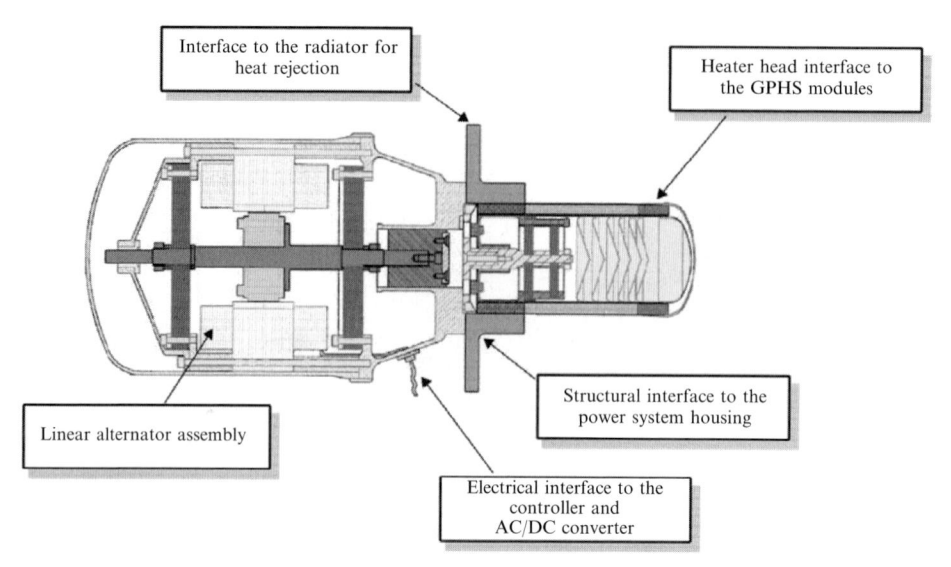

Figure 6.22. A Stirling cycle mechanical electricity generator. The Stirling piston is connected to a linear alternator to convert the reciprocating action into electricity. The hot side of the Stirling cylinder is connected to the RTG heat source, and the cold side to a radiator. Courtesy NASA.

Figure 6.22. The hot side of the cylinder and piston is thermally linked to the heat generator and the cold end to the radiators. The reciprocating motion is converted into electricity by the linear alternator. A current design uses a 500-W thermal power source containing 600 g of plutonium dioxide; the input temperature for the Stirling engine is 650° C, and the radiator temperature is 80° C. The electrical power produced, after conversion to DC, and conditioning, is 55 W. It is likely that this kind of power generation will become standard for use with RTGs, because of the high cost of plutonium, and the safety aspects. An improvement in efficiency, of a factor seven, means a reduction in the quantity of plutonium used by the same factor.

6.8.4 Nuclear fission power generators

It is clear, from the above that RTGs, will never reach the kind of power-to-mass ratio that is required for high power electric propulsion. Neither will solar cells achieve the necessary high absolute power levels. Nuclear fission has long been thought to be the only viable solution to this problem. The available energy in uranium fission is about 70 times greater than from the radioactive decay of plutonium, and the power output is completely controllable; plutonium on the other hand has a specific rate of heat generation that cannot be controlled. It takes 80 years to extract the energy from half the Pu^{238} atoms in an RTG, while for uranium fission the energy can be extracted at any desired rate. At the same time, uranium is a relatively cheap, natural material, while plutonium is an artificial

Sec. 6.8] Electrical power generation 185

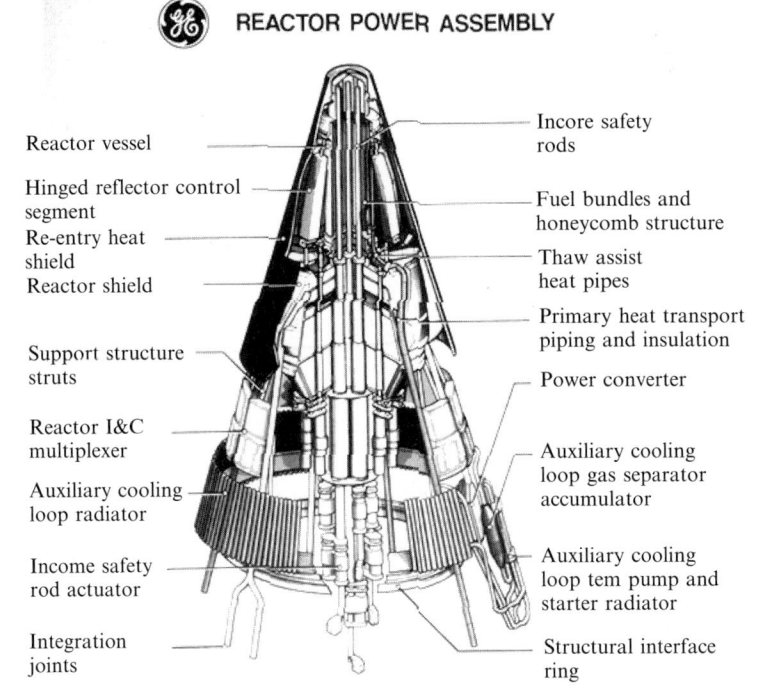

REACTOR POWER ASSEMBLY

Labels (left):
- Reactor vessel
- Hinged reflector control segment
- Re-entry heat shield
- Reactor shield
- Support structure struts
- Reactor I&C multiplexer
- Auxiliary cooling loop radiator
- Income safety rod actuator
- Integration joints

Labels (right):
- Incore safety rods
- Fuel bundles and honeycomb structure
- Thaw assist heat pipes
- Primary heat transport piping and insulation
- Power converter
- Auxiliary cooling loop gas separator accumulator
- Auxiliary cooling loop tem pump and starter radiator
- Structural interface ring

Figure 6.23. An early United States designed nuclear fission power generator. The small fission core is in the nose and the reflectors are placed around it. The rest of the volume is taken up by the power converters and the radiators to dump the waste heat to space.
Courtesy NASA.

element, created in fast-breeder reactors at enormous cost. Uranium is non-poisonous and not radioactive in the pure state, so the safety aspects of its use are much less challenging. It is not often appreciated that small nuclear reactors can be made, typically less than half a metre in diameter; this excludes radiation shielding, of course, and the protective details outlined above for RTGs. Full details of uranium fission reactors are given in Chapter 7, here it is only necessary to look at the electrical power generation aspects.

The thermal power output of the reactor is determined by the instantaneous neutron flux in the core, and this is controlled by means of external neutron reflectors and absorbers, which control the neutron flux being bounced back into the core. The reactor can be launched in an inert state, and can be switched on once in orbit; its power output can be raised or lowered at will. This underlines an important safety aspect of uranium fission reactors: the material is not radioactive until the device has been operated and therefore an accident during launch will not produce radioactive debris from a virgin reactor core.

Thermal power outputs of practical devices are high. The SNAP 10-A reactor (Figure 6.23) developed in the United States for satellite power applications in

1965, produced 40 kW of heat, the Russian Topaz reactor made 150 kW, and modern space reactor designs produce thermal powers up to 500 kW. Even with the low efficiency of thermoelectric generation, the Topaz reactor produced 10 kW of electrical power from 12 kg of uranium in a reactor with an all-up mass of 320 kg. At 32 kg/kW, this is not far from the mass-to-power ratio for solar panels. The difference is that the power output does not depend on the size of the reactor, only on the neutron flux, thus for high power output, the nuclear fission reactor gives the best mass-to-power ratio.

In devices generating heat at such a high power level, active heat transfer is required, and the usual method is to circulate liquid metal—sodium or lithium—through the core in the closed loop. Liquid metals have high boiling points and high heat capacity; they can also be pumped with alternating magnetic fields, requiring no mechanical contact. The liquid metal passes through a heat exchanger, connected to an electrothermal or mechanical generator, exactly as in the case of an RTG. Typical hot-end temperatures are the same as for a modern RTG, in the range 650 to 1,200° C, while the heat is dispersed into space by radiators operating at 80–100° C. The SNAP nuclear reactor, adapted to a space mission is shown in Figure 6.24, and modern systems will have similar characteristics. Note that by far the largest

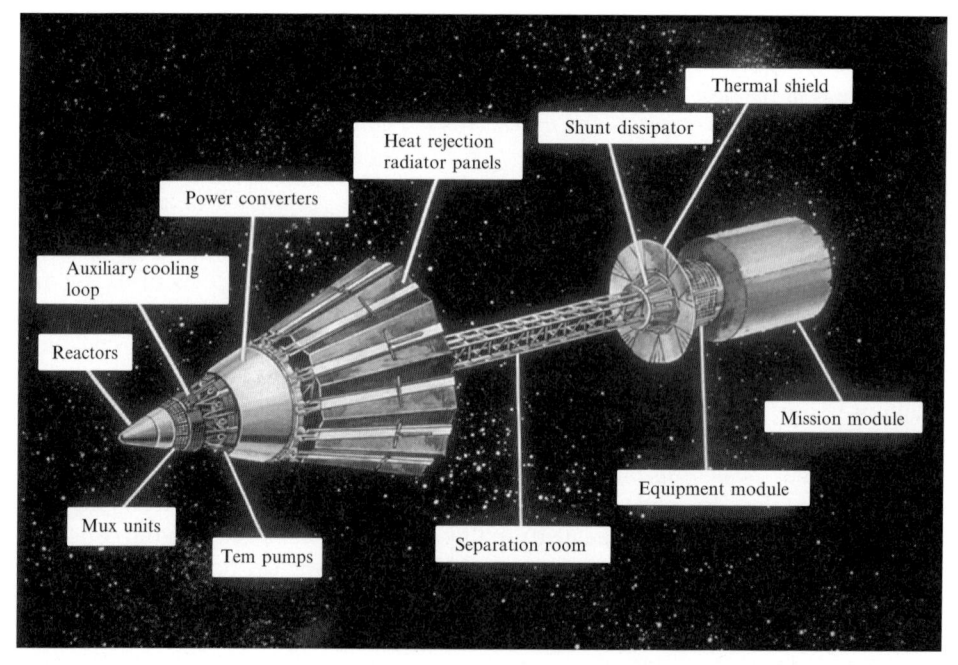

Figure 6.24. An early design for a spacecraft with nuclear electric generation. The reactor is in the nose, and the large radiators dump the waste heat. It is separated from the spacecraft by a long boom to reduce the radiation load on sensitive components.
Courtesy NASA.

component is the radiator, necessary to dump the waste heat to space, while the core itself is very small.

The use of nuclear fission for spacecraft power supply has so far been experimental, and used exclusively for military satellites. It is however now seen as essential for scientific exploration of the outer planets, and their moons. A new mission, called Jupiter Icy Moons Orbiter or JIMO, specifies a fission reactor to provide power for electric propulsion. For such missions, involving visits to several moons in the system, the total delta-V requirement is high, and electric propulsion is *de rigeur*. This will be the first modern use of high-power ion propulsion, coupled with a nuclear fission reactor to provide the electricity. The mission concept is shown in Figure 6.25.

6.9 APPLICATIONS OF ELECTRIC PROPULSION

The advantages of electric thrusters are mainly concerned with their ability to provide a high exhaust velocity, and hence to use propellant very economically. One way to look at this is to consider the following. Rearranging the rocket equation again we find

$$V = v_e \log_e \frac{M_0}{M}$$

$$\frac{M_0}{M} = e^{\frac{V}{v_e}}$$

$$M_0 = M + M_f$$

$$\frac{M + M_f}{M} = e^{\frac{V}{v_e}}$$

$$\frac{M_f}{M} = e^{\frac{V}{v_e}} - 1$$

In this inversion, the ratio of propellant mass to vehicle mass is given in terms of the exhaust and vehicle velocities. This is useful in calculating the quantity of propellant needed for any manoeuvre for a given payload. The ratio M_f/M is the propellant efficiency or *fuel multiplier*, and depends only on the ratio of the vehicle velocity to the exhaust velocity (Figure 6.26). We see that the propellant efficiency depends exponentially on the exhaust velocity, and this is why the high exhaust velocity provided by electric propulsion is so beneficial.

When comparing the performance of electric propulsion devices it is sensible to include a variety of different missions. Here we shall consider three cases: station keeping for a mission lifetime of 10 years, transfer from LEO to GEO, and a nine-month journey to Mars. It is sometimes useful to express the performance in terms of the propellant to total vehicle mass, which is

$$\frac{R-1}{R} = \frac{e^{\frac{V}{v_e}} - 1}{e^{\frac{V}{v_e}}} = 1 - e^{-\frac{V}{v_e}}$$

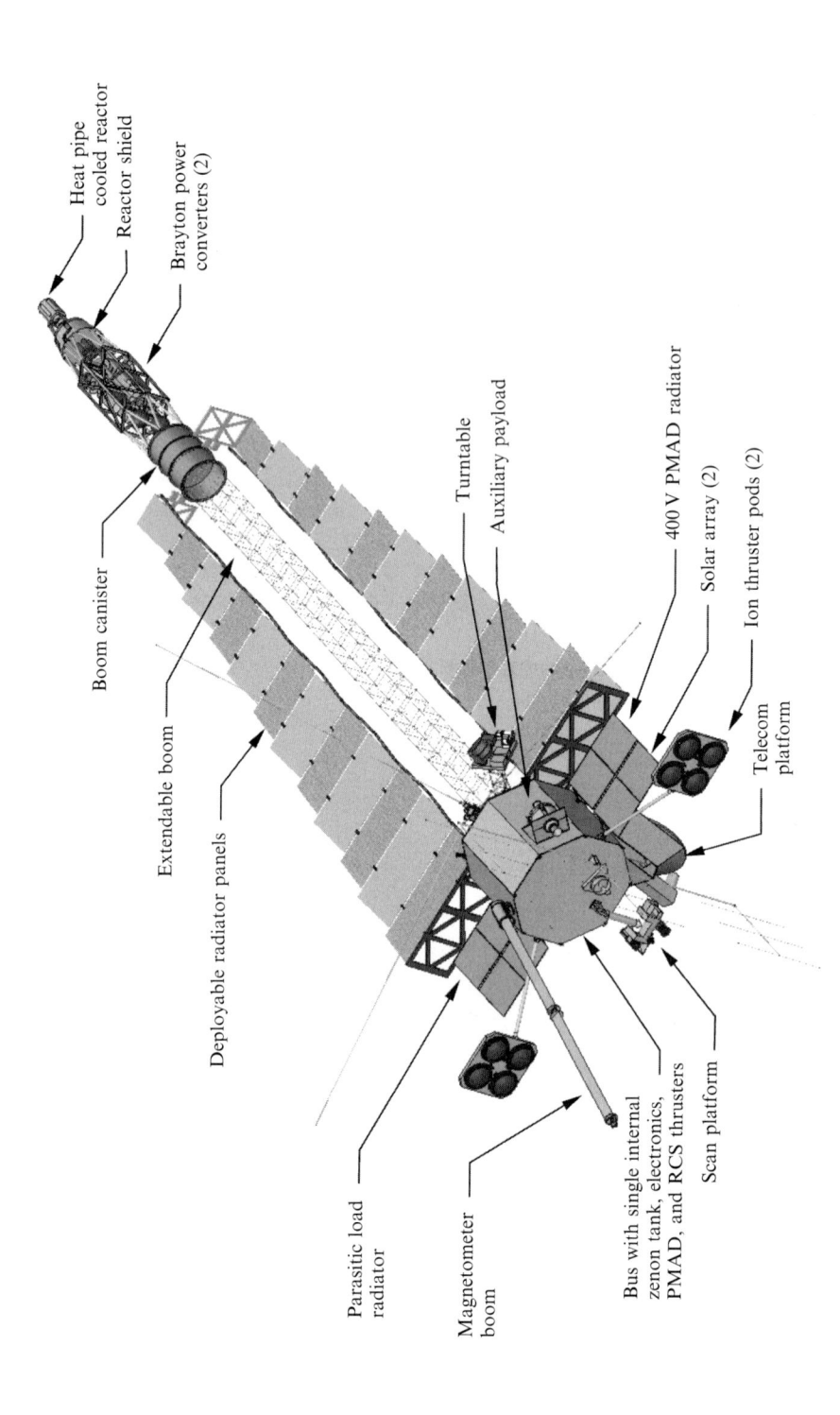

Figure 6.25. The JIMO mission concept, powered by a fission reactor electrical system driving ion thrusters. Note the large radiator area needed to dump the waste heat after electricity generation. This mission to Jupiter's icy moons has a high delta-V requirement, met by nuclear powered electric propulsion. Courtesy NASA.

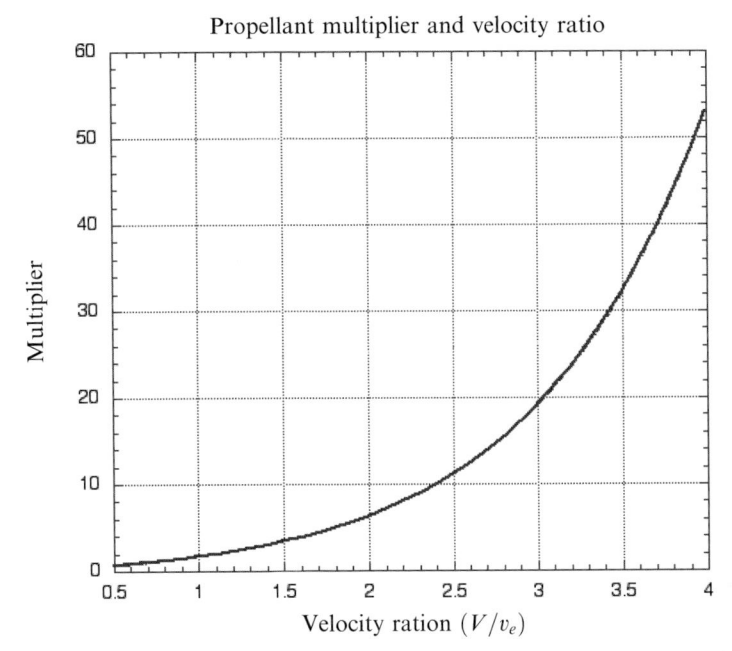

Figure 6.26. The propellant efficiency as a function of the ratio of the vehicle velocity to the exhaust velocity. The efficiency is here described as the *fuel multiplier*, because it allows a direct calculation of the quantity of propellant needed for any manoeuvre, if the exhaust velocity and the payload mass are known.

6.9.1 Station keeping

Here a typical satellite will carry sufficient control propellant to produce a total velocity increment of 500 m/s for station keeping. If this is provided by hydrazine thrusters with an exhaust velocity of 1,000 m/s, the ratio of propellant mass to total vehicle mass is 0.39; 40% of the mass of the satellite is propellant. If a Hall thruster with an exhaust velocity of 15 km/s is used, then the propellant is only 3.3% of the total mass of the satellite, and more than 90% of the propellant mass can therefore be saved. This reduces launch costs (at $20,000–30,000 per kg), or allows a larger satellite to be launched for the same cost. For an electrostatic propulsion system with an exhaust velocity of 30 km/s, the propellant mass drops to 1.6%. Note that once the exhaust velocity becomes much larger than the velocity increment, the saving in propellant mass decreases *pro rata*.

This crude calculation ignores the mass of the power supply, and no reference has yet been made to the power level required by the thruster. The power depends on the available burn time and the required thrust. Clearly, if the thruster requires a longer burn time than the time taken to restore the satellite to its proper orbit, then it will be incapable of carrying out the necessary station keeping. As a rough guide, the total burn time should be less than one half real time. In the case of station keeping, around 50 m/s of velocity change is needed each year, and so the burn time should be less than six months per year. Here we are compounding the many daily short burns of the thruster into a single continuous burn, which does not affect the result. Given such long burn times, the life of the thruster could be the limiting factor. Up to the present, demonstrated lifetimes of the order of one year of operation have been established, which places an upper limit, for a 10-year mission, of a year, or 3.15×10^7 s, to the total burn time.

Using the equation for power supply mass to propellant mass, which is repeated here for convenience,

$$\frac{M_E}{M_P} = \frac{v_e^2}{2\eta\xi t}$$

we find that for the Hall thruster, with efficiency 0.6, the extra solar panel mass (assuming 100 W/kg) is 5.9% of the propellant mass; that is, the extra power supply plus propellant mass is 3.5% of the satellite mass. Using the ion thruster, the factor of two in exhaust velocity requires that the power supply mass has to increase to 35.7% of the propellant mass (assuming 40% thruster efficiency). The propellant mass is half that needed by the Hall thruster, and so the saving is actually larger: the power supply plus propellant mass is 2.2% of the satellite mass.

For station keeping therefore, the savings in propellant mass are very significant—90% being a practical quantity. The choice between a Hall thruster and an ion engine is largely a matter of taste. Arc-jet thrusters and electrothermal thrusters, with their lower exhaust velocities, will provide smaller but still significant propellant savings. It seems likely that the Hall thruster and ion engine will be the station-keeping device of choice for current and future communication satellites.

6.9.2 Low Earth orbit to geostationary orbit

The required velocity increment for an elliptical transfer to geotationary altitude and circularisation of the orbit is theoretically 4.2 km/s. To this should be added the gravity losses associated with the continuous burning of the electric propulsion system, which is of the order 1 km/s, producing a total of 5.2 km/s for this mission. This is 10 times the station keeping requirement. It should also be carried out in a reasonable length of time, so may be expected to be more demanding of the thrusters.

As a comparison chemical thruster we can consider a bi-propellant engine using UDMH and nitrogen tetroxide. It has an exhaust velocity of 3,160 m/s *in vacuo* (see Chapter 3), requiring a propellant fraction of 80%. The same calculation as before produces, for the Hall thruster, a propellant fraction of 29.4%. For the ion engine the propellant fraction is 16%, and so the use of electric propulsion can potentially save about 80% of the propellant. Note that the higher exhaust velocity of the ion engine has a much bigger effect for this mission.

Here the necessary power is determined by the length of time the proprietor of the spacecraft is prepared to wait for it to reach geostationary orbit. Depending on his patience, we can consider six or 12 months as examples. For six months the ratio of power supply to propellant mass is 0.119, and is dependent only on the burn time and the exhaust velocity. The mass of extra solar panels plus propellant is now 33% of the spacecraft mass for the Hall thruster; but for the ion engine the supply-to-propellant ratio is 0.714, and the total mass of the propulsion system is therefore 27% of the spacecraft mass.

These simple arguments demonstrate that the propellant mass can be reduced from 80% of the spacecraft mass to about 30%—a saving of more than half. The difference between the Hall thruster and the ion thrusters is small, because the extra power of the ion engine requires more solar panel area, which is approximately offset by the higher exhaust velocity. For the longer trip-time of a year the power needs are halved, but the propellant requirement remains the same. The ratios are then 31% for the Hall thruster and 22% for the ion engine. This illustrates the advantage of long trip-times for the higher exhaust velocity.

6.9.3 Nine-month one-way mission to Mars

The minimum energy velocity increment for a one-way trip to Mars orbit is about 6 km/s. Intuitively this seems too small compared with that needed for a geostationary orbit, but from Chapter 1 we recall that the escape velocity from LEO is only 3.2 km/s (in addition to the 7.6 km/s LEO velocity). If this is carried out on the correct trajectory, then Mars orbit will be reached. It then remains to effect gravitational capture by Mars, which requires a velocity increment of around 2 km/s. We ignore here the problem of ensuring that Mars is in the correct location along its orbit, for interception to occur. Thus a value of 6 km/s is reasonable. Given the similarity with the geostationary case we can see that the propellant savings will be the same.

6.9.4 Gravity loss and thrust

We have ignored the problem of the gravity loss which occurs when using electric propulsion, because this has been subsumed into the required velocity increment. Gravity loss is a direct consequence of the small thrust. Chemical rockets used in orbit transfers can be very accurately assumed to have no gravity loss, because the burn is so short and is at right angles to the gravitational field. For a low-thrust mission the spacecraft—still thrusting tangentially to the gravitational field—moves in a spiral, gradually increasing its velocity and its distance from the Earth. In this case, gravity loss is significant, as the unburned propellant is being accelerated and moved to a higher altitude throughout the mission.

As might be expected, the important parameter for this situation is the initial thrust-to-weight ratio. As we have seen, chemical rockets typically have thrust-to-weight ratios close to unity, while the thrust associated with electric propulsion is many orders of magnitude lower. The analysis is too complicated to include here, but Figure 6.27 shows the effect (after Sandorff, and quoted in Hill and Peterson). It shows the gravity loss factor.

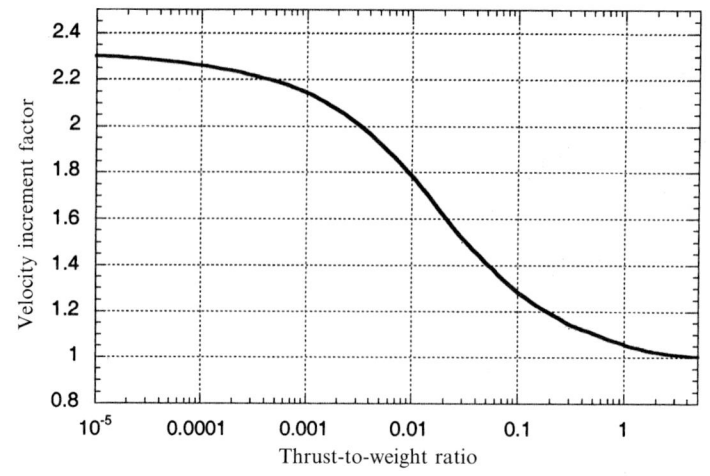

Figure 6.27. Velocity increment loss factor as a function of thrust-to-weight ratio for electric propulsion.

For very small thrust-to-weight ratios this penalty approaches a factor of 2.3 over the chemical rocket. Until now we have not had occasion to consider the thrust explicitly. In terms of the equations given at the beginning of this chapter the thrust can be written as

$$F = \dot{m}v_e = \frac{2\eta\xi M_E}{v_e}$$

remembering that the exhaust stream power is $\eta\xi M_E$, where M_E is the power supply mass. The thrust-to-weight ratio can then be calculated as follows:

$$\frac{M_P}{M_{Total}} = \frac{R-1}{R} = 1 - e^{-\frac{V}{v_e}}$$

$$M_E = M_P \frac{v_e^2}{2\eta\xi t}$$

$$F = \frac{R-1}{R} M_{Total} \frac{v_e}{t}$$

$$\Psi = \frac{F}{gM_{Total}} = \frac{R-1}{R}\frac{v_e}{gt}$$

With these equations we can approximately calculate the thrust-to-weight ratio of the two electric propulsion systems for the Mars mission. The propellant fraction is 0.33 for the Hall thruster, and so the thrust-to-weight ratio is $0.33 \times \frac{v_e}{gt} = 3.6 \ 10^{-5}$. From Figure 6.27 the velocity factor is 2.25, and so the real velocity increment is closer to 12 km/s. The ion engine will have an even lower thrust-to-weight ratio, and the same factor will therefore apply.

This rough calculation shows that for most electric propulsion, or for low-thrust missions, there is a penalty of about a factor of two in the total required velocity increment, due to gravity loss. To see how this affects the preceding results we can recalculate for the new velocity increment. For the Hall thruster the propellant fraction becomes 0.55, and the power supply mass to propellant mass is 0.079. Thus the ratio of total propulsion mass to vehicle mass is 59%. This is still to be compared with the 80% calculated for the chemical rocket, for which there is no penalty for gravity loss.

The ion engine has a higher exhaust velocity, which will make it more efficient for this mission. The propellant fraction is 33%, and the power supply ratio is 0.476. The total propulsion mass to vehicle mass is 49%. Thus there is a significant saving for the Mars mission using the ion engine, the saving with the Hall thruster being somewhat less. Since the key parameter is the ratio of vehicle velocity increment to exhaust velocity, we may expect this difference to increase for a round trip to Mars.

6.10 DEEP SPACE 1 AND THE NSTAR ION ENGINE

We have so far concentrated on the saving in propellant mass for missions starting from LEO. This makes sense for orbit-raising, and ultimately for planetary exploration. Until the International Space Station takes on its proper role as a *station*—a place to prepare spacecraft for interplanetary voyages—all interplanetary spacecraft will be launched from the Earth directly into their transfer orbits. The propellant required for the voyage is therefore part of the *payload* of the launcher. For a SSTO launcher, in the most optimistic case the mass ratio should be about 10, and so the propellant-to-payload ratio is 9. Every kilogramme of propellant needed for the interplanetary voyage therefore requires 9 kg of propellant in the launcher. Since there is a limit to the size of available launchers, this places a severe constraint on the mass of the voyaging spacecraft. The saving of propellant mass through the use of electric propulsion therefore has a major effect in enabling missions which would otherwise need a heavier and more expensive launcher—or, indeed, might be impossible with chemical propulsion alone. The Deep Space 1 mission is propelled by a xenon ion engine, the NSTAR (Figures 6.7 and 6.8 and Plate 12), and is an important milestone in the history of space exploration.

The Deep Space 1 mission was designed as a test-bed for ion propulsion used in interplanetary travel. The objectives—other than to test the propulsion concept—were to execute a fly-by of an asteroid, and if possible to carry out a fly-by of one or two comets. The total velocity increment for this mission is 4.5 km/s, the exhaust velocity of the xenon ions is 30 km/s, and the diameter of the grids is 30 cm. The propellant mass is 81.5 kg, and the spacecraft mass is 500 kg. The maximum power level of the thruster is 1.3 kW, and it can adjust its power automatically to take account of the decreasing solar intensity as the distance increases, and changes in the efficiency of the solar cells.

Figure 6.28. The PPS 1350 Hall thruster mounted on SMART-1. This device has operated faultlessly in space for over a year now as SMART-1 is lifted to the Moon's orbit.
Courtesy ESA.

The asteroid encounter—a fly-by within 16 km—was achieved after 1,800 hours and the use of 10 kg of xenon, and the spacecraft has travelled more than 50 million km from Earth. Most of the remaining 70 kg of propellant was used to place the spacecraft on a trajectory to encounter its first comet in January 2001 (Plate 13).

6.11 SMART-1 AND THE PPS-1350

A recent European example of a mission propelled by electric thrusters is SMART-1 (Plate 15). This is a technological precursor for Bepi-Colombo, an ESA mission to

Mercury. SMART-1 is propelled by an SPT 150 Hall effect thruster (re-designated PPS 1350) (Figure 6.28 and Plate 14), operating at 1,350 W, using 84 kg of xenon propellant, at a thrust of 70 mN, and an exhaust velocity of 15 km/s. In a period of 14 to 18 months, the 370-kg spacecraft spirals out from an initial Earth geostationary orbit, to that of the Moon. Thus in the United States and in Europe, the successful Russian Hall effect technology is being adapted and qualified for the NASA and ESA space programmes.

7

Nuclear propulsion

The idea that nuclear energy could be used for rocket propulsion dates back almost to the beginning of the twentieth century. While Konstantin Tsiolkovsky was writing about the exploration of space, and Robert Goddard was preparing for his first experiments, the aeronautics pioneer Robert Esnault-Pelterie was giving a paper at the French Physics Society in which he identified the release of 'infra-atomic energy' as the only solution to long interplanetary voyages. Typical of those times, he was an engineer who had already developed and built the first all-metal monoplane. Goddard had indeed anticipated this idea in 1906/7, but only in a private journal. This was before the structure of the atom had been fully elucidated; and before Einstein's equation of energy to mass, published in 1905, was well known. The only known process was radioactivity, in substances like radium. Thus, the idea of nuclear energy for space applications grew up alongside the practical development of the chemical rocket. Once a practical demonstration of nuclear energy release had been achieved, in 1942, it was not long before designs of nuclear rockets began to appear. All during the late 1940s and the 1950s, nuclear rocket studies proceeded alongside the studies of large chemical rockets. The early (post Sputnik) ideas for the United States manned lunar programme included the use of nuclear upper stages on the NOVA rocket. In the event, it was the all-chemical Saturn V that gave the United States its unique place in the history of space exploration. Among the reasons for this were the very high thrust and power output needed to escape from Earth's gravity. Nuclear fuel has a very high specific energy (joules per kilogramme), but power levels equivalent, say, to the Saturn V first stage F-1 engines, were not achievable with a nuclear rocket, whilst being held within reasonable mass and size limits. In any case, international treaties would soon ban the use of nuclear rockets in the Earth's atmosphere.

7.1 POWER, THRUST, AND ENERGY

It is easy to see why nuclear rockets held such promise in the pre-Saturn-V days. At a time when humans had yet to enter Earth orbit, the thoughts of the space pioneers were not restricted to that aim, but ranged over the whole gamut of space voyaging, including human missions to the planets. The high specific energy of nuclear fuel made it the obvious choice where *energy* was the main issue; for launching, particularly for the first stages of a multistage vehicle, it is *power* that is most important. For voyages to the planets a spacecraft needs to be given a very high velocity, in excess of 11 km/s; for launching from the Earth's surface, while high velocity (7.6 km/s) is needed, thrust is the main concern for the lower stages, and this is related to power. To see this, consider these equations, borrowed from Chapter 6.

$$P = \tfrac{1}{2}mv_e^2$$

Where P, in this case, represents the power in the exhaust stream (i.e., assuming 100% efficiency).

$$F = mv_e$$

and so

$$F = 2\frac{P}{v_e}$$

In all cases, m is the mass flow rate, in kg/s, with which we are familiar.

From these equations, we can see that thrust depends on the power dissipated in the engine, and is inversely proportional to the exhaust velocity, for a given power output.

Consider now the energy requirement of an interplanetary mission with a departure velocity of 11 km/s. The energy given to the vehicle is just $\tfrac{1}{2}MV^2$ where M and V are the final mass and velocity of the vehicle, respectively. It is helpful here to consider the specific energy of the vehicle (i.e., the energy per unit mass). This is $\tfrac{1}{2}V^2$, and for 11 km/s it is 60.5 MJ/kg. The propulsion energy contained in a kilogramme of hydrogen and oxygen propellant can be derived, approximately, by the following argument:

$$P = \frac{1}{2}mv_e^2 \quad \text{in J/s}$$

$$\frac{P}{m} = \frac{1}{2}v_e^2 \quad \text{in J/kg}$$

remembering that the mass flow rate m has units of kg/s. The maximum exhaust velocity of a practical oxygen and hydrogen engine is about 4,550 m/s, and so the energy per kilogramme is, by substitution, 10.4 MJ/kg. So, about 6 kg of propellant needs to be burnt for every 1 kg of vehicle mass, in order to provide enough energy to set a vehicle off on its interplanetary journey. For comparison the energy contained in 1 kg of pure uranium 235 is 79.3×10^6 MJ; a single kilogramme of uranium 235 could accelerate a spacecraft weighing 1,000 t, to interplanetary velocity, if its energy could be harnessed. In general, efficiency considerations, including the fundamental efficiency limit of reactive devices, restrict this benefit to much lower values, even before considering issues like the mass of shielding required for a nuclear rocket. Nevertheless, the high specific energy of nuclear fuel is a major advantage for high-energy interplanetary missions.

Figure 7.1. Actual test of a nuclear rocket engine at Jackass Flats in Nevada, as part of the NASA NERVA programme. The engine is firing vertically upwards, the storage tanks are for the liquid hydrogen.
Courtesy NASA.

7.2 NUCLEAR FISSION BASICS

While nuclear energy in the form of radioactive decay can be used to provide power for small electric thrusters, this system, based on the radioactive thermal generator or RTG, is very limited in power, and, from the arguments above, thrust. For high-thrust applications, and indeed for high-energy applications, the only practicable form of nuclear energy, is fission. The energy released through fission of a single uranium nucleus is just under 200 MeV, and the rate of fission (i.e., the number of nuclei per second undergoing fission), can be very high indeed. For radioactive decay, the energy release per nucleus is much smaller, and the rate of decay is strictly determined by the half-life, and cannot be controlled. As mentioned above, a considerable amount of development work on nuclear *fission* rocket engines has been done, and it is this process that we shall concentrate on here.

Nuclear fission was discovered, in Germany, by Hahn and Strassmann in 1939, but it was in the United States that the first controlled release of fission energy was established—by Fermi and colleagues in 1942; the first nuclear reactor was built in a squash court at the University of Chicago. The essential process is the absorption of a neutron by a uranium nucleus, which causes the nucleus to split into two nuclei (of mass about half that of uranium), with the release of just under 200 MeV of energy. Most of this is in the form of kinetic energy in the two fission fragments, with a smaller fraction released in gamma-rays. The importance of fission in uranium, is that two or more neutrons are emitted at the same time as the fission of the nucleus occurs. In principle, these neutrons can go on to interact with another uranium nucleus, and cause that to split. In this way, a chain reaction can be set up, with more and more nuclei undergoing fission and more and more neutrons being released to cause yet more fission, and so on. Since the rate at which energy is released depends only on the *neutron flux*, the power output of such a system can be controlled by inserting materials that absorb neutrons. This is the nuclear reactor, used to power electricity generation; it also forms the basis of a nuclear rocket engine.

The energy released in the fission-fragment velocity is very quickly converted into heat, as the fragments slow down in the uranium; so during controlled nuclear fission the uranium becomes very hot—in fact the theoretical limit to the temperature that could be reached is very high indeed. The uranium would melt well before this limit. Thus, once fission energy is being released, the process of making use of this energy is simply that of cooling the uranium, and using the heat extracted to provide power. For the generation of electricity, this can be by any conventional means: some reactors use water as a coolant, which is converted to steam, to use in a turbine driving a generator; others use gas (carbon dioxide), or a liquid metal like sodium, to cool the uranium and carry the heat out of the reactor to power a steam generation system. For a rocket engine, the system is much simpler: the cooling of the uranium is accomplished using the propellant, which passes through the reactor and out through the nozzle, just as in a conventional chemical rocket.

Uranium is a natural material and has properties which make the whole process much more complicated than the simple idea outlined above. There are two main isotopes found in natural uranium: U^{238}, which is the majority constituent, and U^{235}, which forms just 0.72% of the total. Although U^{238} undergoes fission, it is the properties of the much rarer U^{235} that dominate the process. This is because of the complex way neutrons interact with these heavy nuclei. In addition to causing fission, a neutron can be scattered, elastically or inelastically, or it can be absorbed without causing fission. The probability of these different interactions depends, in a complex way, on the energy of the neutron, and which isotope it encounters as it scatters through the uranium. The probabilities of these different processes are expressed as cross sections, and are illustrated in Figure 7.2.

U^{238} is capable of fission, but the probability is low, and falls to zero for incident neutrons with energy less than about 1.5 MeV; inelastic scattering quickly slows the neutrons down to less than this energy, and thereafter they cannot cause fission in U^{238}. Neutrons of *any energy* can cause fission in U^{235} with significantly higher probability. The probability increases rapidly as the neutron energy decreases, and reaches a value some 1,000 times higher than for U^{238} at very low neutron energies. Low-energy neutrons are described as *thermal* because their kinetic energy (much

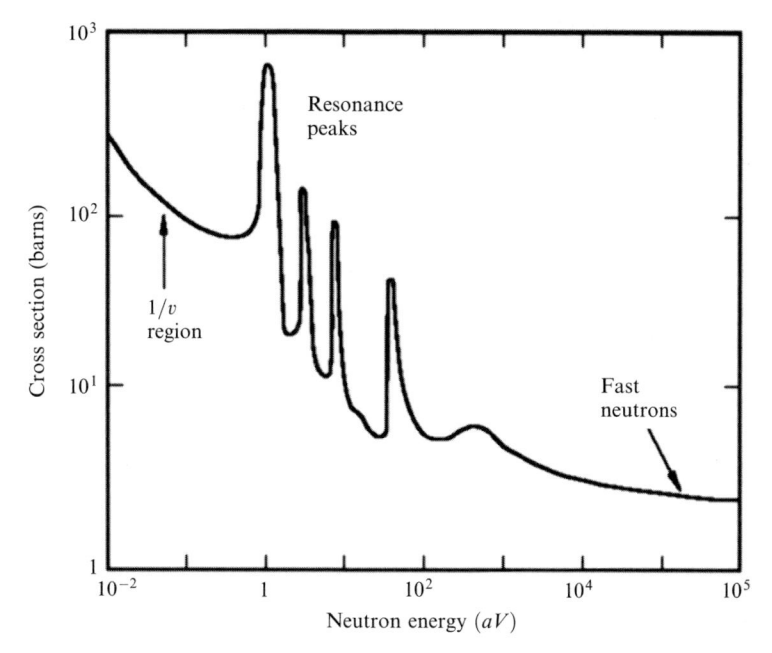

Figure 7.2. Schematic graph of the cross section for neutron interactions in natural uranium. The general smooth curve is for fission in U^{235}, while the resonance peaks are caused by absorption in U^{238} without fission. Fission only occurs in U^{238} for neutrons above 1.5 MeV (off the scale of this graph).

less than 1 eV), is close to that of the thermal motion of the atoms in the uranium matrix. Neutrons only interact with the nucleus—because they have no electric charge, so they can remain free in the matrix, at thermal energies. Inelastic scattering will gradually reduce the energy of the fission neutrons from 200 MeV down to fractions of 1 eV. Thereafter, they can induce fission in U^{235} with high probability.

The problem with natural uranium is twofold: the U^{235} encounters are rare (only 0.72%), so the product of cross section and encounter probability is rather small; resonance absorption occurs in U^{238}, where the probability of loss of the neutron is very high, at one of a band of different intermediate energies (see Figure 7.2). Neutrons losing energy by inelastic scattering in the matrix must pass through this range of energies, where absorption and loss have a very high probability. This means that in natural uranium (mostly U^{238}), very few of the fission neutrons survive down to thermal energies, where they can cause fission in the rare U^{235} nuclei. It is not possible to sustain a chain reaction in pure natural uranium.[1]

---

[1] While present-day natural uranium cannot sustain a nuclear chain reaction, in the past the concentration of U^{235} was higher—because of its shorter half-life. A site in Gabon has been discovered where a natural reactor operated, some 20 million years ago, in a uranium deposit saturated by water, which acted as a moderator.

7.3 A SUSTAINABLE CHAIN REACTION

There are two approaches that can improve the chances for a sustainable reaction. The first and obvious route is to increase the percentage of U^{235} in the matrix. This simply raises the probability of an interaction between a cooling fission neutron and a U^{235} nucleus, until the reaction becomes self-sustaining. Uranium, with enough U^{235} in it to sustain a chain reaction, is called *enriched*, and depending on the intended use, can have 2%, 20%, 50%, or even 90% of U^{235}. The process of enrichment is complicated and costly, since the atoms are only distinguishable by their atomic mass and not charge or chemical nature. Methods, which preferentially select the lighter isotope, are based on diffusion of a gaseous compound of the metal—usually uranium hexafluoride—through filters, or in a centrifuge.

The second approach, is to attempt to slow the neutrons down quickly (i.e., in a very few collisions), so that they reach thermal energies without being lost by resonance absorption in the U^{238}. This process involves a *moderator*, usually carbon or water, that is very good at slowing the neutrons by inelastic scattering, and at the same time does not absorb them. The moderator can be mixed intimately with the uranium atoms, in a *homogeneous* reactor, or the uranium and moderator can be in separate blocks, the *heterogeneous* reactor. The latter is more effective in sustaining the chain reaction with uranium of low enrichment; it can even allow the use of natural uranium. In the homogeneous reactor, the neutrons simply have more collisions with moderator nuclei than with U^{238} nuclei, so the probability of loss is reduced. In the heterogeneous reactor, a further improvement in the reaction takes place. The uranium is in separate blocks—typically cylindrical rods, separated by blocks of moderator. Cooling neutrons, in the energy range where resonance absorption occurs, cannot 'see' every uranium atom in the reactor; they cannot penetrate very deeply into the fuel rod because they are absorbed in the first few millimetres. The neutrons that penetrate to the central region are exclusively those that cannot be lost to resonance absorption, but can cause fission in the relatively rare U^{235} nuclei. This means that more U^{238} can be included in the reactor, without the corresponding loss of neutrons. It is thus possible to build a reactor containing exclusively natural uranium, using the heterogeneous system. This is the system that is used for most nuclear power stations. To sustain a chain reaction in pure uranium (i.e., without moderator), requires it to be highly enriched, perhaps more than 90% U^{235}. Progressive use of moderator allows the use of lower enrichment, down to natural uranium. It will perhaps be obvious that the *size* of the reactor increases, as more moderator is used. It is size, more than anything else, which is the critical parameter for space reactors, whether they are to be used to generate electricity or as rocket engines. The need to keep the reactor dimensions small will require the use of enriched uranium. Plutonium can be used in the same way as enriched uranium, but is so poisonous, and radioactive, that safety issues would add considerably to the complexity of a reactor that had to be launched.

7.4 CALCULATING THE CRITICALITY

The reactor size is complicated to calculate, and only the general principles will be outlined here. The key requirement, in a moderated reactor, is to allow sufficient distance for the neutrons to slow down to thermal energies. This will occur predominantly in the moderator, and for the reason given above, it is more efficient, in the case of uranium of low enrichment, to have the uranium concentrated in *fuel rods* rather than dispersed throughout the moderator. Therefore, the size of the reactor is really dominated by the dimensions of the moderator. Leakage of neutrons, from the reactor, decreases the flux available to generate fission; a large reactor will have a lower leakage than a small one. The remaining factor determining size is cooling. The heat generated by fission must be removed efficiently, both to generate thrust or electricity and to prevent the reactor core from overheating. Channels must be provided to allow the flow of propellant through the reactor, or for heat transfer using liquid metals, for power reactors; these will increase its size. As far as shape goes, while a sphere has obvious advantages because of its high ratio of volume to surface area, and hence low leakage, it has significant engineering difficulties, and the best shape for the reactor core is a cylinder with a height approximately equal to its diameter—the favoured ratio is $R/H = 0.55$.

To approach the design of a space nuclear reactor a little more rigorously we have to consider *criticality* and the so-called 'four factor formula':

$$k_\infty = \eta \varepsilon p f$$

The *multiplication factor* sometimes called the *reproduction constant* is denoted by k_∞, and is the effective number of neutrons, per fission, that survive all the loss mechanisms, and cause fission in another nucleus. A moment's thought will show that, for $k_\infty < 1$, no chain reaction is possible, and for $k_\infty > 1$, the chain reaction will grow continuously. Clearly the condition $k_\infty = 1$ is the critical level, and k_∞ will need to be controlled at 1 for a steady production of heat in the reactor. The subscript 'infinity' in k_∞ refers to a reactor of infinite size (i.e., one where the neutrons cannot leak out through the sides). It is necessary to calculate the criticality for an infinite reactor, before going on to consider one of finite dimensions. The four factors that define k_∞ are given below.

The first factor is η, the number of neutrons that emerge from fission of a nucleus, per incident neutron. This is sometimes called the *fuel utilisation factor*. While the fission of a U^{235} nucleus produces, on average, 2.44 neutrons, the number actually available, per incident neutron, is reduced, because some neutrons absorbed by a nucleus do not cause fission, but instead produce other isotopes of uranium. The value of η for pure U^{235} is 2.07, for thermal neutrons. For U^{238}, fission only happens with high-energy neutrons and this can usually be ignored. Thus, the value of η depends only on the U^{235}, and for the dilute mixture in natural uranium (0.72%), η is 1.335; for 2% enriched uranium, η is 1.726. It is clear that η has to be considerably greater than unity, to allow for loss mechanisms in the reactor.

The number of neutrons that cause fission in U^{238}, which we already know to be small, is expressed by ε, the *fast fission factor*: ε is the probability that a neutron, produced by fission, slows down below the threshold for fast fission (about 1.5 MeV), without causing fast fission (i.e., it is the probability that the neutron is available for further processes). As indicated above, except in special cases, ε is close to unity.

The number of neutrons per fission that are available for further processes is then $\eta\varepsilon$. These must escape resonance capture in U^{238} nuclei and the *resonance escape probability*, p, is the probability that the neutron avoids this capture, and hence survives down to thermal energies. The value of p will depend on the fraction of U^{238} in the fuel, on its distribution in the reactor core, and the amount and type of the moderator. Its calculation is complicated, and will not be attempted here. Suffice it to indicate that, if the neutrons lose energy quickly in the moderator, then their probability of capture will be small; if they lose energy slowly, as in for example an un-moderated reactor, then they are much more likely to encounter a U^{238} nucleus while they have the appropriate range of energies to be absorbed. For a graphite-moderated natural uranium reactor, the value of p ranges between 0.6 and 0.8, depending on the ratio of moderator to fuel. It will of course always be less than unity, unless pure U^{235} is used.

The fourth factor f is called the *thermal utilisation factor*—note the difference between this and *fuel utilisation factor η*. This fourth factor is the fraction of thermal neutrons that are absorbed in fuel nuclei, and not in other components of the reactor, like the moderator, or structure, etc. It will depend again on the distribution of the fuel and moderator, and will always be less than unity.

This now brings us, full circle (see Figure 7.3), back to the neutrons that will be absorbed in the fuel nuclei, which appeared in the definition of η. We can now see why k_∞ should be equal to unity: there must be at least one neutron from the original fission, which survives fast fission and resonance capture, and then encounters another uranium nucleus. From the fact that p and f are both less than one, and appear as a product in the formula, it will be clear that the product of the other factors, η and ε, must be considerably greater than one, for a sustainable chain reaction to occur. It may also be obvious that p, the probability that neutrons will escape resonance absorption, will increase if the moderator-to-fuel ratio increases, while f will decrease, because the probability of absorption by moderator atoms will increase. Even for an optimum ratio of fuel to moderator, which maximises the product pf, the value of the product is only 0.55 for graphite and natural uranium. A quick calculation, using the values above for the other factors, gives a value of 0.734 for k_∞, far below the necessary threshold for a chain reaction. This may seem contrary to the statement that graphite and natural uranium can be formed into a critical nuclear reactor. The difference here is that we have not used a value of the resonance escape probability that takes into account the shielding of most of the uranium from capturable neutrons. This shielding occurs when the fuel is distributed in discrete rods; it raises the value of p to a level where fission becomes sustainable. Another way to make a reactor capable of criticality is to enrich the uranium. The increased fraction of U^{235} raises the value of η, because essentially all the fission neutrons come from U^{235}, and at the same time, by increasing the probability of an encounter with a U^{235} nucleus rather than a U^{238} nucleus, f increases. Even quite

The reactor dimensions and neutron leakage

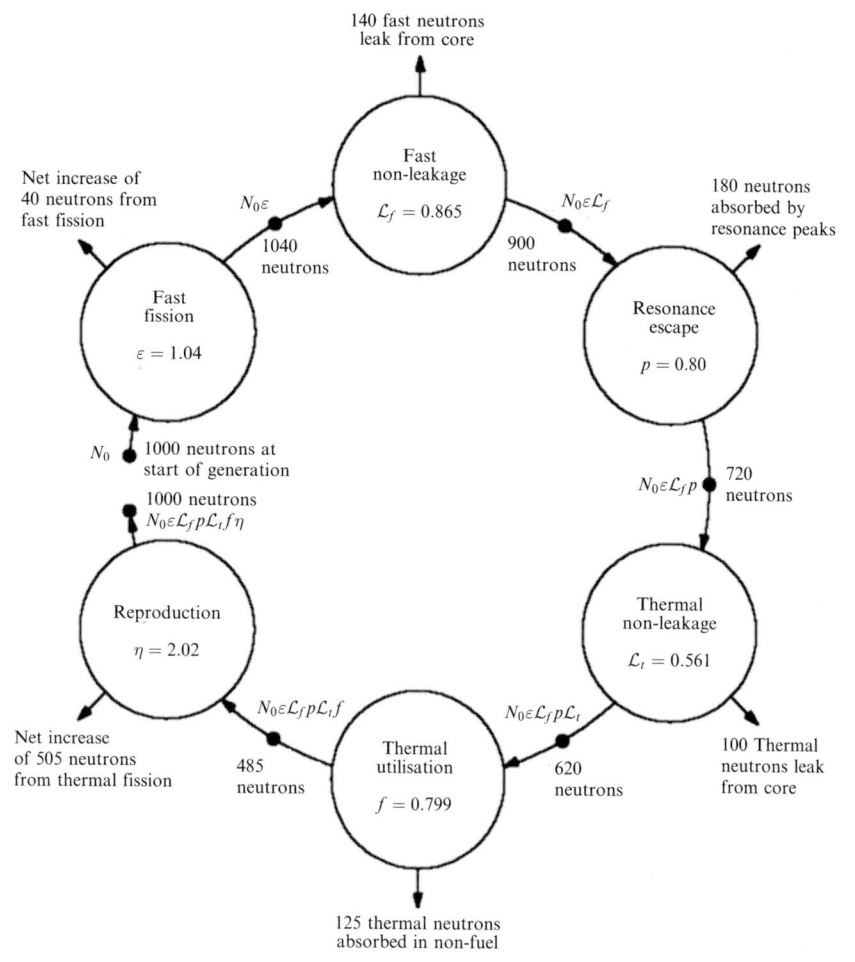

Figure 7.3. The fission chain. One thousand neutrons begin their journey at N_0 in the centre-left of the diagram. After passing through the processes in the circles, they have generated a net surplus of 505 neutrons which, together with the original 1,000 neutrons, go on to cause further fission. Note, the leakage probability $\mathcal{L}$ is included here.

small degrees of enrichment can have a significant effect on these two parameters, and hence on the criticality.

7.5 THE REACTOR DIMENSIONS AND NEUTRON LEAKAGE

The considerations above have implicitly assumed that the reactor is very large—in fact, infinite, so that leakage of the neutrons out of the core is negligible. For any reactor of finite dimensions, an additional loss of neutrons will occur through the periphery, a factor not yet taken into account; it will reduce the value of k

considerably. More neutrons will need to be provided as the reactor size decreases; this can be accomplished by enrichment. The calculation of size is important, but complicated. As the size of the reactor decreases, the leakage of neutrons from the core increases. At the same time, less space is available for moderator, so the resonance escape probability decreases, but the thermal utilisation increases. These factors work in opposite directions so that the net loss of neutrons cannot be guessed; it has to be calculated properly. The overall loss of neutrons, by whatever process, can be made good by increasing the degree of enrichment. For very small reactors highly enriched uranium may be necessary, with very little or no moderator, and perhaps 90% U^{235}.

The key determinants of size are the leakage of neutrons from the core, and the ability of small regions of moderator to thermalise the neutrons. These depend on two properties of neutrons in the core, the *diffusion length*, and the *slowing-down length*. The diffusion length represents the way scattering in the moderator reduces the neutron flux, as the distance from the source of neutrons increases. It is about 52 cm in graphite. The slowing-down length expresses the mean distance travelled by neutrons, through the moderator, before they reach thermal energies—for graphite it is 19 cm.

The simplest way of looking at the nature of the diffusion length L_r is to regard it as the constant in the expression for the rate of decrease in neutron flux with distance from a planar source of neutrons (e.g., a plate of uranium fuel surrounded by moderator). This formula is:

$$N = N_0 e^{\frac{-r}{L_r}}$$

where N is the number of neutrons crossing a unit volume of the material, situated at a distance r from the source. From this formula, setting L_r equal to 0.5 m it can be seen that at a distance 1 m from the source of neutrons the flux has decreased approximately to one-tenth. This shows that neutrons, emerging from a fuel rod, do not diffuse away very rapidly, and the flux remains quite high for tens of centimetres. In the real situation, neutrons are being produced in fuel rods throughout the reactor and the number of neutrons depends on the multiplication factor—neutrons not only diffuse away, but are created throughout the reactor by fission. The neutron flux also varies with time, depending whether the reactor is sub-critical—when the flux will decrease, or super critical—when it will increase. This much more complicated situation cannot be treated here, but some general indications, and approximate formulae, can be given. Such formulae really apply only to a homogeneous reactor in which fuel and moderator are mixed intimately (i.e., the fuel is not in separate rods). As we shall see this is much closer to the conditions for a space reactor, using high enrichment of the uranium, than for ground-based electrical generation reactors using natural or low-enrichment uranium.

The crucial link between the geometry of the reactor and the criticality is given by a constant called the *buckling factor B*. The buckling can be calculated from solutions of the full, time-dependent, diffusion equation for neutrons, using boundary conditions set by the shape of the reactor. For a cylindrical reactor the geometric (i.e., non-time-dependent) value of the buckling is given by:

$$B^2 = \left(\frac{\pi}{L'}\right)^2 + \left(\frac{2.405}{R'}\right)^2$$

where L' and R' are the height and radius of the core, suitably increased to allow for neutron diffusion out of the core; for most cases this increase is only a few centimetres and can be ignored for the present purposes.

The same factor can also be calculated from the time dependent part of the diffusion equation, which includes the fission aspects for the neutron flux as:

$$B^2 = \frac{k_\infty - 1}{L_r^2 + L_s^2}$$

where L_r and L_s are the diffusion and slowing-down lengths respectively. When the reactor is just critical, the two values of B are the same. By equating the two expressions for B, the critical values of L and R can be related to the fission properties of the reactor, and the critical dimensions determined. The minimum volume, taking into account the geometric formula above, is given by:

$$R = \frac{2.405}{B}\sqrt{\frac{3}{2}}$$

$$L = \frac{\pi}{B}\sqrt{3}$$

This is obtained by calculating the volume, in terms of L, R, and B, and minimising it by setting the derivative to zero. The ratio of radius to height for a cylinder of minimum critical volume is 0.55.

Two examples can be given. The first uses uranium, enriched to 2%, with graphite moderator, in the proportion, 300 carbon atoms to one uranium atom. For these fuel and moderator properties, the four factors are: $\eta = 1.73$, $\varepsilon = 1.0$, $p = 0.66$, and $f = 0.923$. This gives a value of 1.054 for k_∞. An infinite reactor with this composition would be super-critical. A reactor of finite size can obviously be made, using this material, which is just critical. Substitution in the formula for B, using the values for diffusion length and slowing-down length already given, results in:

$$B^2 = \frac{0.054}{52^2 + 19^2}$$

$$B = 0.004\,20$$

The height and radius can then be calculated, using the formulae above, as 13 and 7 m respectively. This is a large reactor core, suitable for ground use; even with a ratio of 300 to 1—moderator to fuel, it contains many tonnes of uranium. Note that it is the critical size that is determined here; the critical mass follows from these dimensions and the fraction of uranium in the core. Note also, that the total energy contained in such a reactor is very large, sufficient to provide power for national use, for many years. The total energy is proportional to the amount of fissile material contained in the core, but the power output depends on the neutron flux in the core.

Such a reactor is clearly unsuitable for flight in space: something much smaller is required. To reduce the size, the amount of fissile material, but most of all the amount of moderator, must be reduced, while still keeping the reactor critical. The moderator is present primarily to prevent resonance capture in U^{238} nuclei, by rapidly slowing the neutrons to thermal energies. If moderator is to be removed, then some of the U^{238} must be removed to compensate (i.e., the uranium must be enriched). The enrichment will have two effects: it will reduce the probability of a collision between a cooling neutron and a U^{238} nucleus, and it will increase the probability of a cooling neutron generating fission in the uranium by collision with a U^{238} nucleus. Neutrons with energy less than 1.5 MeV cannot cause fission in U^{238}, while neutrons of any energy can cause fission in U^{235}. Thus η, the number of neutrons that emerge from fission of a nucleus, per incident neutron, will increase; it cannot exceed 2.07, the value for pure U^{235}. At the same time, the resonance-escape probability, p, will approach unity, as the fraction of U^{238} nuclei decreases. The thermal utilisation factor, f, will also approach unity, so that the value of k_∞ will be dominated by the value of η. A glance at the formulae involving the buckling, B, shows that B is proportional to the square root of $k_\infty - 1$, and that the radius of the core, for instance, is inversely proportional to B. The increase in $k_\infty - 1$ from 0.054 to 1.07 for pure U^{235} is a factor of 20, and so the radius might decrease by a factor of 4.5 to 1.5 m. This is still a large reactor, 3 m diameter by about 3 m high, containing several tonnes of uranium.

The error here lies in the assumption that the diffusion and slowing-down lengths remain the same, although much of the moderator has now been removed. When the neutrons spend a significant amount of their time scattering in the uranium itself, the average diffusion length becomes characteristic of uranium, rather than carbon. At the same time, thermalisation of the neutrons is reduced by the absence of moderator. However thermalisation is no longer important, because fission can occur for neutrons of any energy in U^{235} and resonance capture is no longer a problem. The diffusion length for thermal neutrons in uranium is very small by comparison with carbon: rough values for natural uranium (1.5 cm), and U^{235} (0.5 cm), are to be compared with 52 cm in carbon. It is wrong however to use the diffusion lengths for thermal neutrons in this calculation, because the absorption cross section is much smaller for non-thermal neutrons, and these will be the majority of those causing fission in pure U^{235}. Accurate figures are hard to come by for pure U^{235}, but a reasonable assumption, that the absorption and scattering cross sections are equal, at about 10 barns,[2] gives for L_r a value of 1.2 cm for pure U^{235}. At the same time there is no longer need to include the slowing-down length, because neutrons no longer need to be fully thermal before they cause fission, and the

---

[2] A *barn* is the unit used for cross sections and is equal to 10^{-24} cm^2. It used to be said that for a neutron, an area of 10^{-24} cm^2 looked like a barn door.

formula for B becomes:

$$B^2 = \frac{k_\infty - 1}{L_r^2}$$

Substituting the values for pure U^{235}, gives for B, 0.85 and for R a value of 3.47 cm. The volume of this cylinder is then 264 cm^3, and the mass is 4.95 kg. This demonstrates the usual folklore about pure U^{235} (i.e., that a few kilograms of the pure material, in a 'grapefruit sized' sphere, can become critical).

Pure U^{235} is unlikely to be available, and a real space reactor will use enriched uranium, nevertheless containing between 50 and 90% U^{235}. As the percentage of U^{235} decreases, then k_∞ will decrease, and more moderator will need to be introduced to make sure that resonance absorption does not remove too many cooling neutrons. The diffusion length will increase, as neutrons spend more time in the moderator. All of these factors will decrease B, and thus increase the critical size of the reactor core. An optimised mixture of U^{235}, U^{238}, and moderator, can give rise to a core size that is small enough to be used in space, and still provide enough power, see Figure 7.4. A kilogramme of U^{235} contains 79 million MJ, enough energy to boost a 1,000-t spacecraft to interplanetary velocity; the power output (in MW), is dependent on the neutron flux in the reactor core, not on its size or total energy content. Therefore, the main issues in deciding on the reactor core size are likely to be related to the availability of enriched uranium, and engineering challenges, rather than on energy content. In other words, there will always be enough total energy in any practical nuclear rocket core.

7.6 CONTROL

Control of the neutron flux, and hence of the power output of the reactor, is essential if it is to be useful at all. If the multiplication factor, k, is less than unity, then the neutron flux will quickly drop to zero, and there will be no power output. If k is just greater than unity, then the neutron flux will increase indefinitely, and the power output with it, leading to meltdown. There must be a sub-system, in the reactor core, that can control the neutron flux. This consists of a number of *control rods* made of material having a very high absorption cross section for neutrons. The rods move in channels in the core and can be withdrawn, or fully inserted, or suspended at some intermediate position. When fully inserted, neutrons are absorbed, to the point where the reactor goes sub-critical, and fission stops; when fully withdrawn the reactor is super critical and the neutron flux increases indefinitely. At an intermediate position, the neutrons absorbed are just sufficient to hold the reactor at the critical point. The control rods can be connected to a neutron flux sensor, with a feedback mechanism, to hold the reactor in any condition. On start up, the rods are withdrawn, so that $k > 1$, the neutron flux, and the power output, will then increase to the desired level. When this is reached, the rods are partially inserted to return k to the critical point; here the neutron flux will remain constant as will the power level. If a different power level is required, then the rods are withdrawn for

Figure 7.4. The NRX-NERVA nuclear rocket engine at the test stand. Note the engineers standing near it emphasising that an unfired nuclear engine is safe. From comparison with the human figures, we see that the core cannot be much more than 50 cm in diameter, allowing for the pressure vessel and the reflector in the complete engine shown here.
Courtesy NASA.

more power, or inserted, for less, and then positioned at the critical point to maintain power at the new level. Note that the critical point will always be the same, where $k = 1$, the power level will depend on the neutron flux that was reached before k was returned to 1. Shutdown is achieved by fully inserting the rods.

7.7 REFLECTION

For a bare reactor core, a neutron leaving the reactor will never return, and be lost to the fission chain reaction. The calculations made above assume that this is so. However, a smaller reactor core can still sustain the chain reaction if it is surrounded by pure moderator material. Neutrons diffusing out of the core proper, can diffuse back again, after spending some time scattering off the nuclei in the external moderator. Of course, not all the neutrons entering this external moderator will return to the core; but some will, and so help to sustain the chain reaction. A core fitted with this external moderator—called a *reflector*, can be held critical for a smaller load of fissile material, an obvious advantage. Nearly all reactor cores are fitted with a reflector in this way, to reduce the amount of expensive U^{235} needed to maintain criticality. It will perhaps be obvious that a thickness of reflector about equal to the diffusion length will have the optimum effect, and increasing it much beyond this will have little additional benefit. The reflector also has a beneficial effect on the density distribution of neutrons in the core. For a bare core, the neutron flux near the edge will be reduced, because of neutrons leaking out; this reduces the amount of fission going on near the edges. The reflector will return some of these neutrons, and so make the flux distribution in the reactor more even. Since the local power density depends on the neutron flux, this means a more even distribution of power density throughout the core.

While for ground-based reactors the reflector is simply a passive element in the construction, it has a much more active role in space reactors, as a *control element*. Since the neutron flux in the core depends partly on the neutrons scattered back by the reflector, ability to change the efficiency of reflection will enable control of the neutron flux. This reduces the necessity for internal control rods, which are inconvenient in a space reactor. Because space reactor cores are small, a variable external reflector is a much simpler way of controlling the neutron flux in the core, and hence the power level, making internal control rods unnecessary.

7.8 PROMPT AND DELAYED NEUTRONS

The control of the neutron flux would be very difficult if it were not for delayed neutrons. Since fission is a nuclear process, the release of the neutrons takes place on a very short timescale indeed—compatible with the nuclear dimensions. The time between one fission and the next, therefore, depends on the journey time of the neutrons—until they strike another fissile nucleus. This means that the timescale for increasing the neutron flux *throughout* the reactor depends on the neutron transport time. The neutrons travel in a convoluted path, scattering off the moderator and fuel nuclei; nevertheless, the transit time, depending on the amount of moderator present, cannot be longer than a few milliseconds, and much shorter for an un-moderated reactor where faster neutrons mediate the fission. Thus, an increase in reflection, or the withdrawal of a control rod, would be accompanied by an increase in power output with a characteristic rise time of, at most, milliseconds—almost instantaneous. It would be very difficult to control the neutron flux by mechanical movements of control elements, in this case, because the movements could never be as fast as the changes in the neutron flux they induce. Fortunately, however, about 1% of the fission neutrons are delayed. In fact, they are the result of the formation of unstable intermediate nuclei—mainly isotopes of iodine and bromine, which then undergo radioactive decay. The half-lives of these isotopes range from 56 s to 200 ms, with a mean delay of 12 s. The rarity of delayed neutrons compared with prompt ones results in a weighted mean delay of about 80 ms. Because the delay time appears in the exponent of the function describing the growth of fission, this factor of 80 increase in the absolute delay has a strong effect on the rate at which the neutron flux, and hence the fission intensity grows. The formula, given here without proof, is:

$$n = n_0 e^{\frac{(k_\infty - 1)t}{\tau}}$$

where n_0 is the initial neutron flux, and τ is the weighted mean delay time. For $\tau = 1$ ms, and assuming that the multiplication factor (initially 1.0) increases by 1%, the neutron flux rises, in 1 s, by a factor of 22,000. For $\tau = 80$ ms, and the same change in multiplication factor, the flux increases by a factor of 0.125, a much slower rate. This can easily be controlled by mechanical movements of the control elements. It is the existence of these delayed neutrons that makes the controlled release of nuclear energy possible.

7.9 THERMAL STABILITY

There is another factor that makes the controlled release of fission energy easier than it might otherwise be. This is the sensitivity of the multiplication factor k to temperature. In most reactor configurations, k decreases as the temperature rises. This is partly caused by the thermal expansion of the reactor core component materials: if the density of the uranium decreases then the probability of a neutron meeting a fissile nucleus decreases; likewise, the expansion of the moderator, increases the mean distance between moderating collisions, and hence reduces the efficiency of moderation. Voids in the core—the cooling channels for instance—also get bigger and reduce the moderating effect. The probability of resonance capture of neutrons by U^{238} nuclei also increases. This is caused by the Doppler effect. The increased vibratory motion of the atoms of U^{238} broadens the narrow resonance peaks, so that capture can take place over a wider range of neutron energy. All of this contributes to the temperature stability of a nuclear reactor. If k becomes greater than one, and the temperature begins to rise, k will decrease, because of these loss mechanisms, returning the reactor core to the critical point where no net multiplication of neutrons occurs. If k drops below one, and the temperature drops, the increased efficiency of moderation, and improved neutron survival probability, push the multiplication factor back towards the critical level. Thus, a reactor will tend to stabilise automatically at the new power level, after an increase, or decrease, in k.

It is important to realise, in this context, that there are two factors at work, which govern the power output. For any stable state of the core, the value of k is one. The power level depends on the neutron flux in the core, which is stable only if k equals one. To increase the power output, k is allowed to become greater than one. Once the desired power level is reached, k is returned to a value of one, and the reactor continues to produce energy at the new power level. A decrease is accomplished in the same way, by decreasing the value of k to a value less than one, causing a drop in the neutron flux, k being again returned to one, when the desired (lower) flux level is reached. The natural thermal stability of the reactor, described here, helps in this process. It reduces the need for constant small changes in the control elements, in order to keep the power level stable.

7.10 THE PRINCIPLE OF NUCLEAR THERMAL PROPULSION

Using the nuclear physics outlined above we can now establish the essential parameters of a nuclear thermal rocket engine. The engine comprises a nuclear reactor, as described above, with the propellant used as the coolant for the core. The heat generated by fission is carried away by the propellant, and the hot propellant is expanded through a nozzle, in exactly the same way as for a chemical rocket. The core contains highly enriched uranium, mixed with a quantity of moderator that is a compromise between physical size, and the cost of the uranium. A very small engine containing 90% enriched uranium would be very costly, and perhaps difficult to control; lower enrichment, and more moderator, will increase the size of the engine, but the fuel will be less costly, and control will be easier. The general scheme is outlined in Figure 7.5.

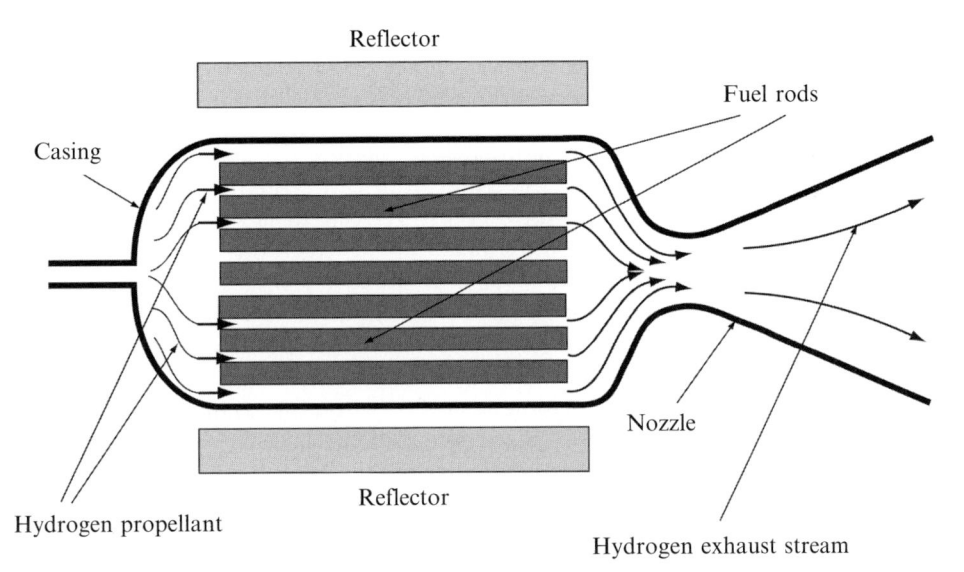

Figure 7.5. The principle of nuclear thermal propulsion. Hydrogen propellant enters the engine from the left, and is heated as it passes down the channels in the fuel rods. The hot gas then expands down the nozzle to generate a high velocity exhaust stream. The rate of fission and hence heat production is controlled by the reflector.

Table 7.1. Melting/sublimation points of some common constituents of nuclear rocket cores.

Type of material	Material	Temperature (K)
Fuel metal	Uranium (U)	1,400
Fuel compounds	Uranium nitride (UN)	3,160
	Uranium dioxide (UO_2)	3,075
	Uranium carbide (UC_2)	2,670
Refractory metals	Tungsten (W)	3,650
	Rhenium (Re)	3,440
	Tantalum (Ta)	3,270
	Molybdenum (Mo)	2,870
Refractory non-metals	Carbon (C)	3,990 (sublimation)
	Hafnium carbide (HfC)	4,160
	Tantalum carbide (TaC)	4,150
	Niobium carbide (NbC)	3,770
	Zirconium carbide (ZrC)	3,450

The nuclear thermal engine has conceptual similarities to the electrothermal engine, or resistojet, outlined in Chapter 6. In both cases the propellant is heated by contact with a hot solid. In the electrothermal case, this is the heating element, and in the nuclear thermal case, it is the hot core of the reactor. The heat transfer problems are the same. The propellant cannot be heated to a higher temperature than that of the heating element—nuclear fuel or electrically powered heater—and it is the mechanical integrity of the element at high temperatures that will limit the propellant temperature. The thrust developed by the engine is, in both cases, generated by expansion in a nozzle, and so the exhaust velocity will again be governed by the temperature of the propellant entering the nozzle, and molecular weight of the exhaust. Assuming that the propellant with the lowest molecular weight, hydrogen, will be used, the exhaust velocity achievable with a nuclear thermal engine will be limited by the temperature at which the fuel elements in the core start to disintegrate. It is not limited, as in the case of a chemical engine, by the maximum temperature obtainable from the reaction; for nuclear fission, this is in the tens of millions of degrees range. Thus, it matters very much which materials make up the core of the reactor; and the high temperature properties of the fuel and the moderator will be the main determinant of performance. Table 7.1 indicates the high temperature properties of different forms and compounds of uranium, and of materials likely to be found in a reactor core.

7.11 THE FUEL ELEMENTS

Uranium metal itself melts at 1,400 K. Comparing this with typical chemical combustion chamber temperatures of 3,200 K, shows immediately that uranium metal cannot be used as a nuclear fuel. It would be useless in a rocket engine, and is dangerous in a ground-based reactor, because an accidental rise in temperature could cause the fuel to melt and accumulate in the bottom of the reactor, in an uncontrolled critical state. The most common compound of uranium to be used as a nuclear fuel is uranium dioxide, UO_2. It is a stable compound with a melting point of 3,075 K; its most important chemical property, from the rocket engine point of view, is its stability in hydrogen, up to its melting point. Uranium carbide is another stable compound—melting point 2,670 K, as is uranium nitride—melting point 3,160 K. Any of these can be used as nuclear fuels, because the interaction is between neutrons and uranium nuclei; it does not matter how they are combined chemically with other elements. The only issue is one of density. Any compound of uranium contains fewer uranium nuclei per cubic centimetre than the pure metal, and so the dimensions of the reactor have to increase proportionally, so that the multiplication factor can be maintained at the critical value. This is generally not a serious problem and most, if not all, designs for nuclear thermal rocket engines use uranium compounds as fuel.

It is fortunate that graphite, the most common moderating material, has very good high temperature properties, with a sublimation point—it does not melt at any reasonable pressure—of 3,990 K. It also has very good structural and dimensional properties at high temperature. It was thus natural that the first nuclear rocket engines made use of graphite, both as moderator, and to form the primary structure containing the fuel. However, it quickly became clear that it has a serious disadvantage in its chemical properties. At high temperature, it reacts chemically with hydrogen, to form hydrocarbons. This results in erosion of the fuel elements by the hot propellant. By itself, this may not matter, provided the period of operation of the engine is short. The main difficulty occurs when the engine is required to operate for a significant length of time, or to be used for several manoeuvres; the damage to the fuel elements may then be sufficient to cause destruction of one or more of them. Given the enormous energy release in a nuclear engine, loose fragments of fuel element, passing through the nozzle are likely to cause a major failure. Indeed, in test firings during the NERVA programme such damage and failure was observed. The other danger from this chemical erosion of the graphite is the entry into the exhaust stream of particles of fuel. The uranium itself is only very mildly radioactive. In the pure state, U^{238} has a half-life of 4.5×10^9 y, and U^{235} has a half-life of 7.5×10^8 y. Once fission has begun, however, the fuel elements contain fission fragments, which are highly radioactive, having very short half-lives—the intensity of radiation depends inversely on the half-life. The presence of fission fragments in the exhaust stream makes it dangerously radioactive, and this has major consequences for ground, or atmospheric, testing of nuclear engines.

It became clear that some kind of protective coating was needed on the fuel elements so that the carbon did not come into contact with the hot hydrogen. This protection was accomplished by coating all the exposed surfaces of the graphite–uranium oxide matrix, with a stable, refractory substance. The neutron cross sections for this coating must be compatible with use in a reactor core: it should not be a strong absorber of neutrons and should preferably have moderating properties. Niobium or zirconium carbides were used, because they are refractory, and benign in the neutron environment, and they do not react chemically with hydrogen.

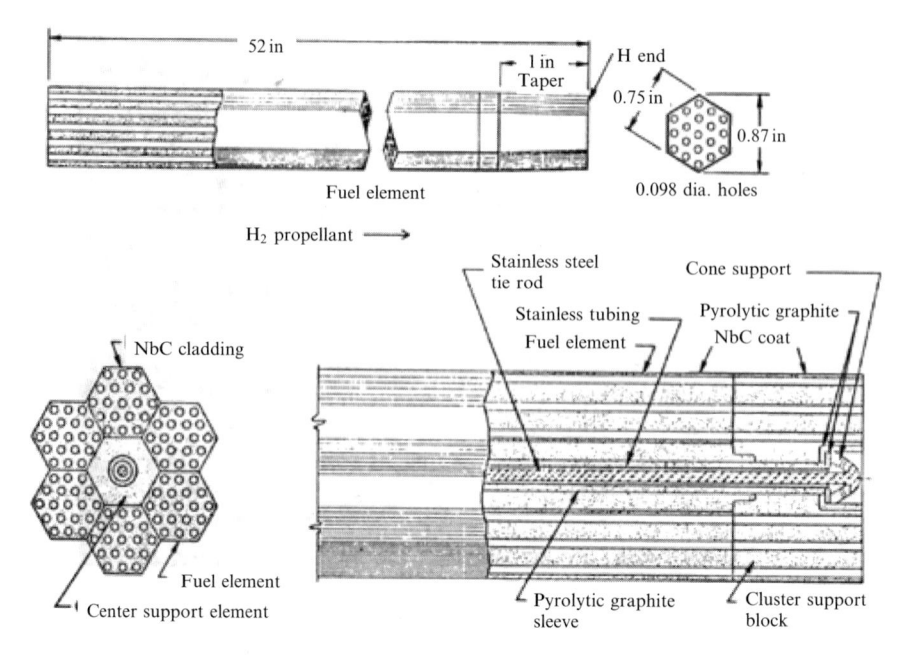

Figure 7.6. Fuel element assembly from the KIWI reactor core. The fuel is enriched uranium oxide spherules embedded in graphite. Each rod has 19 holes for the hydrogen to flow down. A cluster of six rods are held together by a stainless steel tie rod and the elements are coated with niobium carbide. A number of units is stacked together to make the complete core.
Courtesy NASA.

Zirconium carbide has the smaller neutron absorption cross section, while niobium carbide has the higher service temperature. With the carbide coating, the fuel elements survived long enough for a number of test firings, but in almost all cases the coatings cracked eventually, which led to erosion and damage during long firings. Nevertheless, most of the accumulated experience with nuclear rocket engines is with coated graphite.

A fuel element from the 1960s KIWI programme is shown in Figure 7.6. The enriched uranium is in the form of small spheres of uranium oxide dispersed through a graphite matrix. The graphite–fuel matrix is formed into hexagonal rods, 52 inches by 0.75 inches, each with 19 holes drilled lengthwise, through which the hydrogen flows. The surfaces are coated with niobium carbide, and six rods are held in a fuel assembly, locked by a stainless steel tie rod. A number of assemblies are then mounted together to form the reactor core.

7.12 EXHAUST VELOCITY OF A NUCLEAR THERMAL ROCKET

The KIWI reactor was a conservative design but operated very successfully (Figure 7.7): the maximum power output was 937 MW and the hydrogen outlet

Figure 7.7. The KIWA A-Prime reactor on its test stand. This was the second nuclear engine to be tested, and the first to make use of fuel elements coated with niobium carbide to reduce hydrogen erosion. It operated for 307 s at 85 MW.
Courtesy NASA.

temperature was 2,330 K. The effective exhaust velocity of an engine with this outlet temperature, and using hydrogen as propellant, can be computed using expressions from Chapter 2:

$$v_e = C_F c^*$$

The thrust coefficient, C_F is dependent purely on the nozzle properties, and for a well-designed nozzle, used in vacuum, takes a value of about 1.85, the characteristic velocity, c^*, does depend on the temperature and molecular weight of the propellant entering the nozzle; the formula is repeated below for convenience.

$$c^* = \left\{ \gamma \left(\frac{2}{\gamma+1} \right)^{(\gamma+1)/(\gamma-1)} \frac{\mathfrak{M}}{RT_c} \right\}^{-1/2}$$

A reasonable value for γ is 1.2, and with the appropriate substitutions for this and for R, the gas constant, the formula becomes:

$$c^* = 1.54 \sqrt{\frac{8.13 \times 10^3 T_c}{\mathfrak{M}}}$$

For $\mathfrak{M} = 2$, the molecular weight of hydrogen, and T_c equal to the above figure (2,330 K), the characteristic velocity becomes 4,739 m/s, and the effective exhaust velocity becomes 8,768 m/s. This is nearly twice that achievable with liquid hydrogen and liquid oxygen, showing that decoupling the power input to the engine from the nature and flow rate of the propellant allows much higher exhaust velocities than when they are coupled, as in the chemical rocket engine.

The above calculation also illustrates an important practical fact about the nuclear thermal engine: the high performance depends much more on its ability to use hydrogen, alone, as the propellant, than it does on the nuclear source of energy. In this form, the nuclear engine is useful because of its high thrust—related to the high power input from nuclear fission, coupled with its high exhaust velocity—about twice that achievable with a chemical engine. In fact, the nuclear rocket engine has the thrust of a chemical engine, combined with the high exhaust velocity of an electrothermal engine. It cannot achieve the very high exhaust velocities of ion or Hall-effect engines; but these have very low thrust indeed, compared with chemical or nuclear engines.

7.13 INCREASING THE OPERATING TEMPERATURE

Improved performance of the nuclear engine, in terms of the exhaust velocity, is dependent solely on raising the operating temperature of the fuel elements; there is more than adequate power available, from fission, to generate useful thrust. For heat transfer to work, there must be a positive temperature difference between the fuel rod surface and the hydrogen propellant, and between the centre of the fuel rod and the surface. Heat is generated uniformly in each fuel rod, and carried away by the hydrogen, flowing through the holes, and over the surface. Graphite has a relatively high thermal conductivity; this, together with the uniform generation of heat within an individual fuel rod, ensures that the temperature differences and therefore the thermal stresses within the rod are kept relatively small. The heat transfer to the hydrogen propellant is complicated to analyse and beyond the scope of this book. Some general ideas can be given, however. Heat is most easily transferred from solid surfaces to gas if the flow is turbulent, and so the gas velocity down the tubes has to be rather high. This requires a significant pressure difference down the channels (i.e., many small-diameter channels are preferred over a few large-diameter channels). The temperature difference between the hydrogen and the graphite is high at the inlet, and decreases down the channel, as the hydrogen heats up. This means that the equilibrium temperature of the fuel elements will increase from the inlet end to the outlet end, as the heat transfer becomes less efficient, remembering that heat flow into the gas depends on the temperature difference. This means that the highest temperature will be experienced *internally* in the fuel matrix at the outlet end of the core. This temperature must be lower than the maximum service temperature of the material. For graphite, this is about 3,800 K; the *surface* temperature of the rods will depend on the power output of the reactor, but may well be 200–300 degrees lower than the internal temperature. Since everywhere else the temperature must be lower still, the effective mean temperature of the surfaces heating the hydrogen, taken over the whole reactor, will not be much above 3,300 K. The hydrogen gas will achieve a temperature of about 3,000 K at most. A typical case from the NERVA programme, using an optimised graphite matrix, has the rod surface temperature at 3,200 K, and the hydrogen inlet and outlet temperatures as 140 K and 2,800 K, respectively.

The only way to increase the outlet temperature, and hence the exhaust velocity, is to increase the average surface temperature of the fuel elements—assuming that the flow of hydrogen, and the configuration of the cooling channels, have been optimised. There are areas where improvements can be made. In the first place, the conductivity of the matrix, and its service temperature, can be improved, so that the local surface temperature can be higher, and the internal temperature of the rods can be allowed to be higher still. In addition, the distribution of fissile material, and the neutron flux, from place to place in the core, can be optimised. Less heat can be extracted at the outlet end, because the hydrogen here is not much cooler than the surfaces of the fuel rods. If the overall neutron flux is controlled to keep the fission power density at the outlet end low enough to prevent overheating, the power output elsewhere will be unnecessarily held down. This will cause a corresponding reduction in the mean power output and temperature. By tailoring the fissile material density to decrease axially down the fuel rods, the neutron flux can be kept high in the input region where most of the heat is transferred. This is of course no help if the temperature of the gas is limited by the service temperature of the output ends of the fuel rods, so the two improvements must go hand in hand.

Indications from the NERVA programme are that the temperature difference between the gas and the fuel elements at the outlet, can be reduced to 100 K or so, and that higher fuel element temperatures can be achieved by abandoning graphite as the matrix, and replacing it with a mixture of carbides, including uranium carbide. The main reason for this appears to be the thermal instability of the niobium or zirconium carbide coating of the graphite fuel matrix. The graphite itself appears to have suffered cracking and rupture once the protective surface had been eroded away. Using a carbide matrix, the carbon is chemically combined and not subject to hydrogen attack, at the same time the moderating properties of the carbon are retained. Reactor cores using carbide matrices allow a higher exit temperature, even if the melting points of these substances are lower than pure graphite. The superior chemical resistance of carbides and their ability to withstand thermal shock allows the use of higher service temperatures than with graphite. The performance of engines based on some of these types of fuel matrix are shown in Table 7.2.

For uranium enriched to less than 50% U^{235}, a degree of moderation is required, and graphite, or carbon in a chemically combined form as carbides, is essential to provide enough thermal neutrons, and to keep the resonance capture loss within bounds. If the uranium is enriched up to 90%, then the fission is essentially mediated by so-called epithermal neutrons, with energies around 200 keV. This is really 'fast fission', and as mentioned above, the reactor core can be made rather small, with a high power density. The fuel for such reactors (no moderator is required) is formed into a combination of ceramic and refractory metal, so-called *cermet*. The uranium is in the form of particles of uranium oxide, already a ceramic material with a melting

Table 7.2. Complete nuclear thermal rocket engine schemes based on the NERVA programme. NRX XE and NERVA 1 are based on engines developed during the NERVA programme, the later columns refer to evolutions of those engines, based on sub-systems that were tested during the programme, but not evolved into complete engines. For instance the all carbide fuel elements were tested in the Nuclear Furnace programme.

Parameters	NRX XE	NERVA 1	New designs based on NERVA		
Fuel rods	UO_2 beads embedded in graphite	UO_2 beads ZrC coat, embedded in graphite	$UC_2 + ZrC + C$ composite	$UC_2 + ZrC$ all carbide	$UC_2 + ZrC + NbC$ all carbide
Moderator	Graphite	Graphite + ZrH	Graphite + ZrH	Graphite + ZrH	Graphite + ZrH
Reactor vessel	Aluminum	High-strength steel	High-strength steel	High-strength steel	High-strength steel
Pressure (bar)	30	67	67	67	67
Nozzle expansion ratio	100:1	500:1	500:1	500:1	500:1
I_{sp} (s)	710	890	925	1,020	1,080
Chamber temperature (K)	2,270	2,500	2,700	3,100	3,300
Thrust (kN)	250	334	334	334	334
Reactor power (MW)	1,120	1,520	1,613	1,787	1,877
Engine availability (yr)	1969	1972	?	?	?
Reactor mass (kg)	3,159	5,476	5,853	6,579	?
Nozzle, pumps, etc., mass (kg)	3,225	2,559	2,559	2,624	?
Internal shield mass (kg)	1,316	1,524	1,517	1,517	?
External shield mass (kg)	None	4,537	4,674	4,967	

point of 3,075 K, and a refractory metal like tungsten (melting point 3,650 K). The metal and uranium oxide are sintered together to form fuel elements that allow fast fission to occur, and at the same time have very good high temperature strength and chemical resistance. In this way, although the ultimate melting points are lower than for graphite, the improved engineering properties of the cermet elements allow higher operational temperatures. This kind of engine, using highly enriched uranium, is smaller and for many applications, this is an advantage. The thrust-to-weight ratio of the engine is improved and this improves the mass ratio of any nuclear rocket powered spacecraft.

It is not certain that the fast-fission engine will be the device of choice for the next generation of nuclear rockets because of the high cost of the fuel and security issues relating to this material. The carbide systems, using uranium of lower enrichment, seems to be both more affordable and to offer similar operational temperatures, at the expense of a more bulky engine. The form of the fuel elements has evolved to improve the hydrogen flow through the reactor and the efficiency of heat transfer. From the Russian experience comes the twisted-ribbon fuel element, which minimises temperature gradients within the element, and between the surface and the gas; more recent developments in the United States use stacked grids of fuel matrix material that again improve the flow and heat transfer. The destruction of fuel elements in the KIWI and other graphite matrix engines was partly due to the turbulent flow necessary for good heat transfer, coupled with the large amounts of energy being released. The advent of computers capable of executing detailed hydrodynamic calculations, has enabled the optimisation of flow and heat transfer in these extreme conditions. In general we may expect that a propellant temperature in excess of 3,000 K will be achieved, and an exhaust velocity close to 10 km/s, using modern carbide fuel matrices and uranium enriched to 20–50% U^{235}.

7.14 THE NUCLEAR THERMAL ROCKET ENGINE

Having considered the basic process of nuclear fission, the determinants of size, and the nature of the fuel elements, we can go on to consider how these elements are put together to make a functioning rocket engine. Since a nuclear thermal rocket engine is essentially a thermal engine, with the heat provided by nuclear fission rather than chemical reaction, many of the components will be similar to those found in, for instance, a liquid hydrogen–liquid oxygen engine. Several very specific engineering details are unique to the fission engine. These are related to the source of energy—the reactor core and its control, and to the radiation produced; these have several very significant consequences.

7.14.1 Radiation and its management

So far, we have ignored the aspect of nuclear fission that is most commonly associated with it: the radiation it produces, both during operation, and after use. To deal with one aspect immediately, it is important to re-iterate that uranium itself is not to any real extent radioactive. The half-lives of the two major component isotopes are of the order of 100 million years, and so pure uranium is no more dangerous to handle than, for example, granite road-stone. As found in mineral deposits, uranium contains quantities of daughter decay products that are significantly radioactive, such as radium and the gas radon; the purified element, uranium, has these removed. The fission rocket engine will therefore be perfectly safe and non-radioactive, so long as it has not been fired. This immediately makes clear that such an engine must be launched in a virgin state, and that its first firing has to be in space, in such an orbit that return to Earth by any conceivable error or accident has a very low probability indeed. The engine will be quite safe to launch in this state, much safer than an RTG, containing radioactive plutonium. The onset of fission requires such a degree of order in the configuration of the components of the engine, that it is difficult to conceive a kind of launcher accident, which could randomly create the conditions for fission to commence.

Radiation created during the operation of the engine, and after its last use, does have very significant consequences. These may be divided conveniently into those that pose engineering problems, and those that may endanger human life, if not properly managed. A reactor producing 100 MW undergoes about 3.0×10^{19} fissions per second, each releasing about 200 MeV. Much of this energy is more or less immediately converted into heat, by interactions with the fuel matrix. The fission fragments, carrying 168 MeV, stop immediately; alpha and beta particles—helium nuclei and electrons—are also very easily stopped. The neutrons, carrying away some 5 MeV, are penetrating; they survive in the reactor until they are captured in U^{238} nuclei, or cause fission in U^{235} nuclei. The energy they lose while slowing down is converted into heat. There are, in general more than two neutrons produced per fission, so that the number in the core is very high indeed. In addition to neutrons, each fission produces gamma-rays carrying a total energy of 12 MeV, and in general, gamma-rays are also penetrating. So, of the 200 MeV released per fission, some 5 MeV is carried away from the original site by neutrons, and some 12 MeV by gamma-rays. This means that everything in the core, inside the radiator, is bathed in a huge flux of this radiation, while the reactor is operating. This is intentional, because the high flux of neutrons is essential for the chain reaction; but neutrons and gamma rays may also lose energy in the structure of the reactor, the fuel tie rods, and the elements that hold it all together. This energy appears as heat, and so all elements of the reactor are heated by the radiation produced by fission. Cooling has therefore to be provided, not only for the fuel elements themselves, but also for the structural elements. This ensures the integrity of the structure, but also of course helps in transferring the maximum amount of heat from the fission process into the hydrogen propellant stream; in a sense this is analogous to regenerative cooling in the chemical rocket engine.

The reflector and the casing of the engine also require to be cooled. The reflector is placed inside the casing, and is, by definition, exposed to a high flux of neutrons; its job is to return these to the core. The reflector is also exposed to a high flux of gamma-rays, by its proximity to the core. Large amounts of heat are dissipated in the reflector, and it has to be cooled actively, both to keep it within its service temperature, and to return this heat to the propellant. A significant flux of neutrons and a high flux of gamma-rays penetrate the reflector, and impact on the casing. This also is heated by the radiation, and has to be cooled to help maintain its integrity. Vast experience of the effects of such radiation on metals and other structural materials has been built up in the nuclear power industry, and most effects are slight for the total dose experienced in the short firing of a nuclear engine. Beyond the casing, there is a high flux of both neutrons and gamma-rays; these are dangerous to humans, and in fact to electronics; both will need protection during the firing.

The necessary radiation shield, to protect the spacecraft, and any humans, is relatively easy to provide. It is usual for this to be made up of one or more discs of high-density material, mounted on the forward end of the engine. There are no sensitive spacecraft components aft of, or beside, the engine, and it is assumed that any humans will be safely inside their cabin, well forward of the engine, during firing. Geometrically it is most efficient to place the shielding disc as close as possible to the core. Because the radiation spreads out, a smaller thickness, but a much greater area, would be needed if it were some distance from the core; the area required increases with the square of the separation. The radiation absorbed will heat the shield, and it is convenient and efficient to place the first forward shield inside the casing, where it can be cooled using the propellant, with all the obvious advantages this brings. Because it is inside the casing and the same diameter, it cannot cover radiation from the after portion of the core, emerging at an acute angle; at the same time it cannot be made thick enough to stop all the radiation without making the casing unreasonably long. For this reason an additional external shield is also mounted. It is of a larger diameter, and creates a radiation 'shadow' within which the spacecraft and any crew can be safe. A further function of this external shield is to attenuate the gamma-ray flux produced by neutron capture in the internal shield. It will be obvious that there is no intention to shield the reactor core in any direction other than forward; there is no harm in radiation released into unpopulated regions near the reactor.

Neutrons and gamma-rays generate different shielding requirements, gamma-rays are best stopped by high-Z materials in which they have a high pair-production cross section; the pairs generate an electromagnetic cascade which creates further gamma-rays; these can be stopped provided the shield is thick enough. Neutrons are difficult to stop, as they have no electromagnetic interaction with matter: they must either be absorbed by neutron capture, or reduced in energy by scattering until they are harmless. The high density of metals helps somewhat with the neutron absorption, but for a high absorption cross section, boron, cadmium, or one of the rare earths, like hafnium, is required. Boron and cadmium have moderately high thermal absorption cross sections, but to slow neutrons down, hydrogen is the best material. In the form of lithium or zirconium hydride it is relatively dense and can be formed into a shield. Of course, the hydrogen propellant can be used, and in most designs for nuclear powered spacecraft, the propellant tank forms part of the shielding, being shaped to create a neutron shadow for the sensitive components placed forward of the tank. This implies that not all the hydrogen can be expended during firing, with corresponding limitation on the mass ratio. The configuration of primary shielding in a NERVA engine is shown in the cutaway drawing in Figure 7.8. Note that the turbo-pumps, the valves to control the hydrogen flow, and the motors to operate the control drums, are behind the external shield. Radiation does not have a very strong effect on most metals, but organic compounds, used in seals and insulators in the pumps, valves, and motors, need some protection.

The reactor core will be radioactive after use, and must be disposed of safely, without coming into contact with humans, or planetary atmospheres and surfaces. This is best achieved by saving some of the hydrogen to use for a last burn that will

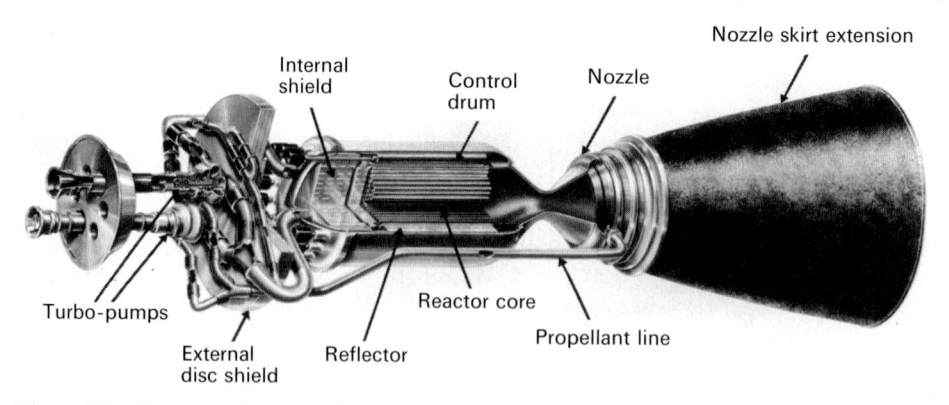

Figure 7.8. Cutaway drawing of a NERVA nuclear rocket engine. The similarities with a chemical engine are obvious, including the high expansion ratio nozzle. The pipe labelled 'propellant line' carries hot gas bled from the nozzle to drive two turbo-pumps. The internal and external shields can be seen as well as the core and the control drums.
Courtesy NASA.

place the engine in a safe orbit which will not encounter planets or humans for millions of years—time for the radiation to die away.

7.14.2 Propellant flow and cooling

The propellant delivery system will be very similar to that used for a cryogenic propellant chemical engine, with the same problems, and solutions. Once at the inlet to the engine, the propellant flow is very different. There are no injectors and mixing, but there is the need to cool several components of the engine, and above all to make sure that the power output of the reactor core is matched by the rate with which heat is extracted by the propellant and exhausted down the nozzle.

The reflector needs to be cooled, as does to some extent the casing. This is done by passing the hydrogen propellant through channels in the reflector. The enlarged view of the NERVA propellant manifold, shown in Figure 7.9, demonstrates the flow pattern. There are two turbo-pumps, each connected to a propellant line from the tank. One pump delivers propellant to the plenum directly above the internal shield, there are cooling channels in the shield to match the channels in the fuel elements, and so this plenum feeds hydrogen at high pressure through the shield, then down the length of the fuel elements and out to the nozzle. The second turbo-pump delivers hydrogen to the outer plenum. This is connected to channels running down the outside of the reflector, to the cavity at the aft end, thus cooling the chamber wall. The gas then flows forward again, through channels in the reflector, which are at a higher temperature than the walls. The hot hydrogen then enters the region directly below the shield and into the channels in the fuel elements.

In this way, the components of the reactor chamber are kept below their service temperature limit, and much of the heat carried out of the reactor core by radiation

Sec. 7.14]   **The nuclear thermal rocket engine** 225

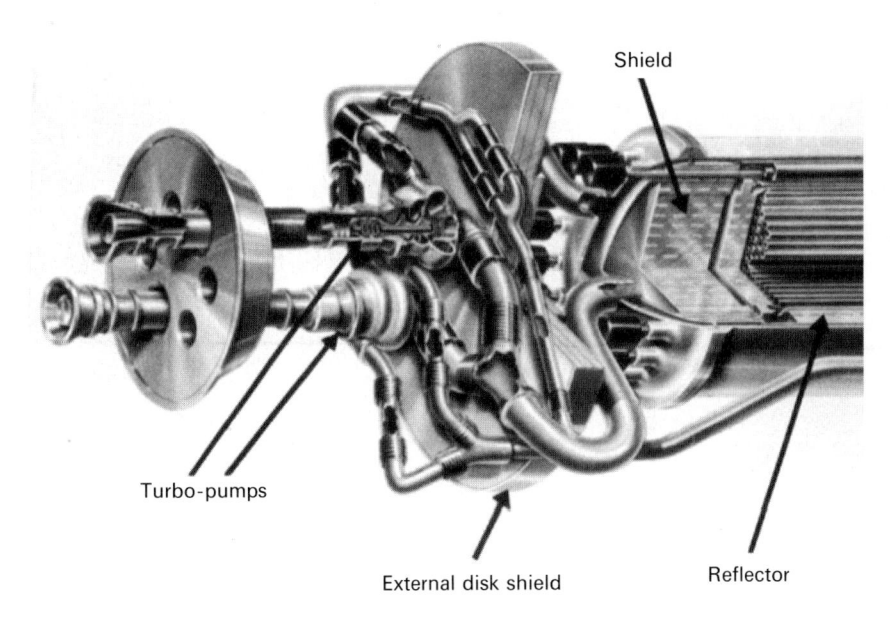

Figure 7.9. Close-up of the propellant delivery part of the NERVA engine showing how the two turbo-pumps deliver propellant, one to cool the reflector and pressure vessel, and the other to cool the reactor core.
Courtesy NASA.

is returned to the propellant stream. This is a kind of regenerative cooling, as used in chemical engines. The use of two turbo-pumps allows separate control of the component cooling function and of the propellant flow through the core. As mentioned above, it is vital to match the flow of the propellant to the energy release in the core. At the same time heating by radiation is a secondary effect, which can benefit from separate control. There is a separate cooling flow for the nozzle, not shown in the diagram.

As there is only one propellant, there is no possibility of driving the turbo-pumps using a gas generator, as in a chemical rocket engine. Instead, heat from the reactor core has to be used in some way. There are two possible approaches. In the *topping cycle*, some of the gas created in the cooling channels is bled back to the turbo-pumps in order to drive them. For low-thrust engines this will work quite well, but for a high-thrust engine with its demands on the turbo-pumps, more energy is needed, and hot hydrogen has to be bled out of the chamber itself, close to the nozzle throat; this is called the *hot bleed* cycle. This line is shown Figure 7.10; it carries hydrogen at high temperature and pressure up to the turbines driving the pumps. There is some stabilising feedback here as well, in that, if the pressure and temperature in the throat rises, the power developed in the turbo-pumps increases, providing more cooling in the core, and reducing the pressure and temperature in the throat. A drop in temperature and pressure in the throat will have the opposite effect.

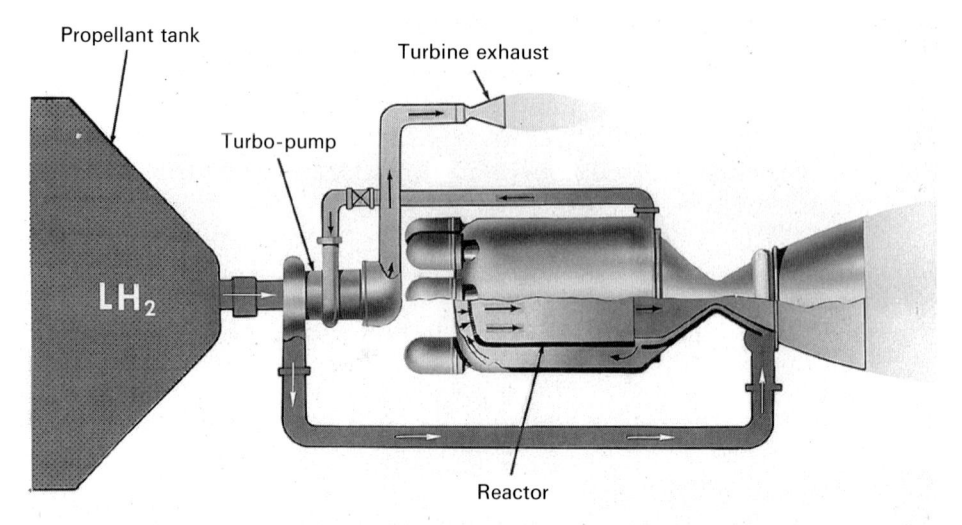

Figure 7.10. The hot bleed cycle. Hot gas extracted from the reactor chamber is used to drive the turbo-pumps and is then exhausted through a small auxiliary nozzle. This design is different from the more detailed drawing in Figure 7.8.
Courtesy NASA.

Bleeding hot gas from the nozzle throat reduces the thrust, and some of this is recovered by feeding the turbine exhausts to small nozzles.

7.14.3 The control drums

Control of fission, using materials with high neutron capture cross section, has been described in general terms. For the rocket engine, control rods are inconvenient, and drums or cylinders are preferred. These are solid cylinders, divided axially into two halves: one semi-cylinder comprises an efficient neutron absorber, while the other is made of moderator. The control drums are embedded in the reflector, and can be rotated by electric or hydraulic actuators situated at the forward end—outside the chamber, via a long drive shaft. The reflector has a high flux of neutrons in it, many of which need to be returned to the core by scattering in the moderating material of the reflector. If all the drums are turned so that the neutron absorbers are towards the core, then the efficiency of moderation and reflection is greatly reduced, and fewer neutrons are returned to the core. The criticality of the core is designed to be dependent on these reflected neutrons; the chain reaction cannot proceed without them. The loss of neutrons in the control drums is enough to make the core sub-critical, and, as we have seen, this will cause the reactor to shut down. If the drums are rotated so that the moderating material is towards the core then the neutrons are not absorbed, and are returned to the core. The core then becomes super-critical and the power output rises. For intermediate positions of the control drums, the criticality can be maintained at unity, with a steady output of power. The drums, shown in the drawing (Figure 7.9), are closely packed around the core, and embedded in the reflector. They are set at the correct rotation angle by means of a feedback loop, fed by neutron sensors that measure the neutron flux in the core; as mentioned above, this flux accurately reflects the state of fission and energy release in the core. For launch from Earth, these drums would be locked in the shut-down position, with perhaps an additional block of neutron poison secured in the engine, only to be removed once the engine is safely in orbit.

7.14.4 Start-up and shut-down

The start-up sequence for a nuclear engine has characteristics similar to those of a cryogenic chemical rocket. The whole distribution system has to be cooled down so that the cold hydrogen does not cause thermal shock in the components. This is done by bleeding a little of the cold gas from the hydrogen tank through the system, forced by the natural pressure in the hydrogen tank. Once the system has been cooled, the control drums can be rotated to the point where k is greater than one. The power output of the reactor will rise, over a period of tens of seconds. It is important that the cooling by the propellant keeps pace with this; it must be controlled using the neutron flux as the monitoring parameter. There is a difficulty here because the pressure in the reactor chamber may not be high enough at this point to drive the turbo-pumps. Auxiliary power for the pumps may need to be provided during start-up. This may be electrical, or by use of a small gas generator using hydrogen and a small tank of oxygen. Once the pressure in the reactor chamber is sufficient to provide pressure to the turbo-pumps, the main system can take over. From this point onwards, the main control loop can function, maintaining the power output and the propellant flow at the correct levels.

Note that this system is not self-regulating like a chemical rocket system. In the latter the power developed in the chamber is directly related to the chemical energy provided by the propellant, and hence to the flow of the propellant. The heat generated, and the flow of propellant through the cooling channels of the nozzle and combustion chamber, rise and fall together, as the flow varies. In the nuclear engine, the power output and the propellant flow are strictly speaking unrelated—this is why a high exhaust velocity can be attained. The power output rises with neutron flux, in response to the position of the control drums; the hydrogen flow is actively controlled in response to the neutron flux, in order to match the power output. Fail-safe systems must be included in the control loop to make sure that the reactor does not overheat in response to an unexpected drop in propellant flow.

Once the engine is in stable operation the thrust can be varied by altering the set-point of the control loop. This is again different from the situation with a chemical engine, where throttling is a rather difficult activity because of the potential for instability in combustion when the flow rate changes. In other respects, the performance of the nuclear engine, while thrusting, is similar to that of chemical thrusters. The shut-down is however markedly different.

An aspect of the fission process that has so far not been described is the formation of intermediate, short half-life nuclei that contribute to the heat produced by fission.

About 1% of the neutrons produced by fission are delayed, as mentioned above, and these enable relatively easy control of the fission process by increasing the characteristic time from milliseconds to fractions of a second. So when the reactor is shut down, the fission process, and hence the power output, decays first by the mean characteristic time, over several tens of seconds; but some fission power will continue to be produced according to the half-life of the longest lived of these delayed neutron reactions, which has a half-life of 80 s. So, fission heating will go on for several half-lives after shut down. This is a significant effect, about 0.05% of the full power output. At a much lower level, radioactive decay, producing beta and gamma-rays, will continue long after shut-down. These are the 'fission products' that make spent nuclear fuel so dangerous in the initial hours after shut-down; heating from the decay of these products is still significant.

These two delayed processes make the shut-down of the nuclear engine complicated. The fuel rods must continue to be cooled by hydrogen flow for some time after shut-down. For many minutes, because of the delayed neutrons, the power output is still in the megawatt region for, for example, a 1,000 MW engine; and the power output will remain in the hundreds of kilowatt region for several days, because of gamma and beta decay of fission products. The hydrogen used to cool the fuel rods during this process will continue to generate thrust, at a decreasing rate. This thrust will contribute to the final velocity of the space vehicle; a fact that must be taken into account in the computations. Not all the propellant can be used in the main thrusting operation: some must be retained for this cooling operation. It will generate thrust less efficiently, and so the overall efficiency of the nuclear engine is reduced from the theoretical value. Note again that the shut-down has to be managed actively. A chemical rocket shuts down when the fuel tank is empty; a nuclear rocket requires a controlled shut-down over several days, and the fuel tank cannot be allowed to empty prematurely. This reduces the available mass ratio and again reduces the overall efficiency from the theoretical maximum.

7.14.5 The nozzle and thrust generation

The nuclear engine is a thermodynamic engine, just like a chemical engine; the only difference is that heat is generated by chemical reactions, in the latter, as opposed to nuclear fission. This means that in general terms the nozzle of a nuclear engine is identical in form and function to that of a chemical engine. The only difference lies in the length of the nozzle and the corresponding expansion ratio. The combustion chamber and nozzle of a chemical engine are rather low in weight, compared with the rest of the vehicle, and contribute only in a minor way to the dry mass of the vehicle—this, it will be recalled, determines the mass ratio. The nuclear engine is rather heavy, because of the shielding and the mass of the core, and its reflector. Also, a rather large pressure vessel is needed to contain the core and reflector. For the light chemical engine, the length of the nozzle has a significant impact on the dry mass; for the heavy nuclear engine, the effect is much smaller. This provides an opportunity to increase the expansion ratio of the nozzle from about 80 (the value typical of a chemical engine for vacuum use) to 300 or so. This pushes the thrust coefficient much closer to its theoretical maximum, and this will contribute to an increase in exhaust velocity. Given that the nuclear engine is used to provide a combination of high exhaust velocity with high thrust, a long nozzle, with a high expansion ratio, is a positive advantage, and has little impact on the dry mass of the engine and hence the mass ratio.

7.15 POTENTIAL APPLICATIONS OF NUCLEAR ENGINES

The hierarchy of rocket engines is one based on the two characteristics of thrust and exhaust velocity. The rocket equation does not contain the thrust as a parameter, and its prediction of vehicle velocity does not require the thrust to be known. The rocket equation expresses the efficiency with which a certain quantity of propellant can give the desired velocity to a required mass of payload. Referring back to Chapter 6, where the propellant efficiency was introduced:

$$\frac{M_F}{M} = e^{\frac{v}{v_e}} - 1$$

we recall that the efficiency of propellant use is independent of the thrust for manoeuvres in space. For missions involving a large value of delta-V, any increase in exhaust velocity produces a dramatic reduction in propellant usage. For interplanetary missions, given that every kilogramme of propellant has to be launched into Earth orbit prior to use, high exhaust velocity is essential. This is why electric and nuclear propulsion, both disposing of high exhaust velocity, have been considered for such missions. For example, it requires some 11 km/s to inject a vehicle into Mars transfer orbit. A chemical rocket, using liquid hydrogen and liquid oxygen, gives 4.55 km/s of exhaust velocity (a ratio of 2.4), requiring 10 kg of propellant per 1 kg of payload. A nuclear engine can give an exhaust velocity of 9.0 km/s (a ratio of 1.22), requiring 2.4 kg of propellant per 1 kg of payload. An ion thruster, giving 20 km/s (a ratio of 0.55), requires only 0.73 kg of propellant per 1 kg of payload.

The advantage of the nuclear rocket is intermediate between the chemical and the ion thruster, when only exhaust velocity is considered. But, if the thrust is taken into account, then the well-known disadvantage of ion engines is dominant. An ion engine can only generate thrust of a fraction of a Newton, while a nuclear engine can generate a thrust of hundreds of kilo-Newtons. The time taken to achieve a given velocity does depend very strongly on the thrust. To see this, consider the following, beginning with the expression for the time to achieve a given vehicle velocity, taken from Chapter 5

$$t = \frac{M_0}{m}\left(1 - \frac{M}{M_0}\right)$$

where m is the mass flow rate. Using the usual expression for thrust, $F = mv_e$, and making the substitution we find:

$$t = \frac{M_0 v_e}{F}\left(1 - \frac{M}{M_0}\right)$$

and we see that the time taken depends both on the exhaust velocity and on the thrust, for a given mass ratio. As an example, consider a 10-t spacecraft to be sent into Mars transfer orbit by being given a delta-V of 11 km/s. In the electric thruster case, assume a generous thrust of 10 N from a cluster of ion engines, an exhaust velocity of 20 km/s, and a required propellant mass of 7.3 t. The time taken to achieve the necessary velocity, given by the formula, is:

$$\frac{17,300 \times 20,000}{10}\left(1 - \frac{10,000}{17,300}\right) = 14.6 \times 10^6 \text{ s}$$

This is 170 days.

For the nuclear engine, with a lower propellant efficiency, but a typical thrust of 300 kN, the result is:

$$\frac{34,000 \times 9,000}{300,000}\left(1 - \frac{10,000}{34,000}\right) = 7,200 \text{ s}$$

This is just two hours.

The above calculation shows both the advantage and the disadvantage of the nuclear engine compared with the ion engine. The propellant mass difference is large: 24 t—nuclear, compared with 7.3 t—ion; but the duration of the burn necessary to enter the transfer orbit is vast, with two hours for the nuclear engine, compared with 170 days for the ion engine.

The future role of the nuclear thermal engine is derived from the advantages illustrated here. In terms of propellant mass, the nuclear engine requires 24 t of propellant in the above example, but the best chemical engine requires 100 t. The nuclear engine has sufficient thrust to execute an interplanetary transfer manoeuvre in a reasonable time. The high exhaust velocity of the ion engine gives a huge propellant saving—only 7.3 t required, but the time taken to execute the manoeuvre is unacceptably long.

7.16 OPERATIONAL ISSUES WITH THE NUCLEAR ENGINE

As noted above, while the start-up and shut-down operations for a nuclear engine are more complicated than for a chemical engine, they are well understood and pose no real challenges. The major operational challenges for a nuclear engine are associated with radiation, and danger to human life. While it was acceptable in the 1960s to operate a nuclear engine at a remote test site, with the exhaust stream entering the atmosphere, it is not acceptable now. The radioactivity of the exhaust stream is minimal for a well-designed engine. The hydrogen will not become radioactive, but there may be fuel rod fragments if there is any erosion. Major efforts have been made to prevent erosion, because of its detrimental effect on the longevity of the engine, and we may assume that any new engine will be designed not to allow erosion of the fuel rods. Nevertheless, it is quite clear that a nuclear engine cannot be operated in the atmosphere. This limits its use to Earth orbit and beyond. Assuming for the moment that the engine can safely be launched into Earth orbit, then the main challenge is to prevent the engine from becoming a hazard after use; as noted above, the engine is not radioactive until it has been operated. Using a nuclear engine in low Earth orbit may be regarded as hazardous, because there is a possibility of re-entry, and contamination of the atmosphere. The firing of the engine must therefore be confined to high Earth orbit, and to interplanetary transfer manoeuvres. In such orbits and manoeuvres, there is a very small probability of accidental re-entry because of the energy required to de-orbit. The nuclear thermal engine is therefore ideal for interplanetary transfer, and is unlikely to find application in near Earth manoeuvres.

7.17 INTERPLANETARY TRANSFER MANOEUVRES

The high delta-V requirements of interplanetary manoeuvres can best be achieved with a nuclear thermal engine. In the example given above, a saving in propellant mass of a factor of four is found. This, combined with relatively high thrust, is ideal. The kinds of missions enabled by this technology will include missions to the Moon, Mars, and Venus, and especially missions to the outer planets where the delta-V requirements are very high indeed. The subject of interplanetary transfer is beyond the scope of this book, and is dealt with in *Expedition Mars*. Here a brief summary will be given, to illuminate the role of the nuclear thermal engine.

Interplanetary transfers are based on elliptical orbits. To make the journey to Mars for example, the lowest energy, and hence the smallest delta-V is obtained when the spacecraft journeys along an ellipse that touches Earth's orbit at perihelion and the orbit of Mars at aphelion. To enter this ellipse, say from low Earth orbit, it is necessary to use an hyperbolic escape orbit. The spacecraft is initially in a circular orbit, with an eccentricity of zero, and the velocity has to be increased to make the eccentricity greater than one. This results in the spacecraft escaping from the Earth's gravitational influence, with some residual velocity; an eccentricity of precisely unity corresponds to an escape with a zero residual velocity. The residual velocity, added to the Earth's orbital velocity, is the necessary velocity for entry into the transfer ellipse. The formulae for this calculation are given below, and are based on the orbit equation given in Chapter 1.

$$\frac{r}{r_0} = \frac{1+\varepsilon}{1+\varepsilon \cos\theta}$$

$$\frac{r_{Mars}}{r_{Earth}} = \frac{1+\varepsilon}{1-\varepsilon}$$

$$\varepsilon = \frac{\left(\frac{r_{Mars}}{r_{Earth}}\right) - 1}{\left(\frac{r_{Mars}}{r_{Earth}}\right) + 1}$$

$$V_0 = \sqrt{\frac{GM_\odot}{r_0}(1+\varepsilon)} = 32.74 \text{ km/s}$$

Having calculated the eccentricity, the necessary velocity with respect to the sun is given by the third formula. The velocity of the Earth in its orbit is 29.8 km/s and so the required residual velocity is only 2.94 km/s. This is the required residual velocity, but the total velocity necessary to escape from Earth with this residual velocity is 11.16 km/s and is calculated using the kinetic energy equation below.

$$\tfrac{1}{2}mv^2 = \tfrac{1}{2}mv_0^2 - \tfrac{1}{2}mv_{esc}^2$$

$$v^2 = v_0^2 - v_{esc}^2$$

Here the final energy is equated to the initial energy, minus the kinetic energy necessary for escape with a zero residual velocity. Thus, the injection velocity is given by:

$$v_0 = \sqrt{v^2 + 2v_{circ}^2}$$

and the eccentricity of the hyperbolic orbit by:

$$\varepsilon = \frac{v_0^2}{v_{circ}^2} - 1$$

All of these can be derived quite simply from the orbit equation given in Chapter 1 and Appendix 1 and the above kinetic energy equation.

The ideal role then for the nuclear engine in a Mars mission is to give this delta-V to the departing mission. It would be in the form of a booster, comprising a hydrogen tank and a nuclear propulsion system that is fired up once the vehicle is at a safe distance from the Earth (Plate 19). When the delta-V manoeuvre is complete, the booster is separated from the coasting Mars vehicle and fired up again, using the remaining propellant, to place it in a safe solar orbit, which will not intersect with the Earth or Mars for a million years or so. The dangerously radioactive spent engine is thus safely disposed of.

7.18 FASTER INTERPLANETARY JOURNEYS

The minimum energy transfer to Mars takes 258 days. Given that the nuclear engine has a higher capability than a chemical engine, it can be considered as a means to shorten this journey. The delta-V values, given above, place the spacecraft into the Hohmann transfer orbit illustrated in Figure 7.11 where the various periods of Mars, Earth, and the spacecraft are shown. If more velocity can be given to the spacecraft, then the eccentricity of the transfer orbit can be increased, according to the formulae above, and the journey will be shorter, with the spacecraft orbit, cutting the Mars orbit, at an earlier point. This is shown in Figure 7.12.

The journey time in this case will be smaller, both because of the reduction in distance, and because the average speed of the spacecraft will be higher according to Kepler's law. The calculation of this time is beyond the scope of this book—it requires a determination of the area bounded by r_0, r, and the transfer trajectory, by integration, and then division of the result by the constant areal velocity derived from Kepler's law. The result is given in Figure 7.13.

Sec. 7.18] Faster interplanetary journeys 233

Thus, using a nuclear engine, a considerable shortening of the journey can be achieved by giving additional velocity to the spacecraft.

For arrival and capture at Mars, the nuclear engine could also be used, provided it can then be disposed of into a safe solar orbit. This requires that the vehicle be shielded throughout the journey, to a level that would be safe, even for continuous exposure during a long voyage. At the same time, the hydrogen would have to be stored without evaporation. This poses an additional technical challenge, because, up to the present-day, hydrogen and oxygen have been used in vented tanks, for all

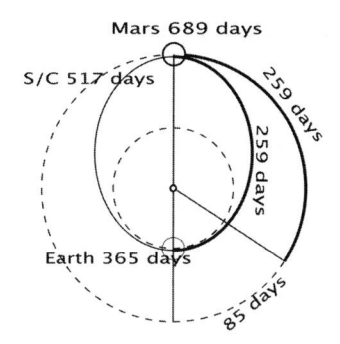

Figure 7.11. The Earth–Mars minimum energy transfer orbit. The elliptical orbit called a Hohmann orbit after its inventor, touches the orbit of Earth at its inner end and the orbit of Mars at its outer end. The spacecraft does a complete 180 degree turn in its journey. The required delta-V for this orbit is the minimum possible.

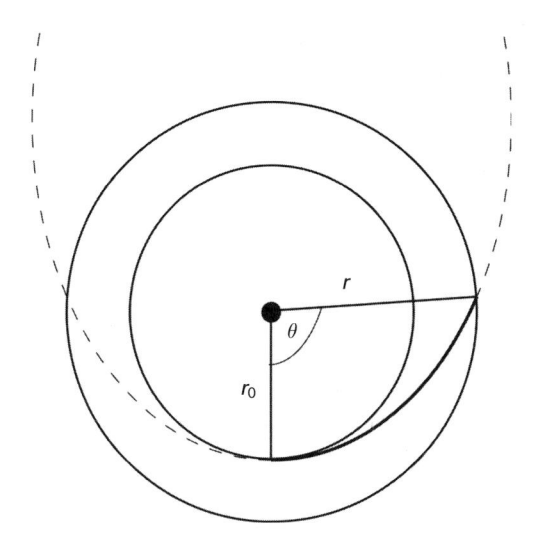

Figure 7.12. A short flight to Mars. Here the spacecraft uses additional velocity to enter an elliptical orbit with much greater eccentricity than the Hohmann orbit. It cuts the orbit of Mars earlier, and the journey is quicker both because the distance is smaller, and because the mean velocity over this short segment is greater.

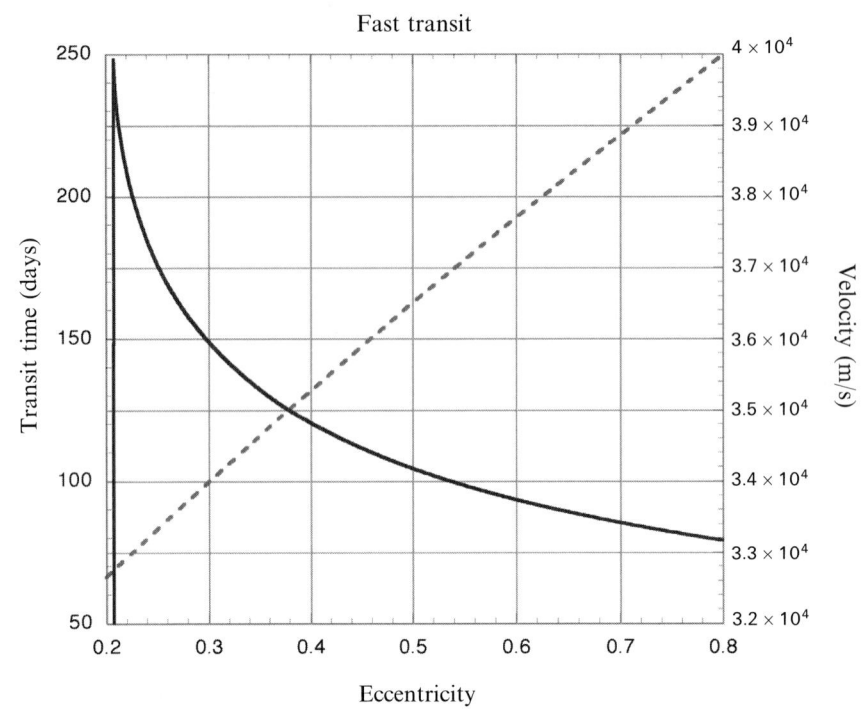

Figure 7.13. The transit time to Mars as a function of initial delta-V and orbit eccentricity. The journey can be shortened considerably for a small increase in delta-V, but it is very difficult, requiring a high delta-V requirement to reduce it much below 100 days. Transit time is the solid curve and left-hand scale; velocity is the dashed curve and right-hand scale.

vehicles; these gases cannot be stored in liquid form, under pressure, at normal temperatures. Some consideration of this problem will be given in the next section.

For the present, assuming hydrogen can be stored for a long voyage, the nuclear engine can be seen as an enabling technology for deep space missions such as visits to Mars and also to the Jovian and Saturnian moons. Here the many manoeuvres necessary to explore the moons could be provided by the nuclear engine, at a much smaller fuel cost than with conventional chemical engines. The journey time could also be reduced, as indicated above.

7.19 HYDROGEN STORAGE

Liquid hydrogen has a critical temperature (above which it cannot be held liquid by pressure alone), of 241°C, and a boiling point of −253°C. To prevent evaporation, it must be refrigerated and kept below its boiling point. In space the only heat-input is from the sun, and the temperature of dark space is −270°C. So, storage comes down to insulation, to reduce the heat input from the sunlit side, and efficient extraction of heat to deep space, on the shaded side of the tank. Multi-layer insulation, the normal type of spacecraft insulation, can be used to reduce the heat input, while combined with a sunshield. When used with up to a hundred layers, it has been established that the heat leakage into the fuel tanks can be reduced to about 3.0 W/t of liquid hydrogen. This is based on spherical tanks in sunlight, radiating to space, with radius scaling with the mass of hydrogen contained. Extraction of these 3 W would be sufficient to keep the hydrogen below its boiling point. This does not seem much, but when the inefficiency of refrigerator systems at low temperature is considered, it is a significant challenge. The method would be to have a cooling loop in the tank, connected to a normal mechanical refrigerator, with the heat disposed from the hot end through a radiator on the shaded side of the tank. The power to pump the heat from the interior of the hydrogen tank to the radiator has to be provided electrically. When the inefficiency of this process is included in the calculation, a value of 524 W of electrical power has to be provided per tonne of liquid hydrogen.

This could be provided by solar panels, at a mass of about 24 kg/kW. But a more efficient solution would be to operate the nuclear engine core at a very low power level during the voyage, to provide electricity. Running the core at low power levels is a simple matter. As we have seen, a reactor can run at any power level with the multiplication factor set to unity. It is simply a matter of allowing the power level to rise, with a given excess multiplication factor, and returning the value to unity when the desired power level is reached. There would however need to be a separate cooling loop in the engine core to carry the heat away, and an electrical generation system, possibly a Stirling engine or a Brayton turbine, driving a generator. Designs have been elaborated for such *bimodal* nuclear thermal engines, which use the nuclear core to provide all the power necessary for the vehicle. The cooling loop is generally specified to be a liquid metal system, similar to those used in ground-based power generation systems. The main issue is how to prevent the cooling loop from being destroyed by the high power output when the engine is being used to provide thrust. It needs to be insulated from the rest of the core, but at the same time be able to extract heat from it. Different schemes are proposed, including vacuum insulated heat-pipes, and the use of only a small section of the core, provided with its own control of criticality.

7.20 DEVELOPMENT STATUS OF NUCLEAR THERMAL ENGINES

Nuclear propulsion has been studied, on paper, by all the major space agencies. In addition, much experimental work has been done, both in the United States, and in Russia (USSR). This included ground testing of nuclear thermal rocket engines. The latter was stopped in the early 1970s when the atmospheric test ban treaty came into force; it was no longer acceptable to release even mildly contaminated hydrogen exhaust into the atmosphere. Non-fissile aspects of nuclear engines have however continued to be studied: including thermal testing, where electrical heat generation is used instead of fission, in an otherwise completely representative engine core; the design of fuel elements; and various control aspects. Experience with practical engines (i.e., those that could really be fired and used to generate thrust), was confined to the period ending in 1972.

The programme in the United States ran from 1955 to 1972, at a cost of 1.4 billion dollars, a huge investment. The initial programme was called KIWI—a flightless bird—and was confined to reactor core and pressure vessel developments; the later programme was called NERVA (Nuclear Engine for Rocket Vehicle Applications). NERVA included flight components, like turbo-pumps and nozzles, and was to all intents and purposes a flyable engine, although it was never actually launched. The NERVA programme made significant design improvements, and at the end, there were several designs that were not tested, but which provide the basis for a re-starting of the nuclear engine programme in our times.

The first nuclear engine test was conducted in July 1959 with KIWI A (Figure 7.14) containing rectangular plates of uncoated UO_2 with a graphite moderator. The

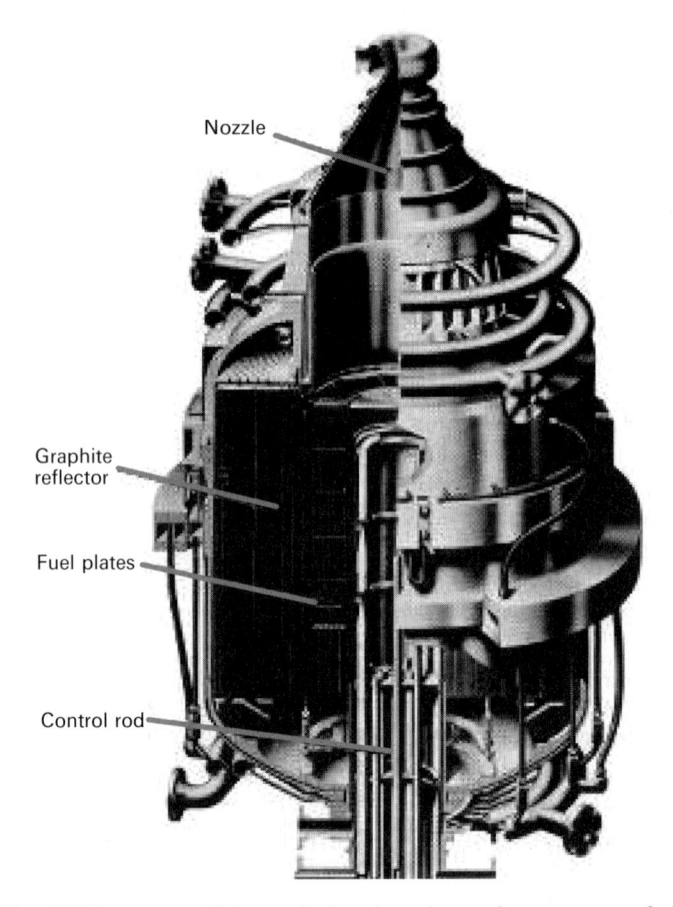

Figure 7.14. The KIWI reactor. This small drawing shows the structure of the core and reflector; there was a nozzle as shown, but no expansion of the emerging hot gas or other flight-like sub-systems. It was intended only for ground testing, hence 'KIWI'.
Courtesy NASA.

power level achieved was 70 MW, and the core temperature was 2,683 K. Immediately identified problems were structural damage to the core, due to high vibration levels, and erosion of the carbon in the fuel elements by chemical reaction with the hydrogen propellant. Re-design of the core flow patterns, and the fuel elements, eliminated vibration damage, but erosion was never fully conquered. The details of engines tested are given in Table 7.3.

There were five series of engines (Figure 7.15). The KIWI series, which comprised essentially the reactor core and pressure vessel, cooled by hydrogen, was not intended for flight; it was used to develop the reactor core characteristics. The Phoebus series, which had all the properties of a flight engine (nozzle, etc.), was designed to meet the perceived need for a nuclear booster for a human Mars mission. Design goals were a thrust of 1,100 MN, exhaust velocity of 8,500 m/s, and a power level of 5,000 MW. Moving on from the core designs in the KIWI reactors, the Phoebus runs tested rocket performance, including the use of an expansion nozzle, over a wide range of power and propellant flow levels. The fuel elements were uranium carbide in a graphite matrix coated with molybdenum and niobium carbide. These were successful in resisting erosion and cracking in the hot hydrogen. These engines also demonstrated cooling of the support structure and reflector by hydrogen flow in a regenerative mode. The PEEWEE reactor was designed to test high power density in the core and alternative fuel-element coatings—zirconium carbide being one. Advanced fuels like uranium-zirconium carbide were tested in the Nuclear Furnace series, and the NRX series tested full-sized engines in simulated rocket operations, including vibration and shock loads as well as start–stop sequences, restart, and thrust control. The XE-Prime engine was tested firing vertically downwards in a simulated space vacuum. It was the nearest thing to a test flight of a nuclear engine that was achieved during the NERVA programme (Plate 17).

The conclusion from all this is that a nuclear engine came very close to a flight test in the late 1960s, and that elements for a flight reactor had been developed to a point where the known technical issues (e.g., fuel rod erosion) could be overcome. Then the programme was stopped. Similar developments were carried out in the Soviet Union until quite late, but they were also eventually stopped, leaving the flight qualification of a nuclear thermal engine still to be done.

As mentioned at the beginning of this chapter, the use of fission as a power source in space is now again being actively canvassed; it will be developed as part of the Jupiter Icy Moons mission, JIMO. At the same time the new Bush initiative, to explore Mars with a human crew, indicates the need for a nuclear thermal rocket engine as the main booster for interplanetary transfer. Thus, there is renewed interest in nuclear fission technology for space. The fission reactor as an electrical power source is covered in Chapter 6. Here we look briefly at possible developments that could re-open the nuclear thermal rocket engine programme.

There are two possible development routes. One is to build directly on the NERVA programme, and develop an engine that uses carbide fuel rods, essentially to the NERVA design, and bring to fruition the programme that was abandoned in 1972. This has the merit of making use of all the NERVA developments, and especially the results of the tests that took place then—they would not be allowed

Table 7.3. The tests carried out for the NERVA programme up to 1972. The results of these tests could be used as the basis for a revived nuclear engine programme. Note that some engine firings exceeded the necessary duration for an in-space delta-V manoeuvre.

Name	Fuel	Moderator	Coating	Temperature	Power	Comment
KIWI A (1 July, 1959)	UO$_2$ plates	Graphite	None	2,683 K	70 MW	Vibration damage and erosion. 300 s of operation
KIWI A Prime (8 July, 1960)	UO$_2$ particles embedded in graphite cylinders	Graphite	Niobium carbide coated cooling channels (4)		85 MW	Some structural damage. 307 s of operation
KIWI A3 (October, 1960)	UO$_2$ particles embedded in graphite cylinders	Graphite	Thicker niobium carbide coating		100 MW	Some core damage, blistering of coating 259 s of operation
KIWI B1A (December, 1961)	UO$_2$ particles embedded in graphite cylinders	Graphite	More cooling channels (7)		300 MW	Planned 1,100 MW. Test terminated after 30 s due to a fire
KIWI B1B (September, 1962)	UO$_2$ particles embedded in graphite cylinders	Graphite	Niobium carbide		900 MW	Some fuel elements ejected through nozzle; test terminated
KIWI B4A (November, 1962)	UO$_2$ embedded in extruded hexagonal graphite blocks with 19 cooling channels	Graphite	Niobium carbide			Terminated before full power because of core disintegration
KIWI B4D	UO$_2$ particles embedded in graphite cylinders	Graphite	Niobium carbide			Vibration eliminated; test terminated due to rupture of nozzle cooling tube
KIWI B4E	Uranium carbide particles coated with 25 μm pyrolytic graphite, embedded in graphite	Graphite	Niobium carbide	1,980 K	937 MW	Operated for 8 min at full power

Reactor (Date)	Fuel	Moderator	Cladding	Temperature	Power	Notes
KIWI TNT (January, 1965)	UO$_2$ embedded in extruded hexagonal graphite blocks with 19 cooling channels					Reactor deliberately destroyed to test fast excursion models
Phoebus 1A (June, 1965)	Uranium carbide embedded in graphite	Graphite	Niobium carbide and molybdenum carbide	2,370 K	1,090 MW	First 'flight-like' engine. Operated for 10 min at full power
Phoebus 1B (February, 1965)	Uranium carbide embedded in graphite	Graphite	Niobium carbide and molybdenum carbide		1,500 MW	Second test. Operated for 30 min at full power, and 15 min at lower power
Phoebus 2A (June, 1968)	Uranium carbide embedded in graphite	Graphite	Niobium carbide and molybdenum carbide	2,310 K	5,000 MW	Most powerful reactor ever. Operated for 12.5 min at 4,000 MW, including a restart
PEEWEE		Graphite	Zirconium carbide and niobium carbide	2,550 K	503 MW 5,200 MW/m^3	Small test reactor. Exhaust velocity 8,450 m/s
Nuclear Furnace 1 (1972)	Uranium carbide, zirconium carbide, and carbon. Also uranium-zirconium carbide	Water	Zirconium carbide	2,500 K	44 MW 4,500–5,000 MW/m^3	Test reactor to develop thermally stable fuel elements and coatings
NRX A2 (September, 1964)	Uranium carbide particles coated with 25 μm pyrolytic graphite, embedded in graphite	Graphite	Zirconium carbide		1,100 MW	Operated for 5 min at high power. Exhaust velocity 7,600 m/s
NRX A3 (April, 1965, May, 1965)	Uranium carbide particles coated with 25 μm pyrolytic graphite, embedded in graphite	Graphite	Zirconium carbide			3.5 min at full power. 16.5 min at full power. Total 66 min operation

continued

Table 7.3. (cont.)

Name	Fuel	Moderator	Coating	Temperature	Power	Comment
NRX EST (February, 1966)	Uranium carbide particles coated with 25 μm pyrolytic graphite, embedded in graphite	Graphite	Zirconium carbide		1,100–1,200 MW	110 min of operation
NRX A5 (June, 1966)	Uranium carbide particles coated with 25 μm pyrolytic graphite, embedded in graphite	Graphite	Zirconium carbide		1,100 MW	30 min of operation
NRX A6	Uranium carbide particles coated with 25 μm pyrolytic graphite, embedded in graphite	Graphite	Zirconium carbide	2,342 K	1,100 MW	60 min of operation at full power—NERVA design goal
XE prime (11 June, 1969)	Uranium oxide particles embedded in graphite	Graphite	Zirconium carbide		1,100 MW	First to operate firing downward. 115 min of operation

Sec. 7.21] Alternative reactor types 241

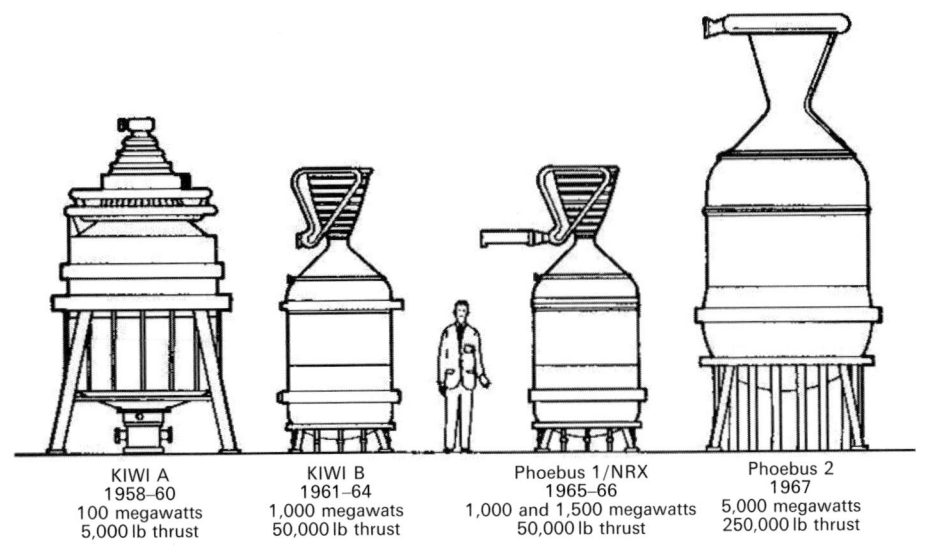

Figure 7.15. The NERVA family of engines.
Courtesy NASA.

KIWI A
1958–60
100 megawatts
5,000 lb thrust

KIWI B
1961–64
1,000 megawats
50,000 lb thrust

Phoebus 1/NRX
1965–66
1,000 and 1,500 megawatts
50,000 lb thrust

Phoebus 2
1967
5,000 megawatts
250,000 lb thrust

today (Plate 18). The technology could be regarded as old fashioned, but we have to remember that it was technology of the 1960s that placed men on the Moon. There is a series of paper designs for modern engines, based on the NERVA programme results. At the same time, using modern computing facilities, and thermal test facilities that simulate the heat of fission using electrical power, new fuel element configurations have been developed that should be stable in use, and have better heat transfer properties. Among these are the twisted ribbon element that originally came from the Russian programme, and the grid matrix designs from the University of Florida. Some of the NERVA based, and newer, designs are described in Table 7.2.

7.21 ALTERNATIVE REACTOR TYPES

There have always been attempts to overcome the fundamental limitation of the solid core reactor by making use of novel configurations. It is difficult to heat the propellant much beyond about 2,500 K, even if the fuel elements themselves can take a much higher temperature, because of the poor heat transfer characteristics of the graphite rods. A way to improve this is to use a particle bed reactor. This has the fuel, in the form of small pellets, held between sieves of refractory metal, called 'frits'. This allows intimate contact between the propellant and the fuel pellets, the former being able to flow through the frits, and over and around the small pellets. The pellets can run much hotter than the solid fuel rods, because there is much less thermal stress across their small diameter, and of course, a much larger surface area is presented to the flowing propellant. The uranium, in carbide, or in so-called 'cermet'[3] form, would have to be highly enriched in the absence of large masses of moderator. But very high temperatures of the exhaust stream are possible, and hence a much higher exhaust velocity, perhaps exceeding 10 km/s.

Other concepts, aiming at still higher temperatures, are too far from achievement, even on the ground, for detailed consideration. Among these are fluid and gaseous core fission reactors, where the fissile material is allowed to melt or indeed become gaseous, so that the temperatures can exceed the service temperature of metals and ceramics. These are all attempts to improve the heat transfer to the propellant from the heated fuel material. They are all limited by the mechanical properties of the matrix holding the uranium. No devices of this kind have been built or even seriously simulated.

There is a more radical approach, which makes direct use of the fission fragments to heat the propellant. These, it will be recalled, have an energy of about 200 MeV; this corresponds to a temperature of hundreds of millions of degrees. A propellant gas, heated directly by these fragments, could have an arbitrarily high temperature. For the fission fragments to escape from the uranium matrix, the plates or fuel elements would have to be very thin (a few tens of microns), because the fragments stop very quickly in matter. Two approaches have been suggested. In the first, very thin foils of enriched uranium, deposited on thin beryllium oxide plates, are suspended in the gas stream, and the fission fragments escape in significant numbers and heat the gas. A moment's thought will show that the product of gas density and width of the channels between the plates must be significantly greater than the product of density and thickness of the fuel plates for this to work. Another approach, pioneered by Professor Carlo Rubia is to use the fissile isotope Americium 242 which has a very high thermal neutron fission cross section. It is deposited in the form of a thin coating on the inside of a rocket combustion chamber. Fission is induced, using an external reflector, and the fission fragments escape into the gas in the combustion chamber and heat it to an arbitrary temperature. The wall of the chamber is cooled using a liquid lithium jacket, and the hot gas can be exhausted through the nozzle to generate thrust. The exhaust velocity could be very high indeed, comparable with ion engine velocities (35 km/s is quoted); the high power output of the fission system couples this high exhaust velocity with a thrust of the order 1.7 MN. This device has yet to be developed beyond paper studies by the Italian Space Agency (ASI), but the idea is quite sound and has many advantages in engineering terms.

Thus, the stage is set for the development of a practical nuclear thermal engine, probably based on the NERVA results. It only awaits a decision to re-start the programme. The major challenges are not technical but socio-political. In essence,

---

[3] *Cermet* is an abbreviation for ceramic–metal composite. The uranium oxide is combined chemically or in sintered form with the refractory metals tungsten or molybdenum, to form a kind of ceramic material that allows fission with un-moderated neutrons, or so-called 'fast fission'.

such devices must be safer than chemical rockets, and must be demonstrated to be so.

7.22 SAFETY ISSUES

Given the natural reluctance of the public to have anything to do with nuclear fission on the ground, it is even less likely that the public would approve a fission core being transported into space on a rocket, and then operated in orbit. There is a vast amount of work to do, both in designing a fail-safe system, and in convincing the public and legislatures that it will be safe to operate. The issues that have to be confronted are: the manufacture and test of the core, its transport to the launch site, the launch itself, and the activation of the core in orbit. We should also include the requirement for safe disposal, either following operation, or following an abort at any stage of the process. These days we also have to consider security issues, given the high enrichment of the uranium fuel.

There is ample experience, and there are ample facilities, for the manufacture of such reactor cores. In essence, they are no different from reactors for nuclear submarines, or indeed for research and certain power applications. Security and safety procedures are fully in place at these facilities and there will be no additional precautions necessary. As far as security and safety goes, the major spacecraft launch ranges also have sufficient precautions in place. Launchers are very dangerous items, and need protection and safe operational procedures. The same could be said for terrestrial transportation of the reactor.

Testing is another matter. Since the engine could release some radioactive material in its exhaust stream, there can be no atmospheric testing. The exhaust will have to be contained, and scrubbed of all radioactive contamination before release into the atmosphere. Such a facility was set up for the all-up Phoebus test, and could be duplicated for ground testing of a new engine. However, the whole safety issue, and particularly the perception of safety, will be much easier if the flight fissile core has never been activated. It will then be inert, in radioactivity terms, and can be launched with no fear of radioactive contamination following an accident. A procedure involving tests of prototypes rather than the actual core will have to be used. This is in fact paralleled in current practice for the testing of certain sensitive spacecraft components, and pyrotechnic actuators, where a pre-launch test of the flight item is sometimes waived, if it can be demonstrated that an identical item has withstood the actual launch loads several times, and functioned correctly afterwards.

With an inert core, safety during launch is simply a matter of preventing the onset of fission under all conceivable accident scenarios. Since the conditions for fission are a highly ordered set of configurations, it is unlikely that they could be duplicated in a random fashion during an accident. The bare core itself cannot become fissile without the reflectors. These could be launched separately and united with the core only in orbit, or they could be locked in a fail-safe manner to make sure they could not, under any conceivable circumstance, configure themselves in such a way as to initiate fission. Neutron poisons could be inserted into the core, to prevent any build-up of neutrons, even should the worst happen. These can be in the form of a plug that can only be removed in orbit. There are many ways of preventing fission during launch or an accident, and it will be a matter of careful analysis to decide exactly how they should best be deployed. It is worth repeating that the material itself is safe, and in a fire or crash, or even a re-entry, the uranium dispersed would not be radioactive.

If the launch can be convincingly demonstrated to be safe, can operation in orbit be similarly demonstrated? It is clear that, once the reactor has been activated, it becomes a container of nuclear waste, with all the real and imagined dangers associated with that product. It must not re-enter, and should preferably be disposed of in a deep space orbit, such that it will not return to the vicinity of Earth until the radioactivity has died away, and it has become inert once more. This can easily be done, and it will be part of the operation procedure of the engine that propellant will be reserved for this purpose. The danger is in a mis-operation or accident in orbit. Here the celestial mechanics have to be such that re-entry cannot happen, or has an extremely low probability of happening. This probably means that the orbit will have to be a high one, not a typical low Earth orbit at 500 km. The permitted manoeuvres will also have to be such that re-entry is not a possibility; in the end it is a matter of getting the celestial mechanics right when planning the manoeuvre.

The safety of the crew of a manned flight is of course paramount. Their safety will in any case depend on the engine operating properly, in that it is needed to place them in the correct orbit; but here we are concerned with radiation safety. While operating, the core will emit a high flux of neutrons and gamma-rays; both of these are penetrating, and require significant shielding to stop them. The engines, described above are all fitted with internal shields. Shielding is most mass-effective close to the source of radiation: because of geometry it can be physically smaller. No shielding is provided in any outward or downward direction, regarding the nozzle as pointing 'down'. We only need to protect the spacecraft, which is forward of the engine. There will probably need to be additional shielding for the base of the spacecraft; and we have already alluded to the use of the partially full hydrogen tanks to protect the crew from neutrons. Gamma-rays need dense, heavy element, shields to stop them; tungsten or molybdenum will need to be incorporated in the spacecraft shield. Another possibility, used in some of the earliest designs by von Braun, but still valid, is to place the engine on a boom, well away from the crew compartment, to allow the inverse square law to do its work. Again, it is a matter for a detailed design and trade-off study to devise the most mass-efficient way to protect the crew.

It is of course important to make sure that the engine in use continues to operate in a safe mode. The fission reaction and the hydrogen flow have to be balanced so that all the heat produced is carried away by the hydrogen. This is a matter for feedback and control loops. The position of the control drums controls the heat produced by fission, and the speed of the turbo-pumps controls the hydrogen flow. These two need to be connected by a fail-safe control loop so that a sudden drop in the hydrogen flow-rate does not cause the fissile core to overheat. In practice, the exit

Sec. 7.22] Safety issues 245

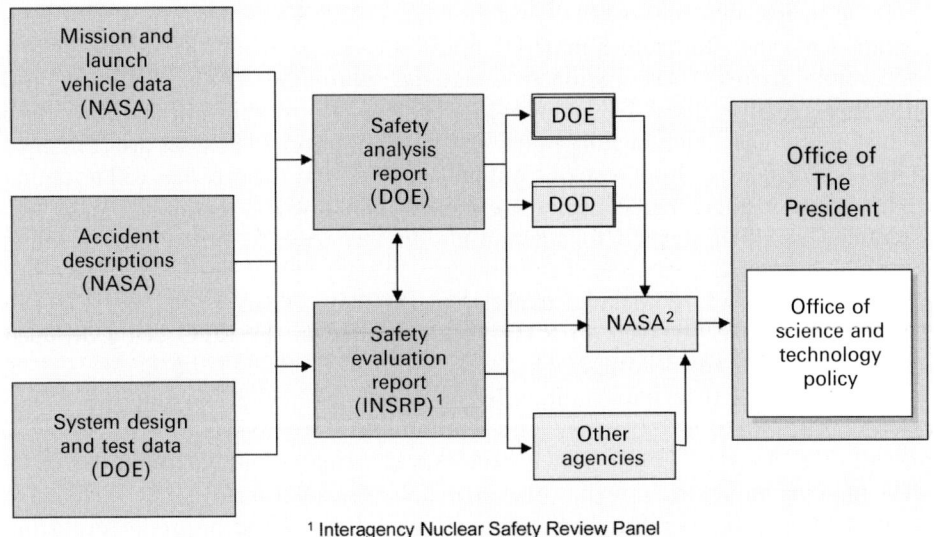

Figure 7.16. The scheme for approval of the use of Radioactive Thermal Generators on spacecraft for launch in the United States. The same scheme could be used to gain approval of the launch of nuclear fisson thermal generators as described in Chapter 6, or for the launch of nuclear thermal rocket engines. There is no corresponding scheme in place for other space agencies.
Courtesy NASA.

[1] Interagency Nuclear Safety Review Panel
[2] Responsible mission agency makes launch recommendation

temperature of the hydrogen needs to be monitored closely as well as the pressure in the chamber. Any increase in the former or decrease in the latter should result in a rotation of the control drums to reduce the fission rate. The rate of fission itself can be monitored using the neutron flux in the core. The many hundreds of hours of operation of engines in the 1960s allowed this system to be brought to a high state of development.

Convincing the public and legislatures of the safety of nuclear thermal engines is a different task from that of ensuring their safety; the technical approaches, outlined above, do the latter, if properly applied. In fact, to secure approval, there has to be complete public visibility of the procedures intended to make the matter safe, and the individuals and organisations involved have to be accountable, and subject to checks and inspections at any time. A legal and organisational model for the control and monitoring process exists in the system used to permit the use of RTGs in space vehicles. This is one of the reasons why their use outside the United States is very difficult: the safeguards and consensus that exists in the United States have yet to be established elsewhere. The organisational model is shown in Figure 7.16; it involves, ultimately, the Office of the President of the United States. Of course, it is the detail of the engineering, and the quality assurance, which makes the devices safe; it is the organisation and the chain of accountability that convinces the legislature, and hopefully the public.

246 Nuclear propulsion   [Ch. 7

The principle that makes RTGs safe for use by NASA is the principle of total containment: the radioactive material must not escape from the device in any conceivable accident. The radioactive material, plutonium oxide is formed into hard, inert, ceramic discs. This material is very resistant to abrasion and formation of dust; it also has a high melting point, and it will not burn. The discs are contained in iridium capsules; iridium is strong but malleable so that the capsules will not burst or rip under impact. The capsules are in turn contained in graphite blocks; the assemblies are then stacked together inside an aero-shell to protect the graphite during high speed re-entry into the atmosphere. This 'belt and braces' approach has prevented release of radioactive material in the many, many, tests and the two accidents that have involved an RTG—both resulted in the RTG being dropped intact into the sea and recovered.

There are two very important differences between the RTG and the fission reactor. The RTG is strongly, and continuously, radioactive. This is what provides the electrical power that it generates. It remains radioactive for over 80 years after manufacture. On the other hand, the fission reactor is not radioactive before first use. It uses enriched uranium, a mixture of the naturally occurring isotopes, U^{238}, which has a half-life of 4.5 billion years, and U^{235}, which has a half-life of 0.7 billion years. Remembering that the activity depends inversely on the half-life, we see that, compared with plutonium, with a half-life of 84 years, the activity of the uranium fuel in a nuclear rocket is negligible. The fissile material itself can be transported quite safely, and an accident that resulted in the release of enriched uranium would not result in any significant radioactive contamination of the human environment. These facts suggest that the conditions for safe use will be very different from those for RTGs. Provided there is no possibility of the reactor becoming fission-active during launch or any conceivable accident, there is no need to contain the fissile material. The fuel rods are designed not to melt, or release dust particles under attack by fast moving hydrogen at pressures of 50 bar and temperatures of 3,000 K, it becomes clear that they would survive most accidents intact.

7.23 NUCLEAR PROPELLED MISSIONS

It seems likely that a nuclear propelled mission will be mounted in the next decade. The proposal in being at the moment, the Jupiter Icy Moons Mission will make use of a fission reactor to provide the electricity necessary for an electric propulsion, however the associated programme will probably include developments related to nuclear thermal thrusters. The 'customer' for these will be the human mission to Mars proposed by President Bush (Plate 19). The need for fuel economy in such a demanding mission, and for a short injection manoeuvre, to minimise crew exposure to the Earth's radiation belts, can only be met by the combination of high exhaust velocity and high thrust that comes with a nuclear thermal engine. Thus, the Mars expedition and the development of the nuclear thermal engine are likely to occur together. The high performance of a nuclear thermal engine would be beneficial for many other planetary exploration applications. Probes to the outer solar system could make the journey in a much shorter time, and carry more mass. Delivery of large landers to the surfaces of bodies in the inner solar system would be much easier than it would be using chemical propulsion for the in-space manoeuvres. If the safety aspects and political acceptance can be demonstrated, then the nuclear thermal engine will take its place in the range of propulsion systems available for the exploration of space in the twenty-first century.

8

Advanced thermal rockets

Launchers and space vehicles are expensive, and only the most pressing economic, strategic or scientific reasons can justify a space mission. Some areas of space activity—such as communications and Earth observation—are justifiable for economic or strategic reasons, while others—such as pure science, human exploration, or even tourism—have a weaker position. If space is to become a routine environment for human activity, either manned or unmanned, then economic means of access are essential. Much effort is being expended towards this end, and in this chapter we examine some of the physical aspects and limitations, and some of the current approaches. Electric propulsion, discussed in Chapter 6, is less important here than improvements to the design and configuration of thermal rockets. We will first review some basic physical principles.

8.1 FUNDAMENTAL PHYSICAL LIMITATIONS

Until now we have treated the rocket in terms of the vehicle velocity that can be obtained, in relation to the mass ratio and engine performance. When investigating means for improvement it is helpful to look at the *efficiency* of a rocket in terms of how well the chemical energy in the fuel is applied to propulsion.

8.1.1 Dynamical factors

Consider first a simple situation: a rocket motor being fired perpendicular to the gravitational field, which would be, for example, injected into orbit by a third stage, or an interplanetary manoeuvre. This is the simplest situation, and clarifies some of the physical limitations. Gravitational losses can be ignored if the burn is short or if the rocket is guided during the burn.

The simple dynamical situation is shown in Figure 8.1: two masses—the vehicle with its propellant, and an energy release leading to their separation. Momentum

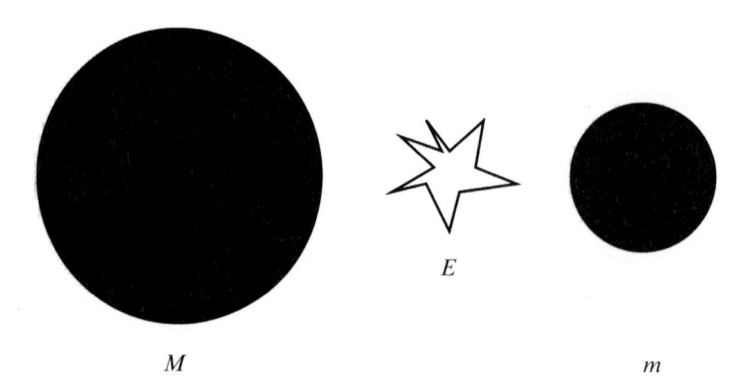

Figure 8.1. Separation of two masses.

and energy are conserved, so we can solve for the velocities and therefore the energy transfer from the propellant to the vehicle.

The quantity of interest is the energy gained by the vehicle, which is

$$E_{Vehicle} = \tfrac{1}{2}mV^2$$

The larger mass is that of the propellant. The energy gained by the vehicle can be expressed in terms of the 'exhaust velocity' (the velocity gained by the propellant) and the ratio of the two masses:

$$V = -\frac{m}{M}v$$

$$E_{Vehicle} = \frac{1}{2}M\left(\frac{m}{M}v\right)^2 = \frac{1}{2}\frac{m^2v^2}{M}$$

$$\frac{E_{Vehicle}}{E} = \frac{1}{2}\frac{m^2v^2}{M} \bigg/ \left(\frac{1}{2}\frac{m^2v^2}{M} + \frac{1}{2}mv^2\right)$$

This determines the efficiency with which energy is converted into vehicle velocity. The ratio M/m is not the mass ratio as we have defined it previously, but rather the ratio of vehicle to propellant mass. Converting it into the more useful mass ratio (the ratio of total mass to vehicle mass), the efficiency is represented by

$$\frac{E_{Vehicle}}{E} = \frac{R-1}{R}$$

This is pure dynamics, and simply shows that the energy transfer depends on the relative masses of the bodies involved. In our terms, the larger R is (the greater the mass of propellant), the more efficient is the energy transfer to the vehicle. For large propellant mass the efficiency would approach 100%.

Unfortunately this dynamical argument applies only when the energy of the propellant is instantaneously released and converted into velocity—for example, in an explosion. In the real situation the efficiency is much lower, because the energy is

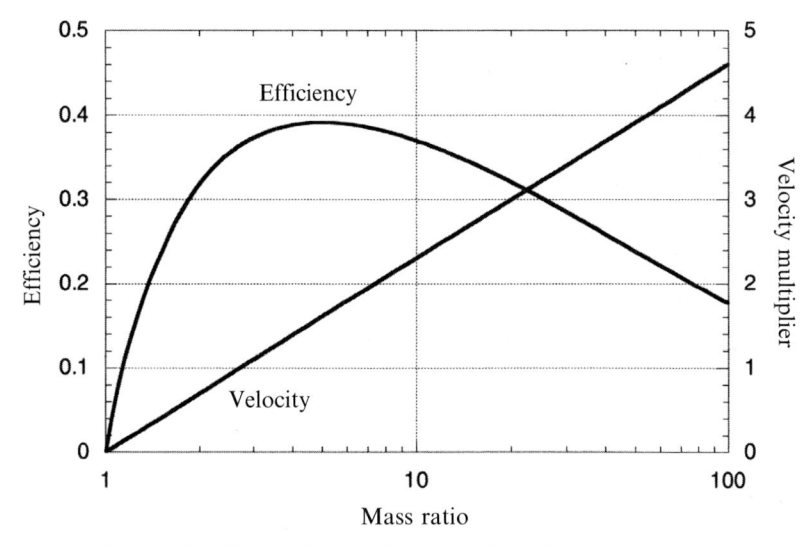

Figure 8.2. Propulsion efficiency as a function of mass ratio.

released gradually; the unburned propellant has to be accelerated with the vehicle, and this energy is wasted. For gradual release an analogous argument can be applied, although instead of deriving the vehicle velocity from the momentum equation, the rocket equation from Chapter 1 is applied. The effective exhaust velocity is the propellant velocity, and the mass ratio is defined in the classical way. The energy of the vehicle is

$$E_{Vehicle} = \tfrac{1}{2} M v_e^2 \log_e^2 R$$

The energy of the propellant is $\tfrac{1}{2} M (R - 1) v_e^2$, and the efficiency is expressed as

$$\frac{E_{Vehicle}}{E} = \frac{\log_e^2 R}{\log_e^2 R + (R - 1)}$$

This equation takes into account the energy expended during acceleration of the unburned propellant. Figure 8.2 shows both the efficiency as a function of the mass ratio, and the value of $\log_e R$, which is the ratio of the vehicle velocity to the exhaust velocity.

The velocity multiplier is the natural log of the mass ratio, and is the factor by which the vehicle velocity exceeds the exhaust velocity. The velocity of the vehicle, for a constant exhaust velocity, always increases with R, but the efficiency now peaks at 40% and a mass ratio of about 5. For larger mass ratios, although the momentum transfer increases, and the vehicle velocity increases, the acceleration of unburned propellant reduces the overall efficiency, and more of the chemical energy is wasted. Thus there is an optimum mass ratio, and for maximum efficiency a rocket should not depart too far from this value. A vehicle with a mass ratio of 10 has only 30% efficiency but, of course, a higher ultimate velocity. This is bought at the cost of extra propellant, which reduces the mass available for the payload and also has to be ferried into orbit.

Although the above arguments have been developed for the case of an orbital change rocket the principles apply to any rocket and, with certain modifications, to the case of a launcher. These simple dynamical arguments indicate that access to space can be less costly through the development of more efficient rockets. While the velocity increment necessary to change orbit—for example, to a martian transfer orbit—can always be met by a higher mass ratio, the cost of transporting this propellant into Earth orbit may be prohibitive. A better method would be to increase the exhaust velocity and return closer to the optimum mass ratio. The use of electric propulsion to do this has been discussed in Chapter 6, but the low thrust of electric propulsion is a limitation. It seems clear that to progress much further with human planetary exploration, higher-thrust rockets of greater efficiency must be designed, such as the nuclear thermal rockets described in Chapter 7.

8.2 IMPROVING EFFICIENCY

To achieve high vehicle velocity with optimum mass ratio requires an appropriately high exhaust velocity. The other factor in the efficiency equation is the so-called 'dry mass' of the vehicle—the mass of the empty vehicle before fuelling. This will be dealt with later.

8.2.1 Exhaust velocity

For most rockets the exhaust velocity *in vacuo* is fairly well optimised, but for launchers working in the atmosphere there is greater scope for optimisation

The effective exhaust velocity, v_e, is essentially the ratio of thrust to mass flow rate; that is, the thrust per kilogramme/second ejected. Using the analysis in Chapter 2 we can use the *characteristic velocity* c^*, and the *thrust coefficient* C_F. The product of these determines the effective exhaust velocity, and their values should be maximised in order to create the most efficient rocket:

$$F = C_F c^* m$$

$$C_F = \frac{F}{p_c A^*}$$

$$c^* = \frac{p_c A^*}{m}$$

$$v_e = C_F c^*$$

Characteristic velocity
The characteristic velocity is a function of the combustion temperature and of the molecular weight of the exhaust gases. These depend mainly on the chemical nature

of the propellants; as described in Chapter 3, high temperature and low molecular weight are required for optimum velocity. Oxygen and hydrogen are the best combination in this respect, although some improvement could be attempted using fluorine and hydrogen, which produces higher heat of combustion and lower molecular weight. The theoretical value of characteristic velocity is higher, but fluorine and its compounds are very corrosive.

Apart from the choice of propellant combination and ratio, an increase in exhaust velocity can be brought about by attention to the heat losses from the combustion. Heat conducted through the walls of the combustion chamber and nozzle, or radiated away from the exhaust stream, reduces the efficiency. This is why regeneratively cooled rocket engines have higher exhaust velocities, and improvements in this area are valuable.

Another area where efficiency can be improved is in the turbo-pumps. The fraction of propellant used in these contributes to the mass of propellant, but not to the thrust, and so greater efficiency here will help. There are inevitable thermodynamic losses in the turbines, and these set a limit to the achievable improvement. However, while there is little that can be done with the waste heat in the exhaust of turbines used in terrestrial applications, this is not so for rocket engines. Expansion of the exhaust through an appropriately oriented nozzle will recover some of this energy as thrust. The SSME and some other rocket engines carry this further: the injection of the hot exhaust into the combustion chamber increases the overall combustion temperature and hence the exhaust velocity.

Some liquid-fuelled rockets do not use turbopumps; they use gas pressure to deliver the propellant to the combustion chamber. This saves engine mass, which improves the mass ratio and, more importantly, the propellant needed to power the pumps. It is not suitable for high-thrust engines, where the required propellant flow rate cannot be met by gas pressure alone without prohibitively thick tank walls. It is, however, appropriate for third-stage engines and orbital transfer engines, where high exhaust velocity and mass ratio, with relatively less thrust, are needed. A good modern example of this is the Ariane 5 upper-stage Aestus engine, which uses helium pressure to deliver the propellants to the chamber.

Thrust coefficient
The value of the thrust coefficient expresses the efficiency of conversion of the thermal energy in the combustion into the kinetic energy of the exhaust gas. It is to improvements in thrust coefficient that most endeavours to produce higher efficiencies are directed, particularly for launchers. The thermodynamic expression for the thrust coefficient, derived in Chapter 2, is

$$C_F = \left\{ \frac{2\gamma}{(\gamma-1)} \left(\frac{2}{\gamma+1} \right)^{\frac{\gamma+1}{\gamma-1}} \left[1 - \left(\frac{p_e}{P_c} \right)^{\frac{\gamma-1}{\gamma}} \right] \right\}^{1/2} + \left(\frac{p_e}{P_c} - \frac{p_a}{P_c} \right) \frac{A_e}{A^*}$$

We can use this expression to help identify where improvements could be made. The value of γ is fixed for a given exhaust temperature and composition, and we shall assume a value of 1.2. This gives for the constant involving γ a value of 2.25. This is

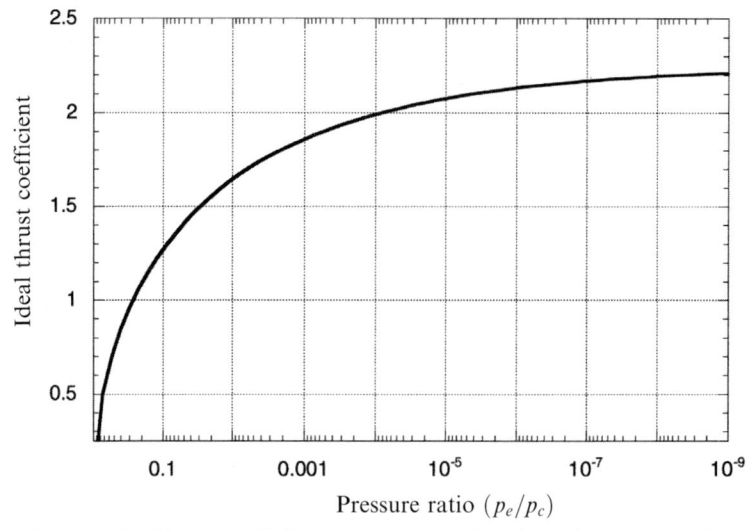

Figure 8.3. Thrust coefficient *in vacuo* as a function of pressure ratio.

the value of thrust coefficient, which a perfect rocket engine having the exit pressure p_e equal to zero would have in a vacuum. The other terms vanish. This is the ideal value, which most rockets do not achieve: the expansion is normally not complete, and the exhaust pressure is finite. There is a good reason for this, since it would require a nozzle of infinite length to produce an exhaust pressure of zero. We should now determine how far this ideal can be approached in practical nozzles.

Figure 8.3 shows the thrust coefficient *in vacuo* as a function of the pressure ratio of the nozzle. It is immediately apparent that the coefficient is a very slow function of the pressure ratio, and that the expansion ratio needs to be very large indeed before a value approaching the ideal can be reached. Even for a pressure ratio of 1 : 10,000, the value of the coefficient is still only 2.0. This implies an exhaust velocity 90% of the ideal, for a typical combustion chamber pressure of 50 bar and exhaust pressure of 5 mbar. This slow behaviour is because the index of the pressure ratio in the above expression is only 0.2 (the fifth root). Smaller values of γ will make this even smaller.

Thus, a large nozzle is needed *in vacuo* to bring the thrust coefficient close to the ideal value. Whether or not this is done depends on whether the extra velocity to be gained is offset by the increase in mass of the nozzle, which will lower the mass ratio. It is not only the mass of the nozzle which needs to be considered, but also the cooling and the extra length required—for example, in a multistage launcher—to accommodate it. (The relationship between nozzle area ratio and pressure ratio is shown in Figure 2.10.) For the above-mentioned pressure ratio of 1 : 10,000 the area ratio is 500, or diameter ratio of 23. This is far too big for most applications. Rocket engines with large expansion ratios have been designed, an early example being the Apollo lunar transfer vehicle main engine. In general, expansion ratios much greater than 80 are seldom used. This limits the pressure ratio to about one part in a 1,000, and a value of thrust coefficient of 1.8, which is equivalent to 80% of the ideal exhaust velocity. To gain the extra 20% of thrust *in vacuo* requires further advances in nozzle materials and thermal design.

As well as changing the expansion ratio to extract a higher exhaust velocity, the shape of the nozzle can be optimised. For a simple conical shape the expanding gas diverges, roughly at the opening angle of the nozzle. Since some of the exhaust is moving at an angle to the axis, some of the momentum is perpendicular to the axis and is cancelled by that on the other side of the nozzle. This reduces the exhaust velocity. It can be corrected by making the nozzle of the correct 'bell' shape, dictated by the equations in Chapter 2. This ensures that all the momentum transfer is axial.

Again, whether or not this is done depends on whether the cost of design and manufacture of a complex nozzle shape, in expensive high performance alloys, can be offset against the increased performance. Again, the bell nozzle is longer for a given expansion ratio, and so weighs more. Examples of bell-shaped nozzles are the SSME and the Apollo engine, in which ultimate performance is required; and, at least for the SSME, there is the possibility of recovery and reuse.

This brief examination of the possibility of improvement in thermal rocket engines for vacuum use is not very encouraging. There is not much prospect of improving on oxygen and hydrogen as propellants, and thermal efficiency is already high. Nozzles are only about 80% efficient, and there is room for development to gain the remaining 20%. Fortunately, electric propulsion offers very significant scope for improving the performance of reaction propulsion *in vacuo*, as discussed in Chapter 6. As we shall see, there is much more scope for improvement of thermal rockets used in atmosphere, and particularly for launchers.

8.3 THERMAL ROCKETS IN ATMOSPHERE, AND THE SINGLE STAGE TO ORBIT

The foregoing discussion of thermal rocket engines in a vacuum is useful in establishing the parameters of importance. As we shall see, there is both scope, and an urgent need, to seek improvements in thermal rocket engines operating in the atmosphere. As described in Chapter 1, the multistage rocket enabled human access to space, but with a very high penalty in terms of complexity, risk and cost. For space activity to expand, this penalty must be removed, and improved thermal rocket engines and launch vehicles are the essential building blocks of such an expansion. These are combined in the concept of the *single stage to orbit,* or SSTO. It is understood that in this concept the vehicle will also return to Earth for reuse. The concept is not yet realised, although significant steps are being taken. The clear promise of such a vehicle is that it can be used over and over again like an aircraft or a ship. It could transform the way in which space is used, by reducing the cost and risk associated with present-day space access. The whole issue of SSTO is multi-faceted, and with many complexities which are beyond the scope of this book. However, the reusable launcher is so important to the future of space activity that some basic ideas should be explored.

In Chapter 1 the multistage rocket was shown to be the only means by which access to space could be gained, given certain (historical and present-day) boundary conditions. The multistage rocket achieves orbit by using separate stages, each with rather modest mass ratio and performance. The combination delivers access to space, but at a uniquely high cost. The whole complex machine is discarded after one flight. A single stage to orbit vehicle, which can return to Earth and be reused, saves both on the number of separate rocket engine systems needed and on the cost of the replacement vehicle for the next launch.

There are many operational considerations, but the central engineering issue is whether sufficiently high performance can be obtained from a single stage, to allow injection of a payload into orbit and return of the vehicle. This requirement translates into a certain total velocity increment, or ΔV, for the whole mission, including the landing; and in addition to the orbital velocity of 7.6 km/s, the potential energy and gravity loss have to be included. In Chapter 1 an approximate figure for the total velocity increment was determined to be 8.7 km/s. To this has to be added the return velocity increment, which raises the estimate to about 9.6 km/s. An accurate calculation of gravity loss is not possible, because it depends on the trajectory details, and what is needed here is a general approach.

8.3.1 Velocity increment for single stage to orbit

The trajectory of most launchers varies throughout the launch, and it is difficult to establish a basic trajectory for calculations. For our purposes, however, it is enough to take a simple approximation—constant pitch trajectory at 45°. This is a crude approximation, especially to modern trajectories with constantly changing pitch angle, but it allows the inclusion of gravity loss in the calculation and provides some insight into the problems to be solved. We shall assume that a rocket leaves the ground at an angle of 45°, and travels at a constant pitch angle until all the fuel is exhausted. It then coasts to orbital altitude, and has enough residual kinetic energy to remain in orbit. This subsumes the velocity increment necessary to circularise the orbit into the total velocity developed by the constant pitch angle burn, which is a reasonable approximation for present purposes.

It is convenient to use energy considerations in this estimate. The kinetic energy imparted to the vehicle should equal the total (kinetic and potential) energy of the vehicle in orbit. For the moment we define this total energy as E. The vehicle velocity at burn-out is given, by equations in Chapter 5, as

$$V = \sqrt{(v_e^2 \log_e^2 R - 2v_e gt \sin\theta \log_e R + g^2 t^2)}$$

where t and g are the burn time and the acceleration due to gravity, respectively. Ideally we should carry out an integration over the path with varying pitch angle, and here we assume a constant pitch angle of 45°, which is a crude approximation.

The burn time t can be expressed as

$$t = \frac{M_0}{m}\left(1 - \frac{1}{R}\right)$$

where m is the mass flow rate, and M_0 is the initial mass of the vehicle and propellant. The kinetic energy of the rocket at burn-out is then

$$\tfrac{1}{2}MV^2 = \tfrac{1}{2}M(v_e^2 \log_e^2 R - 2v_e gt \sin\theta \log R + g^2 t^2)$$

Equating this—the total kinetic energy given to the vehicle by the rocket burn, to the combined kinetic and potential energy of the orbit—we can define an equivalent velocity:

$$V_{Orbit} = \sqrt{\frac{2E}{M}} = \sqrt{V^2 + 2gh}$$

where V is the orbital velocity as defined in Chapter 1, and h is the difference in altitude between the burn-out and orbit. Thus

$$V_{Orbit}^2 = v_e^2 \log_e^2 R - 2v_e gt \sin\theta \log R + g^2 t^2$$

Rearranging this as a quadratic in $v_e \log_e R$, the solution is

$$v_e \log_e R = gt \sin\theta \pm \sqrt{V_{Orbit}^2 - g^2 t^2 (1 - \sin^2\theta)}$$

This refers to the vehicle velocity required at burn-out to ensure that the rocket reaches orbital altitude with sufficient velocity to stay in orbit, with the approximations we have made. It includes the gravity loss, and the contribution to horizontal velocity from the 45° pitch angle. Figure 8.4 shows this function plotted against burn time, together with the value of mass ratio required for different exhaust velocities. The value of $V_{Orbit} = 8.1$ km/s is taken, and includes both the orbital velocity and the velocity equivalent in kinetic energy to the work done in reaching the orbital altitude. It is computed for a 500-km circular orbit

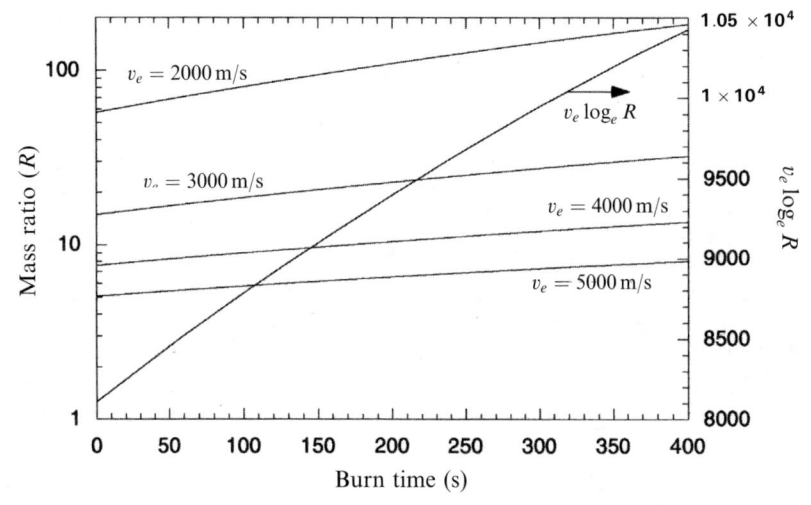

Figure 8.4. Velocity increment and mass ratio necessary to reach orbit, as a function of burn time.

258 Advanced thermal rockets   [Ch. 8

In Figure 8.4 the scale at right and the arrowed curve show the necessary burn-out velocity increment ($v_e \log_e R$) and, as expected, this is a strong function of the burn time; for zero burn time (no gravity loss) it is equal to V_{Orbit}. The effect of gravity loss increases dramatically as the burn time increases, and this gravity loss means that the required velocity gain for a real launch is much greater than the orbital velocity. For a typical burn time of 150 s it is already more than 9 km/s. Thus, for single stage to orbit vehicles we must be considering values in this region. An often-quoted practical value, which includes return to Earth, is 9.6 km/s.

The scale at left indicates the mass ratio required to generate the required burn-out velocity, for different exhaust velocities. For SSTO an exhaust velocity of 2,000–3,000 m/s requires a totally impossible mass ratio. This is of course the case for solid propellants, and explains the absolute requirement for such launchers to be staged. Only when the exhaust velocity reaches 4,000 m/s does the possibility of a single stage emerge; and here the mass ratio is still around 10 for a reasonable burn time. Shorter burn times imply strong acceleration, with the accompanying stress on the (light-weight) vehicle from thrust and atmospheric forces. The above simple calculation indicates the mass ratio which would be needed for SSTO, and the great importance of a high exhaust velocity.

8.3.2 Optimising the exhaust velocity in atmosphere

The effective exhaust velocity is the most important parameter for SSTO, but it depends on the ambient pressure (as explained in Chapter 2). For maximum exhaust velocity the expansion ratio of the nozzle should result in a nozzle exit pressure equal to the ambient pressure. Hitherto we have assumed that the exhaust velocity is constant during the ascent, which in practice cannot be true. A nozzle of fixed expansion ratio can only be optimal for one pressure, which is approximated on a multi-stage rocket by producing different nozzles for each stage. For an SSTO vehicle this cannot be done, and the resulting variation in exhaust velocity needs to be addressed.

The actual, as opposed to the ideal, exhaust velocity of the rocket can be expressed via the thrust coefficient, as defined above. Most rockets have a fixed expansion ratio so that the exhaust pressure p_e is also fixed, while the ambient pressure p_a changes with time. Since the exhaust pressure is optimal for only one ambient pressure, the rocket engine will be inefficient for most of the ascent. Figure 8.5 illustrates this by plotting the thrust coefficient as a function of pressure for different expansion ratios. Also shown is the thrust coefficient for a nozzle in which the expansion ratio adjusts, providing the correct expansion at every altitude. The expansion ratio of a correctly expanded nozzle for each atmospheric pressure is also shown for reference.

Remembering that the exhaust velocity is proportional to the thrust coefficient, it is clear that the curves for constant expansion ratio fall far below the optimised curve for nearly all pressures. For example, the curve for expansion ratio 10 is optimum near sea-level, but is limited at about 1.7 for pressures below 0.1 atmospheres, while the optimised curve continues to a value of 2.2. For a nozzle optimised for high

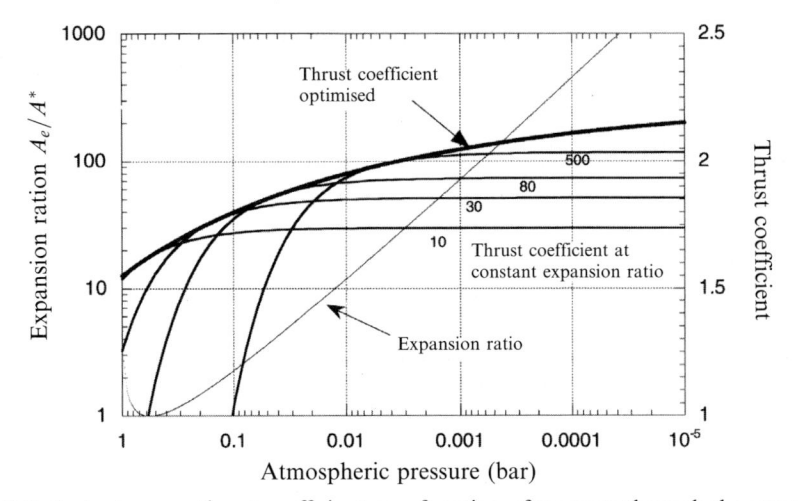

Figure 8.5. Instantaneous thrust coefficient as a function of pressure through the atmosphere for fixed and variable ratios.

altitude—say an expansion ratio of 80—the efficiency is low for pressures above 0.1 atmospheres, and limits at 1.9 for low pressure, which is still well below the optimised curve.

This means that the exhaust velocity—which until now we have assumed to be constant—is not constant, it falls below the ideal value for much of the launch trajectory. Since every ounce of thrust is needed to make SSTO work, this is not an acceptable situation. In fact, it is worse than indicated in Figure 8.5 because the atmospheric pressure drops exponentially with altitude, which renders the leading part of the thrust coefficient curve much steeper than indicated is in Figure 8.5.

The variation of thrust coefficient with altitude—not important for staged rockets—becomes of significant importance for SSTO, since the same rocket engine must be used throughout the flight. This is one of the main issues connected with SSTO development: over the flight, the difference in performance between a fixed expansion ratio engine and one which adapts has a very significant effect on the achieved velocity.

8.3.3 The rocket equation for variable exhaust velocity

The rocket equation with fixed exhaust velocity is no longer appropriate, and variable exhaust velocity has to be included. The differential equation from which the rocket equation is derived can be written for variable exhaust velocity as

$$dV = v_e(R)\frac{dR}{R}$$

where $v_e(R)$ is an assumed relation between the exhaust velocity and the mass ratio (here taken to represent progress through the flight). It can be derived empirically by

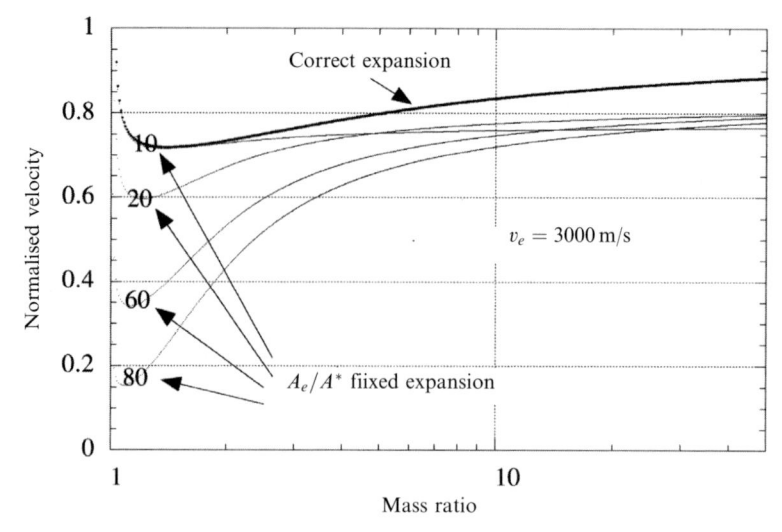

Figure 8.6. Normalised vehicle velocity as a function of mass ratio for fixed and variable expansion.

the following procedure. The altitude as a function of mass ratio can be derived from equations in Chapter 5, and the pressure as a function of altitude from the standard atmosphere. This pressure can then be substituted in the thrust coefficient formula for different expansion conditions. The resultant instantaneous exhaust velocity can then be substituted in the above equation, and a numerical integration performed. The result is shown in Figure 8.6 for an ideal (vacuum) exhaust velocity of 3,000 m/s, and vertical flight, without gravity loss. This is sufficient to illustrate the principle.

The lower curves show (as a function of mass ratio) the ultimate vehicle velocity integrated for variable exhaust velocity normalised to the ideal vehicle velocity. The ideal vehicle velocity is defined here as the velocity the vehicle would have if the exhaust velocity were constant at 3 km/s throughout the flight. In this instance the mass ratio should be taken to indicate time, or progress through the trajectory. The (fixed) expansion ratio of the nozzle labels each curve. While the mass ratio is less than 10—which here indicates the early part of the flight—the smaller expansion ratios are more efficient. This reflects the strong influence of the early acceleration of the rocket in atmosphere. Only for high mass ratio does a larger expansion ratio generate a higher vehicle velocity, reflecting its greater efficiency in near-vacuum conditions. The upper curve shows the vehicle velocity for a nozzle that continuously adjusts for the ambient pressure. It approximates the low, fixed expansion ratio curve for small mass ratios, and elsewhere is always greater than any fixed expansion ratio curve. It also shows that a nozzle optimised for vacuum use achieves very little, in terms of vehicle velocity, low in the atmosphere.

We have not considered thrust here, but Figure 8.4 shows that the thrust coefficient of large expansion ratio nozzles is small, low in the atmosphere, and an SSTO vehicle has to have high thrust at launch. All of this illustrates the advantage of a nozzle with a variable expansion ratio, if it can be realised. The possibility of achieving this, and the other desiderata for SSTO, are dealt with below.

8.4 PRACTICAL APPROACHES TO SSTO

Because of the obvious advantages of SSTO, much current research and development is directed toward solving the challenging technical problems. The key engineering challenges are the achievement of a high mass ratio and a high exhaust velocity. To this may be added the reduction of the required total velocity increment for an SSTO mission by, for instance, using aerobraking for the return journey; the value of 9.6 km/s quoted above includes such an assumption. If aerobraking is not used, then in principle the velocity increment would be double that required to attain orbit. (Gravity always acts in the contrary direction, by slowing the ascending rocket and accelerating the descending rocket.)

8.4.1 High mass ratio

This is equivalent to reducing structural mass for a given mass of propellant. In reality this implies reduction of the mass of everything except the payload and the fuel, and reduction in the mass of the engine and of, say, guidance electronics, rocket telemetry equipment and so on, is important. However, for high mass ratio vehicles the largest structural mass will be in the propellant tanks, and it is here that reductions will be most rewarding. The density of the propellants plays an important part here. Low-density liquids require proportionately larger tanks for the same mass of propellant, and in this respect it is unfortunate that liquid hydrogen has such a low density.

Propellant tanks
Given a certain mass of propellant, and its density, then the tank volume and the surface area of the walls is determinable. The only variable left is the thickness of the tank walls and the density of the wall material. The thickness depends on the stiffness of the material and the mechanical loads it must bear. In general the Young's modulus of the material determines the stiffness for a given thickness, and the density, of course, determines the mass. Thus for the lowest tank mass, materials with a high ratio of the Young modulus to density are required. Since this requirement is common to many design issues in engineering, and particularly in aeronautics, aluminium alloys are the obvious choice. As in aircraft (and spacecraft) construction, a single-skin wall is less mass-efficient than a double-skinned honeycomb wall, and this approach could also be appropriate for tank manufacture. The need for the tank to be hermetic under high mechanical loading, and the weight reduction to be gained by welded joints, which are unsuitable for honeycomb walls, limits the areas in which a classical honeycomb approach can be applied. The common alternative is a ribbed structure with a single thin hermetic skin and welded joints. This is a classical engineering approach, but is bound to be costly because of

the machining complexity. It is nevertheless a safe and conservative design, and may be appropriate for a reusable vehicle.

Any improvement in the efficiency of tanks must lie in the direction of thinner walls, which points towards materials with a very high modulus and low density, such as carbon fibre composites. It is important, however, in reducing the wall thickness to take account of buckling. While a thin wall can easily sustain the internal hydrostatic pressure of the propellant under several g, any small inward displacement of the wall can lead to buckling and collapse, however high the modulus. Ribs and honeycomb structure can prevent this at the cost of extra mass, although buckling can also be prevented by pressurising the tank in such a way that the outward forces are always high, preventing any inward displacement of the thin wall. This can easily be appreciated from the behaviour of a plastic carbonated drink bottle: when sealed and under pressure the bottle is very stiff, but collapses easily once the cap is removed and the internal pressure is released. This approach is used in several current rocket vehicles. Of course, the walls need to have a greater hoop stiffness, but often the mass of the additional skin thickness is offset by the reduced need for ribs or other external support.

Liquid oxygen
Many of the substances used as propellants are more or less corrosive, but liquid oxygen presents special problems. It reacts with many organic substances, with potentially dangerous results. For this reason, metal tanks have so far been used. Certainly the use of CFRP (carbon fibre reinforced plastic) with liquid oxygen is potentially dangerous. Intensive research is currently being directed to solving the problem of containing liquid oxygen in non-metallic tanks, because of the great advantage in mass ratio which would result. Aluminium is currently the material most used, but aluminum–lithium alloys show promise in reducing the tank mass, and are under development.

Composite tanks
Metal tanks have to be used for liquid oxygen, but tanks made of composite materials of high stiffness-to-density ratio are actively being designed for new liquid-fuelled rockets. Composites have long been used for the casings of third-stage solid rockets because of the mass ratio advantage they confer. In general, composite tanks are made by winding carbon or glass fibre, impregnated with a plastic material, on to a mandrel of the required shape. The plastic material is then cured under high pressure and temperature to form a lightweight integrated structure. The tensile properties of the structure derive from those of the fibres used, and by the use of different layouts, and combining different fibres, different properties for the composite can be realised. Typically, fibres can be laid so that they run along the direction of greatest stress, and composite walls can be much lighter than equivalent metal walls because of this anisotropy of properties. As with all thin structures, while the hoop stiffness can be very great, prevention of buckling may limit the thinness well before the stiffness limit is reached. This method of construction is well tried for components such as the wings of high-performance

aircraft, and its application to propellant tanks is straightforward. Of course, for cryogenic liquids the thermal contraction of the tank is an issue. It is possible to wind structures of very low expansion coefficient over a small temperature range, but in general, plastic materials have higher expansion coefficients than metals.

Vehicle structures

While early rockets had separate tanks and structure, in modern launchers the tank is integrated with the structure wherever possible. For the Ariane 5 the main cryogenic tank is also the main vehicle structure, being attached through a rear thrust-frame to the single engine, and through a forward skirt and CFRP bulkhead to the upper stage. The tank is cylindrical with hemispherical ends made of 2219 aluminium alloy, and a hemispherical bulkhead separates the oxygen and hydrogen tanks. This approach can also lead to mass savings in SSTO craft, and with the use of composites for tanks, more complex shapes are possible—perhaps making the tank shape conform to the aerodynamic shape of the vehicle.

Another aspect of vehicle structure relates to the stress it has to bear. For an expendable launcher, in general the main stress is axial, since at all times the thrust vector is close to the vehicle axis. This means that the structural mass can be concentrated to take this axial stress, and very little needs to be applied to transverse stress. The degree to which this is achieved can be appreciated from the fact that launchers that go off course very quickly begin to break up due to transverse aerodynamic stresses. For a reusable launcher the stresses of return and landing have also to be taken into account, and this can considerably increase the mass required for the structure. In a vehicle such as the Space Shuttle, which launches vertically and lands horizontally, two load paths have to be accommodated. During launch, thrust is transferred to the vehicle from the engines and boosters in an axial direction, while during aerobraking and landing the forces act on the vehicle body and the undercarriage in a transverse direction. This dual load path increases the structural mass of the vehicle, but it could be avoided by landing the rocket on its tail—the so-called 'Buck Rogers' landing, seen in early space films—which reduces structural loads during landing. In this case, aerobraking loads also need to be axial for a significant mass saving.

These structural issues may seem to involve rather small mass improvements, but it should be remembered that the mass of the dry vehicle is the crucial determinant of the mass ratio and hence the vehicle performance. Small changes in structural mass have a large impact on the overall mass ratio.

8.5 PRACTICAL APPROACHES AND DEVELOPMENTS

Having discussed the problems associated with single stage to orbit vehicles, it is appropriate to examine some of the technological developments and test vehicles which are emerging. This field is very active at present, as the search for lower-cost access to space is pursued by agencies and the aerospace industry. There are two main thrusts: improved engines, both in terms of exhaust velocity and reusability;

and improved vehicles, vehicle structure and vehicle/mission concepts. The Space Shuttle, which first flew in April 1981, remains the only re-usable vehicle in service; only the orbiter itself is re-furbishable, and most of the propellant is carried in expendable tanks and boosters; the combustion chamber segments of the boosters are re-filled with propellant. The cost is also very high because of the high specification of the engines and other subsystems. The prize for success in developing a fully reusable vehicle will be very high, not only for the growing commercial market, but also because human exploration of space will be rejuvenated.

8.5.1 Engines

Since the objective is a completely reusable vehicle, solid propellants are not applicable in this context. Solid vehicles can be refilled with propellant, but nozzles and thermal protection have to be renewed. Thus the main improvements being sought refer to liquid-fuelled engines. There is still interest in improved propellant combinations, including tri-propellants, and in simplification of the propellant supply, in an effort to reduce cost and weight. The most interesting development, however, is in the area of adapting the expansion ratio of the nozzle to the changing ambient pressure during launch. This provides significant advantages in total vehicle performance, as noted above, and is logical, since any SSTO vehicle has to produce its best performance throughout the ascent.

There are two practical ways to adjust the expansion in flight: to make the nozzle shape variable with altitude, or to develop a nozzle with a performance independent of ambient pressure. The former approach is now commonly applied to the upper stages of expendable vehicles; the nozzle is made in two sections, with the extension nested inside the inboard part. This saves on vehicle length and improves the mass ratio, because the long nozzle is compressed. Once the stage is released, a disposable spring (in the case of the Japanese Mu-V rocket) extends the nozzle to its full length. This is relatively easy to do before ignition, and has benefits, but it is more difficult with an already burning rocket, and is being investigated.

Flow separation

Under some circumstances a normal bell-shaped nozzle can adapt to ambient pressure changes. This relies on the phenomenon of flow separation, familiar in aerodynamics.

In Figure 8.7 the bell-shaped nozzle is shown under three ambient pressure conditions. In (a) the nozzle is operating *in vacuo*, and the expansion is correct for this condition. This is the normal situation dealt with in Chapter 2, but for other pressures the nozzle would be shorter with a smaller expansion ratio. When a nozzle is used in an over-expanded situation, in which the ambient pressure is higher than the exit pressure, a shock develops at the edge of the nozzle. Under certain circumstances this can cause separation of the exhaust flow from the nozzle wall, as shown in Figure 8.7(b). When this is uncontrolled it diminishes the thrust and generates turbulence in the exhaust stream. Separation of the flow can, however, be induced deliberately, and if it occurs under controlled conditions the flow

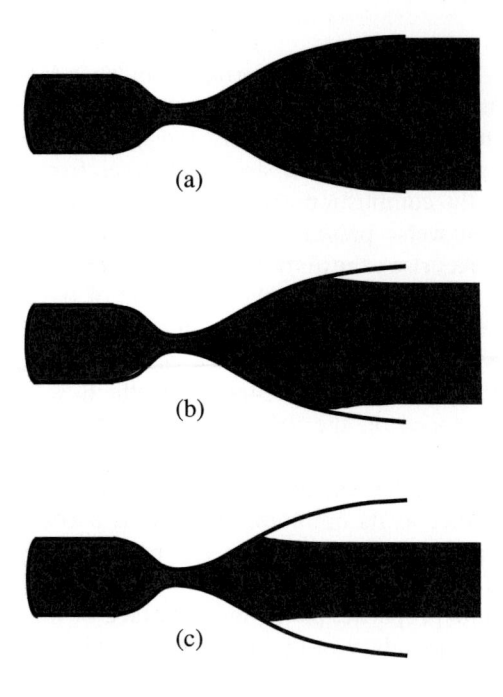

Figure 8.7. Flow separation in a nozzle.

downstream of the separation point can be smooth. In these circumstances the expansion of the exhaust is reduced; it is essentially confined by the atmospheric pressure, and the nozzle behaves as if it were shorter, with a smaller expansion coefficient. The thrust is developed on the portion of the nozzle where the flow is still attached. The atmospheric pressure downstream of the separation point is the same inside and outside the nozzle, and has no effect on the thrust.

The separation can be controlled by small abrupt changes in slope of the nozzle wall. The general effect is to allow the separation point to migrate along the nozzle as the atmospheric pressure changes. At launch, therefore, the high ambient pressure causes flow separation deep in the nozzle, towards the throat, and the effective expansion ratio is small. As the rocket rises through the atmosphere the separation point moves towards the end of the nozzle, as shown in Figure 7.7(b). Finally, when the ambient pressure drops to a low value, the separation point coincides with the end of the nozzle and the expansion is appropriate for a vacuum. This technique is employed to improve the performance of the cryogenic engines on the Space Shuttle and on Ariane 5. They are ignited at launch, but are fitted with bell nozzles with a high expansion ratio. As we saw earlier in this chapter, the sea-level performance of such highly-expanded nozzles is very poor, but flow separation can improve this performance so that the engine provides more thrust in the early stages of the launch.

The plug nozzle

The other approach—a nozzle with a performance independent of altitude—reverts to an early concept: the plug nozzle. Instead of the thrust being developed against the inner surface of a cone containing the exhaust stream, it is developed on the outer surface of a conical plug. The exhaust stream emerges from an annular aperture between the plug and the combustion chamber wall; its outer boundary is now the slipstream, and the transverse pressure acting on the exhaust is that of the local atmosphere. As the rocket rises, the ambient pressure drops, and the outer boundary expands and changes shape. In this way the expansion of the exhaust adjusts to the local conditions, and the performance in terms of exhaust velocity is much closer to ideal, throughout the ascent. In a sense, this use of atmospheric pressure to confine the exhaust and control its expansion occurs also in the flow separation nozzle, but there the thrust is developed on the *inner* surface of the nozzle, not on the *outer* surface of a plug.

It is important to realise that the plug nozzle, illustrated in Figures 8.8 and 8.9, has the same performance at its optimum pressure as a conventional nozzle. The exhaust stream expands in the same way and generates the same thrust. The nozzle can be thought of as a spatial inversion of the conventional nozzle. The outer surface of the plug has the same properties as the inner surface of the conventional nozzle, and the exhaust stream boundary in the plug nozzle is the analogue of the central axis of the exhaust stream in the conventional nozzle. The difference is that the shape of the exhaust stream boundary depends on the ambient pressure. At design pressure the boundary is parallel to the axis (Figure 8.9(b)) and the performance is entirely analogous to that of a conventional nozzle; for higher ambient pressures (Figure

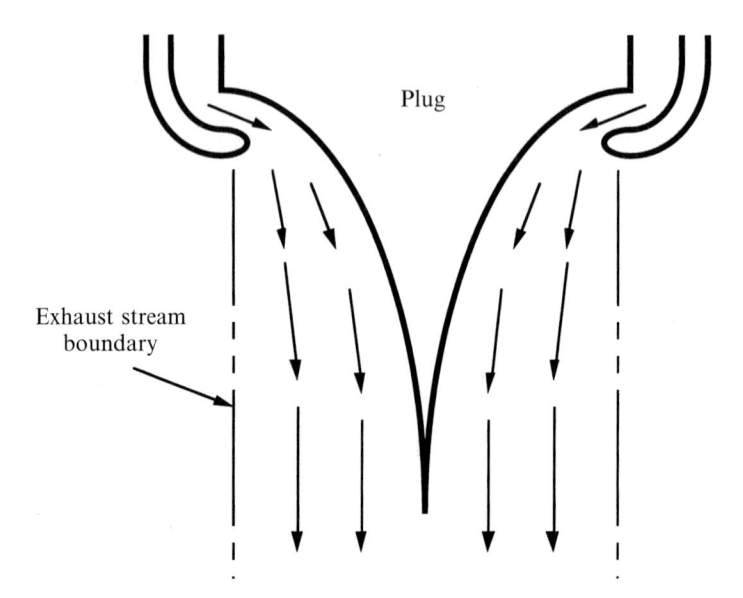

Figure 8.8. Principle of the plug nozzle.

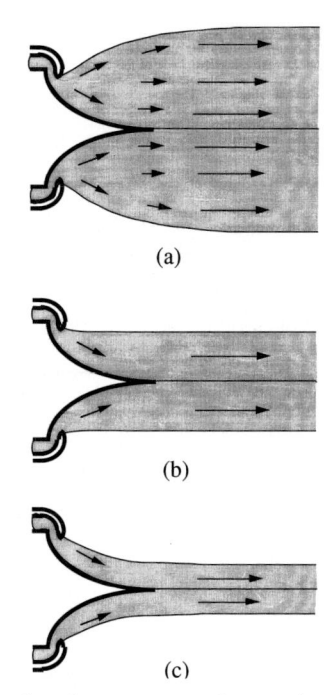

Sec. 8.5] **Practical approaches and developments** 267

Figure 8.9. Plug nozzle exhaust streams for varying atmospheric pressure.

8.9(c)) the boundary curves inwards, raising the effective exhaust pressure; and for lower pressures (Figure 8.9(a)) the boundary curves outwards, allowing the gases to expand to a lower pressure. This explains why the performance away from the design pressure is much better than the conventional nozzle.

In understanding how the plug nozzle works it should be realised that the boundary between the atmosphere and the exhaust stream only has the transverse force of the ambient pressure acting upon it. It does not generate any thrust, which is all generated on the plug itself. The boundary just controls the pressure in the exhaust stream as a function of distance, and thereby the expansion ratio. This concept is very good at automatically matching the expansion ratio to the ambient pressure.

The plug nozzle is a relatively old idea, and several experimental nozzles have been tested, although there are significant technical difficulties. An annular combustion chamber is much more complicated to construct than a simple cylindrical chamber. If a plug is used with a normal cylindrical chamber, then it must be immersed in the hot gases, and it is almost impossible to cool actively. Moreover, the supporting members are themselves immersed in the hot exhaust stream; they should have a small cross section not to impede it, and they have to bear the full thrust developed on the plug, and transfer this to the rocket. For this reason the plug nozzle has not seen much use, despite its advantages. There are, however, developments of this simple idea which show promise, and are being pursued vigorously with particular application to SSTO. The *aerospike nozzle*—and more particularly the linear aerospike nozzle—solve the problems of cooling and support.

The aerospike nozzle
The aerospike nozzle replaces the long-shaped plug with a short truncated cone from the face of which cool gas is injected into the stream. This creates a recirculating flow with an outer boundary which approximates to the plug nozzle shape. The thrust is transferred to the face of the truncated cone by the pressure of this central core of cool gas, and the outer boundary reflects the ambient pressure, as in the plug nozzle. The problem of cooling the plug is solved because its inboard surface is accessible, but the need for an annular combustion chamber remains. One advantage of this nozzle is that if an annular or toroidal combustion chamber is mounted around the rim of the cone, then the propellant distribution equipment can be located within it. This not only enables the cool gas to be injected through the face, but also the walls of the plug to be cooled by one of the propellants. This device is illustrated in Figure 8.10.

The linear aerospike engine is an adaptation in which the cylindrical geometry of the aerospike is transformed to linear geometry; the cylindrically symmetric plug becomes two inclined curved planes joined together at the apex, with a row of ordinary combustion chambers along each outer edge. This solves the problem of complicated combustion chamber geometry, and at the same time provides a large volume between the inclined planes for the turbo-pumps and propellant distribution pipe-work. This type of engine is being developed for a specific SSTO vehicle, and

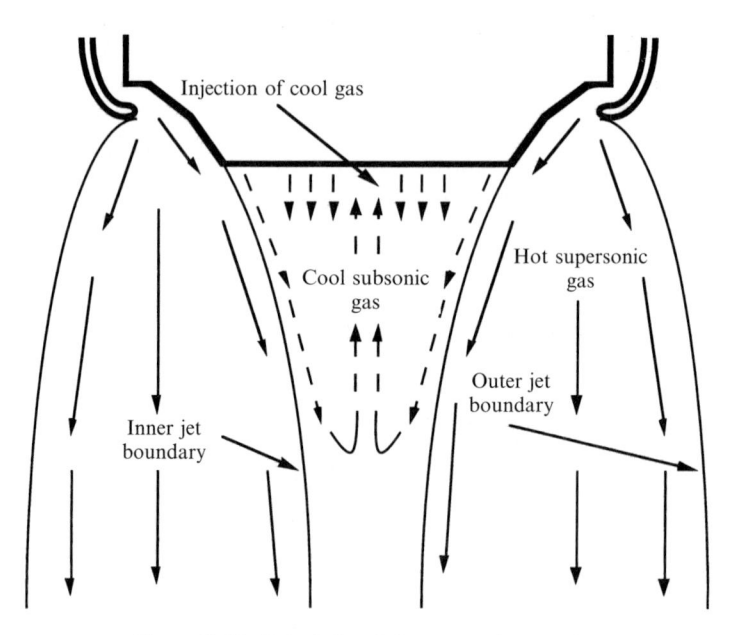

Figure 8.10. Principle of the aerospike nozzle.

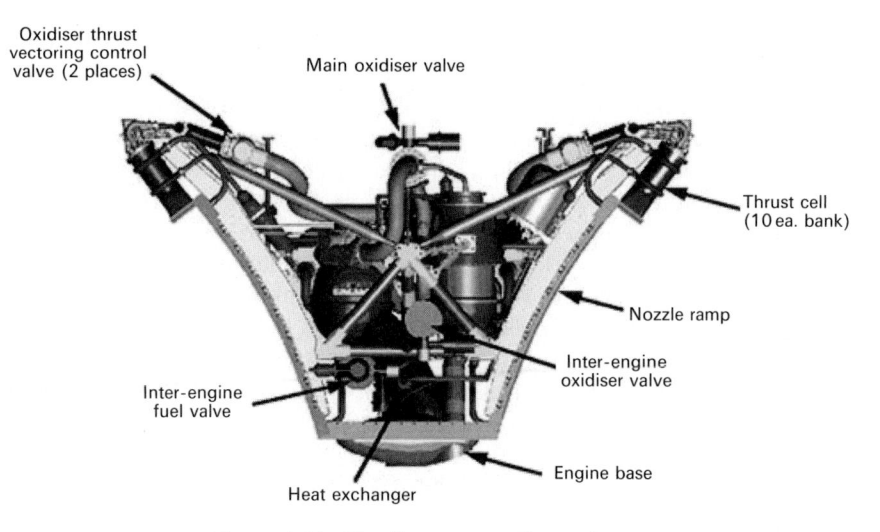

Sec. 8.5]	Practical approaches and developments	269

Figure 8.11. The linear aerospike engine.

has undergone test flights, as well as extensive ground testing. Figure 8.11 shows the layout.

The linear aerospike engine also has the advantage that the total thrust vector can be steered by varying the thrust of one or more of the combustion chambers. Its rectangular cross section is suitable for several new space vehicle concepts which use a lift-generating elliptical cross-section *aerobody*. There is no separate wing to the vehicle, as all the necessary lift is generated by airflow over the body of the vehicle itself. If the whole rear cross section of the vehicle is filled with the exhaust stream, the drag is reduced. This occurs with most rockets.

The linear aerospike engine shows great promise in increasing overall launcher performance, and there is a vigorous development programme. The development of tri-propellant engines and pulse detonation engines is less advanced, but they are briefly considered here.

Tri-propellant engines
Liquid hydrogen as a fuel has two significant disadvantages: it is a very low-density cryogenic liquid, needing high-volume and therefore heavy tanks; and the energy released per kilogramme of fuel burned is quite low, compared with, for example, hydrocarbons (see Chapter 5). In specific impulse terms these disadvantages are outweighed by the low molecular weight of the combustion products, and the exhaust velocity is the highest of any in-service propellant system. However, when considering not just the rocket engine, but the whole vehicle, then the disadvantages of hydrogen must be considered.

One potential improvement is to use hydrocarbon as the fuel and hydrogen as the *working fluid*. In this way the energy necessary for heating the exhaust stream comes from a hydrocarbon fuel, while hydrogen is injected into the chamber simply to lower the molecular weight of the exhaust gases. To appreciate this, consider what happens in a normal combustion chamber, as two parallel processes: the generation of heat by combustion, and the heating of the exhaust products to form a high-pressure working fluid. The expansion of this fluid through the nozzle produces thrust. It is immediately obvious that, while convenient, there is no particular reason why the combustion products should exclusively form the working fluid. Indeed, many liquid hydrogen engines inject excess hydrogen into the combustion chamber simply as a working fluid, and it plays no part in the combustion. Thus a tri-propellant engine—which burns hydrocarbon fuel to heat the third component, hydrogen—is a perfectly valid concept; the extra injectors and supply system require no new technology, and the advantages may be significant. The ratio of hydrocarbon to hydrogen may be changed during the ascent. In particular, hydrocarbon can be used exclusively for the early part of the flight, producing high thrust (a high mass ejection rate) with a low exhaust velocity, and hydrogen can later be added to raise the exhaust velocity for the injection into orbit. Another advantage is that most of the chemical energy needed for the ascent is stored onboard in the form of the denser hydrocarbon fuel, so the overall tank mass is reduced, therefore reducing the dry weight of the vehicle. This is even more advantageous if hydrocarbon is burned exclusively during the early stages of the flight, reducing the volume of hydrogen which has to be carried to altitude. Whether or not such a scheme is advantageous depends very much on the vehicle design parameters, but again this is a type of engine which is being studied for SSTO.

The pulse detonation engine
The pulse detonation engine seeks to improve the thermodynamic efficiency by ensuring that the combustion occurs at constant volume instead of at constant pressure as in a normal engine. Sufficient propellant is admitted to the combustion tube to fill it in vapour form, and it is then ignited at the closed end of the tube. A compression wave travels down the tube, igniting the mixture as it does so. This wave travels at the speed of sound, and the pressure and temperature are such that complete combustion occurs as it passes. This process is known as 'detonation', and occurs because of the positive dependence of reaction rate on temperature and pressure. The most common example is any kind of chemical explosion; for example, gunpowder. The characteristic is that the reaction travels through the mixture at sound speed, and not at a rate limited by the combustion process. The latter is effectively a flame, and travels at a much lower speed. Thus there is no time for the gas in a pulse detonation tube to expand before the reaction is complete; all the energy goes into temperature, and the maximum thermodynamic efficiency is achieved. An engine of this type should, in principle, have a higher exhaust velocity, but the difficulty is to achieve a more or less constant thrust rather than a single pulse.

The solution to this problem goes back to the V1 flying bombs used during the Second World War. These were propelled by a pulse detonation engine, using aircraft fuel, combined with oxygen from the atmosphere. When flying at speed a 'venetian blind' flap valve opened at the leading end of the tube, admitting air. Fuel was injected, and the mixture was ignited by a spark plug. The detonation closed the flap valve, ensuring that the thrust was directed backwards. Once the internal pressure dropped below the dynamic pressure on the valve, it opened again, restarting the cycle. The bomb was given sufficient initial velocity by a solid rocket booster, and then continued using its pulse engine.

When considering this system for a rocket engine with two liquid propellants, a different kind of valve is needed, and systems with a rotating disk which rapidly opens and closes the valves have been tried. An important property of the pulse detonation engine is the fact that the propellants are introduced at ambient pressure, after the gas from the previous pulse has expanded. This means that the propellant delivery system and injectors work at much lower pressures than that of a conventional combustion chamber, and are much simpler. Such engines are at present the subject of experimental programmes.

The rotary rocket engine
SSTO transport requires both a significant improvement in engine and vehicle performance and a reusable engine. The latter implies that the complexity of the engine and the level of stress on individual components need to be reduced; that is, parts should not wear out during a single use. Combustion chambers are generally robust, and can be reused with relatively little refurbishment. However, this is not true of the propellant delivery systems of high performance engines—particularly the turbo-pumps. These endure very high stresses, and have many moving parts and seals which would require frequent replacement. Simplification of the propellant delivery system is also the subject of much attention. The typical high-performance system, such as that of the SSME, uses two gas generators and two turbo-pumps. The simplest system uses helium from a high-pressure storage tank to force the propellants through the pipe-work and injectors. The Aestus engine used in the upper stage of Ariane 5 is a good example. As explained in Chapter 3, this approach is not suitable for high flow rates. Simplified turbo-pump systems can be used in intermediate cases: the gas generator can be omitted, and gas resulting from heating of one propellant used for regenerative cooling is used to drive the turbine. A single turbine can be used for both pumps, but this needs gearing to match the different fuel and oxidiser rates.

A radical solution is the rotary rocket engine, and again this is a rather old concept being revived. It comprises a number of combustion chambers mounted around the rim of a wheel. Radial pipes connect the chambers to a central propellant stem, and pressure induced in these pipes by centrifugal force delivers the propellants to the combustion chambers as the wheel rotates. A small tangential offset of the chamber thrust axes provides the rotary force. High propellant delivery rates are achieved, with no pumps. The main problem with this approach is the design and durability of the central stem and its rotary seals. The number of parts is dramatically reduced, however, and if this one problem can be overcome a much simpler and potentially more durable engine would result. The design of the rotary seals is simplified by the fact that near the stem, propellant pressures are quite low, and they only rise to injection levels near the periphery. This kind of engine is being actively considered for a particular SSTO vehicle.

8.6 VEHICLE DESIGN AND MISSION CONCEPT

The achievable mass ratio and the exhaust velocity are the two parameters whose optimisation will lead to a successful SSTO vehicle; engines determine the exhaust velocity, while the vehicle and mission concept determine the achievable mass ratio. In arriving at a vehicle design and concept, both the ascent and descent have to be considered. Most of what has already been discussed refers to the ascent. In general the descent problem is equally challenging: in theory the task of removing orbital velocity is equal to that of gaining it, which would require twice the velocity increment derived earlier. Fortunately the atmosphere provides a ready means of dissipating kinetic energy, and the manned space programme has relied on this from the beginning. The problem can be divided into three areas: the de-orbiting of the vehicle, the dissipation of most of the kinetic energy, and the landing. Early returns from orbit used parachutes for this last step, fuel for de-orbiting was part of the velocity increment calculation, and a hypersonic re-entry vehicle dissipated the kinetic energy. For an SSTO vehicle, return to Earth is a key requirement; it has a number of possible solutions, each with its own problems. These will emerge as different concepts are examined.

8.6.1 Optimising the ascent

Here a fully laden vehicle, with all the propellant necessary to complete the mission, has to lift off and complete its trajectory into the desired orbit. The technologies of importance here are those which improve the exhaust velocity and the 'dry mass': engine optimisation, structure, choice of propellant, and propellant tanks. The vehicle concepts discussed below generally follow the same lines: nozzles that adjust to the ambient pressure, and the imaginative use of composites and special alloys. Choice of propellant is of course limited, but there is some scope for improvement using tri-propellant systems and in *densification* of propellants. This latter uses very low-temperature storage to increase the propellant density and to lower the vapour pressure. Both of these enable a reduction in tank mass for a given propellant mass.

8.6.2 Optimising the descent

A reusable vehicle has to descend safely. The mass is much lower than during ascent, and the atmosphere can be used in different ways to ease the task. All concepts use aerobraking to remove most of the orbital kinetic energy, but they differ in the way they complete the descent. Those that follow the Space Shuttle heritage make an unpowered landing, essentially behaving like a glider. This requires a runway to land on, and a significant cross-range manoeuvring capability. Without the ability to move substantially off the quasi-ballistic trajectory defined by the re-entry point and the orbital inclination, a vehicle cannot be sure of locating the runway and making an accurate approach. The aerofoils and control surfaces needed to achieve this without power are part of the mass penalty for an aircraft-type landing. A different option is to make a powered landing using the rocket engines to land the rocket on its tail. The early part of the descent is essentially the same, but the vehicle has then to invert so as to move tail first towards the ground. Cross-range performance can be achieved with control surfaces before the inversion, and with the rocket engines afterwards. The engines are then used to bring the vehicle to rest as it touches down. This kind of landing (on the Moon) was of course used for the Apollo missions. No runway is required, and the slow-moving vehicle can be accurately placed down on a small landing pad. The mass penalty for this kind of vehicle is the extra propellant needed for the landing manoeuvres.

8.7 SSTO CONCEPTS

Mission concepts can be divided into those seeking to make a technological breakthrough—either with a new engine, a new vehicle design, or both—and those which rely on existing technology, used in a particular way to achieve the objective. Amongst the former are the Venture Star project, the Delta-Clipper and the Roton project, while the K-1 and Astroliner are among the latter class.

The Venture Star relies on new technology to achieve a reusable vehicle. The linear aerospike engine provides the motive power, burning liquid hydrogen and liquid oxygen. The vehicle and propellant tanks are newly designed, using modern composites and alloys to produce a very high mass ratio, and the de-orbiting and landing relies on the heritage of the Space Shuttle. The vehicle itself is an aerobody which provides hypersonic aerobraking after de-orbiting. At lower speeds and altitudes it develops lift, and can be landed like an aircraft. The aerobody and the propellant tanks form part of an integrated structure, minimising the mass of the vehicle. The high-volume hydrogen tank is made of lightweight composites, as is the aerobody, while the oxygen tank uses a high specific stiffness aluminium–lithium alloy to avoid potential problems with composites and liquid oxygen. Tests of various parts of this concept—including the engine, the tanks and the aerobody— have been carried out, but the attempt to reach orbit is yet to be made.

Ability to land a rocket tail-first is seen by some teams as vital to achieve SSTO, and it is also an easily demonstrated first step. The advantages have been detailed above, and a good example of this approach is the Delta-Clipper.

The technology for a vertical landing is relatively straightforward. It involves an accurate altimeter, and control of the engine thrust by a control loop. The first vertical landing was, of course, on the Moon. Since then, control systems have advanced greatly, and this kind of landing can be taken as mature technology. The DC-X—the precursor to the Delta-Clipper—has successfully taken off and landed several times. The propellant tanks and structure of the Delta-Clipper use very similar technology to the Venture Star: composites and aluminium–lithium alloy. The body of the vehicle is more like a rocket, as the cross section is circular. Integration of tanks and structure result in a tapering shape, with the larger hydrogen tank nearest the tail, which is convenient for stability during re-entry. The engines will eventually be tri-propellant, and are based on the well-tried RL-10 design.

The nozzles adjust for altitude either by using a composite nozzle extender deployed in flight, or a 'dual bell' nozzle. This comprises two bell-shaped nozzle sections: a low expansion ratio section immediately connected to the combustion chamber, and a larger expansion ratio nozzle section joined to its end. Low in the atmosphere, back pressure, and the sharp change of contour, cause flow separation of the exhaust stream from the nozzle wall at the junction of the low- and high-expansion ratio sections. The stream beyond this junction essentially does not 'see' the rest of the nozzle, and is correctly expanded for high ambient pressure. As the pressure decreases the flow separation point moves down the nozzle, so that the larger expansion ratio section comes into play, and the exhaust stream is correctly expanded for high altitude.

A choice between these two vehicle concepts would depend on the precise details of the trade-off of the gliding concept with higher structural mass and less propellant, versus the tail landing concept with lower structural mass and more propellant. Since the objective is to provide cost-effective access to space, other factors will also be important. Common to both concepts is the serviceability of the vehicle, in particular of the engines. These need to be capable of several flights without major refurbishment, and this in turn implies that they should operate below their maximum output. The thrust margin necessary to do this has to go into the calculation. Economic aspects include the launch and recovery infrastructure, and here the need for a long runway rather than a small launch pad may play a part.

Another vehicle concept is the Roton. This is like the Delta-Clipper in that it is launched and landed vertically and has the advantage of a single load path. Ultimately it will use the rotary rocket engine (described earlier) for the ascent, but the descent will be made by using a set of deployable rotor blades on the nose rather than under rocket power. Aerobraking takes place in a tail-first orientation, and heat shielding for the engines is provided by a deflector incorporated in the engine disk. At lower altitude and speed the rotor is deployed, and spins up under aerodynamic forces to brake the vehicle, bringing it to a gentle tail-first landing. This, of course, saves the propellant needed for a normal tail-first landing, and provides a significant mass advantage.

8.7.1 The use of aerodynamic lift for ascent

Much of the propellant burned during ascent is used in the early part of the trajectory. The vehicle itself is very heavy at that time, and drag is a significant force; moreover, gravity loss is at a maximum during near vertical flight. Thus a great deal of the total chemical energy in the propellant tanks is used while the rocket is gaining a modest altitude and velocity. This is more obvious in a staged launcher, with the huge mass flows of the boosters discarded at 60 km altitude or so. Aerodynamic lift is a much more efficient way of gaining altitude, and typical aircraft speeds of 300–400 m/s are a significant fraction of the 7,600 m/s needed for orbit. The Pegasus air-launched expendable rocket makes good use of this to provide a very efficient vehicle.

Using an aerofoil to gain initial altitude and velocity is also attractive for SSTO. The Astroliner is an aircraft-like vehicle which is towed to 6 km altitude and Mach 0.8 (350 m/s) by a modified Boeing 747. The rocket engines are then fired to take the vehicle to a sub-orbital altitude of 100 km, and a velocity of 2.7 km/s. An expendable rocket stage then completes the injection. This is not strictly SSTO, but the main vehicle returns to Earth and makes an unpowered glide landing, which enables reuse of all but the injection rocket. Lift is used both to attain the ignition altitude and during the rocket-powered ascent. Another concept uses lift to transfer the vehicle to a similar altitude and velocity powered by efficient turbo-fan jet engines. The vehicle is loaded with fuel on the ground, but the liquid oxygen is carried to altitude by a tanker which meets with the vehicle. The oxygen is transferred and the rockets are then ignited, taking the vehicle into orbit. Return to Earth is accomplished by the familiar aerobraking and glide landing. These approaches use the lift to carry unburned propellant to a significant altitude and speed without loss to the main vehicle. A further advantage is the low ambient pressure at ignition, which enables a fixed high-expansion ratio nozzle to be used for all the rocket-powered portion of the flight.

The reusable single stage to orbit vehicle will eventually bring about the low-cost access to space which is so important for scientific exploration and commercial expansion. However, it seems likely that this will not be achieved through a single breakthrough in technology but by a development programme in which small technological advances eventually combine to produce the desired result—and this cannot now be very far from being accomplished.

Appendix 1

Orbital motion

Spacecraft move in orbits governed by the local gravitational field and the momentum of the spacecraft. The nature of these orbits is described in Chapter 1, and here the mathematics of the orbital equation are reviewed.

A1.1 RECAPITULATION OF CIRCULAR MOTION

The force F holding the spacecraft in circular motion about a planet is defined by Newton's law of gravitation:

$$F = \frac{mV^2}{r} = mr\omega^2 = mr\dot{\theta}^2 = \frac{GMm}{r^2}$$

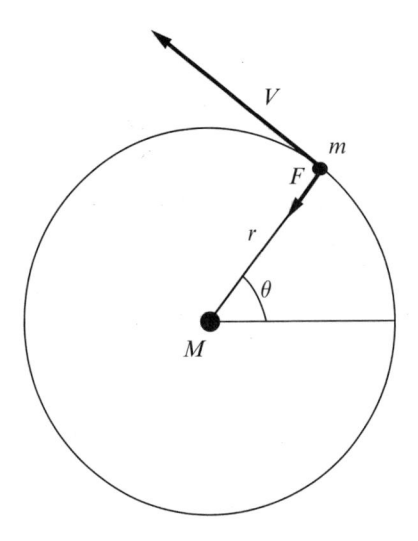

Figure A1.1. Circular motion.

278 Appendix 1

where $\dot{\theta} = \dfrac{d\theta}{dt}$ is the angular velocity in radian s^{-1} (sometimes written as ω), and G is the gravitational constant. Using this relationship the following equations, linking the linear and angular velocities with the gravitational field, can be written as

$$\frac{GMm}{r^2} = \frac{mV^2}{r} \quad \frac{GMm}{r^2} = mr\dot{\theta}^2$$

$$V = \sqrt{\frac{GM}{r}} \quad \dot{\theta} = \omega = \sqrt{\frac{GM}{r^3}}$$

These equations are easy to derive, and include the relationship for the velocity in a circular orbit, quoted in Chapter 1. To deal with the general case of a non-circular orbit, separate differential equations for the radial and transverse motion need to be set up and solved.

A1.2 GENERAL (NON-CIRCULAR) MOTION OF A SPACECRAFT IN A GRAVITATIONAL FIELD

In a circular motion the radial distance of the spacecraft from the centre of the planet is constant. Only the angle θ changes, rotating through 2π for each completed orbit. For general motion both θ and r change with time, and their rates of change have to be derived separately.

The forces acting are the same as in the case of circular motion, but the velocity has to be resolved into two components—a radial component u and a transverse component v.

The radial acceleration a_r is itself comprised of two components:

$$a_r = \frac{du}{dt} - v\frac{d\theta}{dt}$$

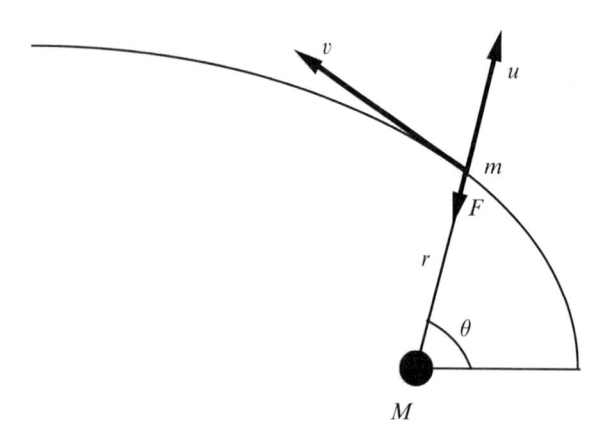

Figure A1.2. Non-circular motion.

The first term is the simple radial acceleration, and the second term is the radial component of the transverse acceleration, which is zero for circular motion.

The transverse acceleration now also has two components: the simple transverse acceleration and the transverse component of the radial acceleration:

$$a_t = \frac{dv}{dt} + u\frac{d\theta}{dt}$$

Expressing u and v in terms of r and θ, and using the notation

$$\dot{r} = \frac{dr}{dt}; \qquad \ddot{r} = \frac{d^2r}{dt^2}$$

$$\dot{\theta} = \frac{d\theta}{dt}; \qquad \ddot{\theta} = \frac{d^2\theta}{dt^2}$$

these equations can be written as

$$a_r = \ddot{r} - r\dot{\theta}^2$$

$$a_t = r\ddot{\theta} + 2\dot{r}\dot{\theta}$$

where

$$v = r\dot{\theta}; \qquad \frac{du}{dt} = \ddot{r}$$

For a spacecraft in orbit, with no rocket firing, the transverse accelerating force is zero, while the radial force is simply the force of gravity acting on the spacecraft. Applying Newton's laws

$$ma_r = -\frac{GMm}{r^2}; \qquad ma_t = 0$$

or, using the above relationship

$$m(\ddot{r} - r\dot{\theta}^2) = -\frac{GMm}{r^2}$$

$$m(r\ddot{\theta} + 2\dot{r}\dot{\theta}) = 0$$

These equations govern the motion of the spacecraft. The first defines the radial acceleration under gravity, and the second defines the motion along the orbit, conserving angular momentum. The angular momentum is constant because the rockets are not firing and there are no loss mechanisms in space.

$mr^2\dot{\theta} = mh$ is the angular momentum, and $h = r^2\dot{\theta}$ is the angular momentum per unit mass. The angular momentum per unit mass is also a constant of the orbit, and represents the momentum given to the spacecraft by the rocket:

$$m(r\ddot{\theta} + 2\dot{r}\dot{\theta}) = 0$$

$$\frac{dh}{dt} = 0 = \frac{d(r^2\dot{\theta})}{dt} = r(r\ddot{\theta} + 2\dot{r}\dot{\theta})$$

$$\dot{r}(r\ddot{\theta} + 2\dot{r}\dot{\theta}) = 0$$

Appendix 1

The two equations can now be written as

$$m(\ddot{r} - r\dot{\theta}^2) = -\frac{GMm}{r^2}$$

$$r^2\dot{\theta} = h$$

We now need expressions for $\dot{r}$ and $\ddot{r}$:

$$\dot{r} = \frac{dr}{dt} = \frac{dr}{d\theta}\frac{d\theta}{dt} = \frac{h}{r^2}\frac{dr}{d\theta}$$

$$\frac{d(1/r)}{d\theta} = \frac{d(1/r)}{dr}\frac{dr}{d\theta} = -\frac{1}{r^2}\frac{dr}{d\theta}$$

$$\dot{r} = -h\frac{d}{d\theta}\left(\frac{1}{r}\right)$$

and

$$\ddot{r} = \frac{d\dot{r}}{dt} = \frac{d\dot{r}}{d\theta}\frac{d\theta}{dt} = \frac{h}{r^2}\frac{d\dot{r}}{d\theta}$$

$$\dot{r} = -h\frac{d}{d\theta}\left(\frac{1}{r}\right)$$

$$\therefore \ddot{r} = -\frac{h^2}{r^2}\frac{d^2}{d\theta^2}\left(\frac{1}{r}\right)$$

A change of variable is now necessary to set up and solve the differential equation of motion:

$$k = \frac{1}{r}$$

Substituting for the radial velocity and radial acceleration,

$$m(\ddot{r} - r\dot{\theta}^2) = -\frac{GMm}{r^2}$$

$$-\frac{h^2}{r^2}\frac{d^2}{d\theta^2}\left(\frac{1}{r}\right) - r\left(\frac{h}{r^2}\right)^2 = -\frac{GM}{r^2}$$

$$h^2\frac{d^2k}{d\theta^2} + h^2k = GM$$

$$\frac{d^2k}{d\theta^2} + k = \frac{GM}{h^2}$$

This is the differential equation for the spacecraft motion in terms of the variable $1/r$ and the (constant) angular momentum h.

The solution to this equation has been quoted in Chapter 1, and is

$$k = \frac{1}{r} = \frac{GM}{h^2} + C\cos\theta$$

The constant C can be evaluated by considering the initial conditions at the point of injection into the orbit. If the initial radius is r_0, the initial velocity is v_0, and $\cos\theta = 1$, then

$$C = \frac{1}{r_0} - \frac{GM}{h^2}$$

$$h = r_0^2 \dot\theta_0 = r_0 v_0$$

$$\therefore C = \frac{1}{r_0} - \frac{GM}{r_0^2 v_0^2} = \frac{1}{r_0}\left(1 - \frac{GM}{r_0 v_0^2}\right)$$

Rearrangement of the equation of motion produces

$$\frac{1}{r} = \frac{GM}{h^2}\left(1 + \frac{Ch^2}{GM}\cos\theta\right)$$

$$\frac{1}{r} = \frac{GM}{h^2}(1 + \varepsilon \cos\theta)$$

$$\varepsilon = \frac{Ch^2}{GM} = \frac{r_0^2 v_0^2}{GM}\frac{1}{r_0}\left(1 - \frac{GM}{r_0 v_0^2}\right) = \left(\frac{r_0 v_0^2}{GM} - 1\right)$$

The constant ε is called the *eccentricity* of the orbit, which depends on the ratio of the specific angular momentum given by the rocket to the gravitational potential of the planet given by GM/r_0. The eccentricity defines the shape of the orbit.

Considering the equation of motion

$$\frac{1}{r} = \frac{GM}{r_0^2 v_0^2}(1 + \varepsilon \cos\theta); \qquad \varepsilon = \left(\frac{r_0 v_0^2}{GM} - 1\right)$$

it can easily be seen how this is. For $\varepsilon > 1$ (the initial velocity squared is much greater than the gravitational potential) the product $\varepsilon \cos\theta$ becomes less than -1 for some value of θ greater than $90°$. At this point the ratio $1/r$ becomes zero, and r becomes equal to infinity. This is the case in which the spacecraft escapes from the gravitational field of the planet with a finite velocity. Such a case could arise for a journey to Mars, for instance, when the requirement is to complete the trip in a short time. These orbits are hyperbolic, and the spacecraft never returns to its starting point.

If $\varepsilon = 1$, then the $\varepsilon \cos\theta$ term equals -1 for $\theta = 180°$, and again the reciprocal of the radius goes to zero and the radius goes to infinity. This is the case where the spacecraft just escapes from the system with zero residual velocity. The value of v_0 is termed the *escape velocity*, as it is the minimum velocity necessary to escape the gravitational influence of the planet.

If $\varepsilon = 0$, then the radius becomes independent of θ and the orbit is circular; between zero and unity, the orbit is an ellipse of the eccentricity defined by ε. If ε is equal to unity or to zero, then it is easy to derive the necessary velocity of the spacecraft at injection:

$$\varepsilon = 0; \qquad r_0 v_0^2 = GM; \qquad v_0 = \sqrt{\frac{GM}{r_0}}$$

$$\varepsilon = 1; \qquad r_0 v_0^2 = 2GM; \qquad v_0 = \sqrt{\frac{2GM}{r_0}}$$

Thus the escape velocity is just 1.414 times greater than the circular velocity.

Appendix 2

Launcher survey

At present there are many types of launch vehicle available, and if past launchers are included the tally runs into several hundreds. Here vehicles are included if they have launched a non-military payload during the last ten years. The list is not exhaustive, and some vehicles used in the last ten years may no longer be available.

A2.1 LAUNCH SITE

The capability of a launcher to deliver a payload into the required orbit depends to some extent on its launch site. The extra velocity increment from the Earth's rotation is significant, and this will depend on the latitude of the site and the inclination of the orbit. Inclination is an important parameter of an orbit, and different missions require different inclinations. For communications satellites zero inclination is normal, because they should be geostationary above the equator. For Earth-observation satellites, highly inclined orbits are preferred, so that most of the Earth's surface passes beneath the orbit. Changing the inclination of an orbit from one defined by the latitude and azimuth of launch requires a large additional velocity increment, and in the vast majority of cases the range of inclinations available via a direct launch from a particular launch site will be of importance.

In addition, all launch ranges have restrictions on the azimuth along which the ground track of a launch can be aligned. This is a safety requirement. Launchers *do* fail, and falling debris must not endanger the lives or the property of people living under the launch trajectory. Also, the discarded boosters and second stages must not fall on inhabited land. Since the azimuth, together with the latitude, determines the inclination, safety requirements will limit the range of inclinations possible at a given site. The easiest solution to range safety is to launch over the ocean, and the majority of launch sites are located on the coast, adjacent to an ocean. Russia, China and Australia have large inland deserts within their borders, and so launch sites can be inland.

A2.2 LAUNCHER CAPABILITY

There are many ways of defining the capability of the rocket, the most important parameter being the velocity increment, which depends on the mass ratio and therefore on the payload. The same payload will require different velocity increments, depending on the orbit. In general, for no inclination change the increment ranges from about 8 km/s for a low Earth orbit to 11.5 km/s for a geostationary orbit. An inclination change of 15° requires an additional 2 km/s at LEO, but considerably less at geostationary altitudes.

In the following table the maximum payload capability for LEO and GTO is given, unless the rocket is not capable of GTO. In many cases a family of launchers is available, with different variants having different payload capabilities. For example, the Ariane 4 can increase its payload capability by adding boosters to the configuration, and other launchers vary the upper stage to give different payload capacities. Another aspect is the ability to restart the upper stage engine. Essentially this allows the perigee of the orbit to be selected independently of the launch site. If no restart is possible, then the time of day of the launch determines the range of perigee locations possible from a given launch site. The launch window may be as short as half an hour, and may be available during only part of the year, due to solar constraints. The additional flexibility of restart is therefore an important parameter. The *vacuum* thrust of each stage is included. For first stages and boosters the thrust will be somewhat lower than the value quoted, due to atmospheric pressure. The launchers are listed alphabetically by country. In some cases only a few payloads have been launched by a particular vehicle in the last 10 years, and in others, large numbers of launches have taken place. Example payloads only are included, and serial numbers have not been given.

A2.3 HEAVY LAUNCHERS

Defining a heavy launcher as one capable of delivering a payload of four metric tons or greater into GTO, we find examples in all the main space nations' portfolios. The Space Shuttle, Ariane 5, Titan IV and Delta IV (heavy) all have a capability in excess of six metric tonnes. In the four–six tonne range are Long March CZ-3B, Ariane 44, H-2, Proton, Zenit and Titan III. This is driven by the need for large communications satellites, and the Proton capability was used also for interplanetary probes.

For building and servicing the International Space Station, and for launching satellite constellations, heavy-lift capability into LEO is important. The Space Shuttle (24T), Delta IV (heavy) (23T) and Proton (22T) head the list, followed by Ariane 5 and Titan IV at about 18 tonnes. Long March CZ-2E, Zenit 3 and Titan III all have a capability between 10 and 14 tonnes. In general there seems to be an increasing demand for heavy launchers, and consequent pressure to increase the capability.

A2.4 MEDIUM LAUNCHERS

There is little demand for payloads smaller than one metric tonne in GTO, and the majority of launch vehicle families have a capability from one to four metric tonnes in GTO. The vast majority of current satellite launches are contained in this category, and several hundred are launched each year worldwide. Examples can be found throughout the tables. A capability of up to four tonnes in GTO is accompanied by LEO capability up to seven tonnes—which again is useful for satellite constellations and for military purposes.

A2.5 SMALL LAUNCHERS

There is a serious world shortage of small launch vehicles. The ISAS programme leads the availability of small launchers with the Mu series, the Pegasus air-launched vehicle is very significant, and the Russian ROKOT is also a contender. Typically, the requirement is for a vehicle capable of placing a few tonnes in LEO. There is no strong commercial or military driver for this capability, and the result is a dearth of small launchers. Manufacturers of small satellites have to share vehicles with larger payload capacity, either as a lightweight passenger with a large spacecraft, or as a group of small satellites. It is difficult to deliver the satellites into different orbits from the same launcher, and so this is far from being convenient.

Table A.2.1. Launchers.

China

Launcher	Boosters	Stage 1	Stage 2	Upper Stage(s)	Payload LEO	Payload GTO	Restart	Launch site	Example Payload
Long March, CZ-2C	Two CZ-2E liquid strap-on boosters; Single YF-20B engine; N_2O_4/UDMH 832 kN × 2	CZ-2C 1 Four YF-20A engines; N_2O_4/UDMH 3,060 kN	CZ-2C 2 Single YF-22A/23A engine; N_2O_4/UDMH 777 kN		3.2T	1T	yes	Jiuquan	Freya FSW RV Iridium
CZ-2D		CZ-4A1 Four YF-20-B engines; N_2O_4/UDMH 3,329 kN	CZ-4A-2 Single YF-25/23 engine; N_2O_4/UDMH 847 kN		3.5T		no	Jiuquan	FSW RV
CZ-2E	Four CZ-2E liquid strap on boosters; Single YF-20B engine; N_2O_4/UDMH 832 kN × 4	CZ-4E1 Four YF-20-B engines; N_2O_4/UDMH 3,329 kN	CZ-4E-2 Single YF-25/23 engine; N_2O_4/UDMH 847 kN		9.5T	3.5T	yes	Jiuquan	Optus Asiasat Echostar
CZ-3		CZ-3-1 Four YF-20-A engines; N_2O_4/UDMH 3,060 kN	CZ-3-2 Single YF-22A/23A engine; N_2O_4/UDMH 777 kN	CZ-3-3 Four YF73 engines; LO_2/LH_2 20 kN	4.8T	1.4T	yes	Xichang	Asiasat Apstar
CZ-3A		CZ-3A1 Four YF-20-B engines; N_2O_4/UDMH 3,329 kN	CZ-3A-2 Single YF-25/23 engine; N_2O_4/UDMH 847 kN	CZ-3A-3 two YF75 engines; LO_2/LH_2 160 kN	7.2T	2.6T	yes	Xichang	KF DFH

CZ-3B	Four CZ-3B liquid strap-on boosters; Single YF-20B engine; N_2O_4/UDMH 832 kN × 4	CZ-3B1 Four YF-20-B engines; N_2O_4/UDMH 3,329 kN	CZ-3B-2 Single YF-25/23 engine; N_2O_4/UDMH 847 kN	CZ-3B-3 two YF75 engines; LO_2/LH_2 160 kN	5T	yes	Xichang	Intelsat Agila Apstar Sinosat
CZ-4		CZ-41 Four YF-20-B engines; N_2O_4/UDMH 3,329 kN	CZ-4-2 Single YF-25/23 engine; N_2O_4/UDMH 847 kN	CZ-4-3 Two YF-40 engines; N_2O_4/UDMH 1,028 kN	4T 1.1T	yes	Taiyuan	China Feng Yun

Europe

Launcher	Boosters	Stage 1	Stage 2	Upper Stage(s)	Payload LEO	Payload GTO	Restart	Launch site	Example Payload
Ariane 40		L220 Four Viking V engines; N_2O_4/UDMH (25) 3,094 kN	L33 Single Viking IVB engine; N_2O_4/UDMH (25) 786 kN	H10 Single HM7-B engine; LH_2/LH_4; 62 kN	4.9T	2.02T	yes	Kourou	Spot ERS Webersat
Ariane 42P	Two solid strap-on PAP boosters; 625 kN × 2	L220 Four Viking V engines; N_2O_4/UDMH (25) 3,094 kN	L33 Single Viking IVB engine; N_2O_4/UDMH (25) 786 kN	H10 Single HM7-B engine; LH_2/LH_4; 62 kN	4.8T	2.74T	yes	Kourou	Satcom Superbird Galaxy Telcom Koreasat
Ariane 44P	Four solid strap-on PAP boosters; 625 kN × 4	L220 Four Viking V engines; N_2O_4/UDMH (25) 3,094 kN	L33 Single Viking IVB engine; N_2O_4/UDMH (25) 786 kN	H10 Single HM7-B engine; LH_2/LH_4; 62 kN	8.3T	3.29T	yes	Kourou	N-Star Intelsat Panamsat ISO
Ariane 42L	Two liquid strap-on boosters; Viking 5C engine; N_2O_4/UDMH (25) 766 kN × 2	L220 Four Viking V engines; N_2O_4/UDMH (25) 3,094 kN	L33 Single Viking IVB engine; N_2O_4/UDMH (25) 786 kN	H10 Single HM7-B engine; LH_2/LH_4; 62 kN	7.4T	3.35T	yes	Kourou	Panamsat Telstar Intelsat PAS
Ariane 44LP	Two liquid and two solid strap-on boosters; Viking 5C; N_2O_4/UDMH (25) 766 kN × 2 625 kN × 2	L220 Four Viking V engines; N_2O_4/UDMH (25) 3,094 kN	L33 Single Viking IVB engine; N_2O_4/UDMH (25) 786 kN	H10 Single HM7-B engine; LH_2/LH_4 62 kN	6.6T	4.06T	yes	Kourou	Skynet JCSAT Hipparcos Meteosat Intelsat Panamsat Immarsat

Ariane 44L	Four liquid strap-on boosters; Viking 5C engine; N_2O_4/UDMH (25) 766 kN × 4	L220 Four Viking V engines; N_2O_4/UDMH (25) 3,094 kN	L33 Single Viking IVB engine; N_2O_4/UDMH (25) 786 kN	H10 Single HM7-B engine; LH_2/LH_4 62 kN	7.7T	4.46T	yes	Kourou	Superbird Intelsat TDF Eutelsat Immarsat Sirius
Ariane 5	Two solid MPS strap-on boosters; 6.700 kN × 2	Main Stage Single Vulcain engine; LO_2/LH_2 1,096 kN	Upper Stage Single Aestus engine; N_2O_4/UDMH 28 kN		18T	6.8T	yes	Kourou	Cluster Maqusat
Ariane 5 ECA	Two solid MPS strap-on boosters; 7.14 kN × 2	Main Stage Single Vulcain engine; LO_2/LH_2 1,370 kN	Upper Stage Single HM7-13 LH_2/LO_2 64 kN			10.5T	yes	Kourou	Qual. flight end 2004

Japan

Launcher	Boosters	Stage 1	Stage 2	Upper Stage(s)	Payload LEO	Payload GTO	Restart	Launch site	Example Payload
ISAS Mu-3-S & Mu-3-S2	Two Strap-on solid; single SB735 engine; 334 kN × 2	Solid Mu-3S Single M-13 engine; 1,287 kN	Solid Mu-3S Single M-23-Mu engine; 534 kN	Solid Mu-3S Single M-3B-Mu engine; 135 kN	0.77T	0.52T	no	Kagoshima	Tenma Ginga Asca
ISAS M-V		Solid M-V Single M14 engine; 3,854 kN	Solid M-V Single M24 engine; 1,270 kN	Solid M-V Single M-34 engine with extendible nozzle; 300 kN	1.8T	1.2T	no (kick-stage motor option)	Kagoshima	Haruka Nozomi Astro-E
NASDA H-1	Nine Castor-2 strap-on solid boosters; single TX-354-3 engine; 264 kN × 9	H-1 Single MB-3-3 engine; LO$_2$/Kerosene; 884 kN	H-1 Single LE-5 engine; LO$_2$/LH$_2$ 105 kN	H-1 Single solid H-1-3 engine; 79 kN	3.2T	1.1T	yes	Tanegashima	Sakura JERS GMS
NASDA H-2	Twin H-2 solid strap-on boosters; 1,570 kN × 2	H-2 Single LE-7 engine; LO$_2$/LH$_2$; 1,099 kN	H-2 Single LE-5EC engine; LO$_2$/LH$_2$ 124 kN		10.5T	4T	yes	Tanegashima	Himawari TRMM Kakahashi ETS

Russia

Launcher	Boosters	Stage 1	Stage 2	Upper Stage(s)	Payload LEO	Payload GTO	Restart	Launch site	Example Payload
Molniya	Four Liquid Boosters; single RD-107 engine with four combustion chambers; LO$_2$/kerosene 1,015 kN × 4	Block A; single RD-108 engine with four combustion chambers; LO$_2$/kerosene 996 kN Fired after booster burn-out	Block I; single RD-0110 engine; LO$_2$/kerosene 304 kN	Block L; single S1.5400A engine; LO$_2$/kerosene 68 kN	1.8T	1.6T	yes	Baikonur Plesetsk	Venera Molniya IRS-C Cosmos
Proton 8K82K		UR500 Polyblock; central oxidiser tank with six fuel tanks around periphery; six RD253 engines mounted on fuel tanks; N$_2$O$_4$/UDMH 10.7 MN	UR200 monoblock; four RD0210 engines; N$_2$O$_4$/UDMH 2,446 kN	UR200 monoblock; single RD0212 engine; N$_2$O$_4$/UDMH 642 kN	20.6T		yes	Baikonur	ISS Zarya Spectrum Mir
Proton 8K82K/ 11S824 11S824F 11S824M		UR500 polyblock; central oxidiser tank with six fuel tanks around periphery; six RD253 engines mounted on fuel tanks; N$_2$O$_4$/UDMH	UR200 monoblock; four RD0210 engines; N$_2$O$_4$/UDMH 2,446 kN	3) UR200 monoblock; single RD0212 engine; N$_2$O$_4$/UDMH 642 kN 4) Block D single RD58M engine; LO$_2$/kerosene 86.7 kN	5.4T to translunar orbit 6.2T to transmartian orbit 4.7T to transvenusian orbit		yes	Baikonur	Zond Lunik Mars Venera Mars-96 Granat Astron

Russia (*cont.*)

Launcher	Boosters	Stage 1	Stage 2	Upper Stage(s)	Payload LEO	Payload GTO	Restart	Launch site	Example Payload
Proton 8K82K/ 11S86 11S86-01		UR500 polyblock; central oxidiser tank with six fuel tanks around periphery; six RD253 engines mounted on fuel tanks; N_2O_4/UDMH 10.7 MN	UR200 monoblock; four RD0210 engines; N_2O_4/UDMH 2,446 kN	3) UR200 monoblock; single RD0212 engine; N_2O_4/UDMH 642 kN 4) Block DM/11S86; single RD58M engine; LO_2/kerosene 86.7 kN		2.1T 2.5T	yes	Baikonur	Gals Ekspress Kupon
Proton 8K82K/ DM1, 2 (different payload adaptors)		UR500 Polyblock; Central oxidiser tank with 6 fuel tanks around periphery; six RD253 engines mounted on fuel tanks; N_2O_4/UDMH 10.7 MN	UR200 monoblock; four RD0210 engines; N_2O_4/UDMH 2,446 kN	3) UR200 monoblock; single RD0212 engine; N_2O_4/UDMH 642 kN 4) Block DM-2, DM-5; single RD58M engine; LO_2/kerosene 86.7 kN	DM2: 5T	DM1: 2.1T	yes	Baikonur	Immarsat Iridium
Proton 8K82K/ DM3, 4		UR500 polyblock; central oxidiser tank with six fuel tanks around	UR200 monoblock; four RD0210 engines; N_2O_4/UDMH 2,446 kN	3) UR200 monoblock; single RD0212 engine; N_2O_4/UDMH 642 kN 4) Block DM-	DM4: 20.6T	DM3: 2.5T DM4: 2.5T	yes	Baikonur	Astra 1F Panamsat 5 Asiasat Echostar Telstar

Launcher	Stage 1	Stage 2	Stage 3	Stage 4	Payload	Launch site	Manned	Satellites
(continued)	periphery; six RD253 engines mounted on fuel tanks; N_2O_4/UDMH 10.7 MN			2M; single RD58S engine; LO_2/kerosene 85.1 kN	2.5T			Raduga
Proton 8K82KM (improved version 1999)	UR500 polyblock; central oxidiser tank with six fuel tanks around periphery; six RD253 engines mounted on fuel tanks; N_2O_4/UDMH 10.74 MN	UR200 monoblock; four RD0210 engines; N_2O_4/UDMH 2,446 kN	3) UR200 monoblock; single RD0212 engine; N_2O_4/UDMH 642 kN 4) Briz-M single S5.98M engine; N_2O_4/UDMH 20 kN		22T	Baikonur		
Rokot	UR100N1 Four RD0232 engines; N_2O_4/UDMH 56.9 kN	UR100N2 single RD0235 engine; N_2O_4/UDMH 14.9 kN	Rokot-3 Briz single S5.98M engine; 17 kN		1.8T	Baikonur	no	Radio-ROSTO Abrixas
Soyuz 11A511U	Four liquid boosters; single RD-107 engine with four combustion chambers; LO_2/densified-kerosene; 1,015 kN.	Block A; single RD-108 engine with four combustion chambers; LO_2/densified-kerosene; 997 kN; fired after booster burn-out	Block I; single RD-0110 engine; LO_2/densified-kerosene; 304 kN		6.8T	Plesetsk Baikonour	yes	Soyuz Cosmos Progress Photon Gamma X-Mir-Inspector

Russia (*cont.*)

Launcher	Boosters	Stage 1	Stage 2	Upper Stage(s)	Payload LEO	Payload GTO	Restart	Launch site	Example Payload
Soyuz 11A511U2	Four liquid boosters; single RD-107 engine with four combustion chambers; LO$_2$/ kerosene; 1,015 kN	Block A; single RD-108 engine with four combustion chambers; LO$_2$/ synthetic kerosene; 997 kN: fired after booster burn-out	Block I; single RD-0110 engine; LO$_2$/ densified-kerosene; 304 kN		7.1T		yes	Baikonour	Progress VBK-Raduga Znamya Soyaz
Tsiklon		R36; single RD251 engine with six combustion chambers; N$_2$O$_4$/UDMH 2,750 kN	R36; single RD252 engine with two combustion chambers; N$_2$O$_4$/UDMH 956 kN	S5M; single RD861 engine; N$_2$O$_4$/UDMH 78 kN	3.6T		yes	Plesetsk	Meteor Ocean Intercosmos Cosmos-constell-ation Gonets
Zenit 3 SL		Zenit 1; single four-chamber RD171 engine; LO$_2$/kerosene 7.9 MN	Zenit 2; single RD 120 engine; LO$_2$/kerosene 8334 kN	Block DM; single RD58S engine; LO$_2$/ kerosene 85.1 kN	13.7T	5.18T	yes	Sea Launch	Demosat Directv-1R

United States

Launcher	Boosters	Stage 1	Stage 2	Upper Stage(s)	Payload LEO	Payload GTO	Restart	Launch site	Example Payload
Athena: current model Athena II		Castor 120 Solid 1934 kN	Orbus 21D Solid 194 kN	OAM hydrazine 2 kN	1.6T		yes	Vandenburg, Canaveral	Lunar Prospector, ROCSAT
Atlas Centaur: current models 2, 2A and 2AS	Atlas aft-skirt; MA-5A twin nozzle; LO_2/kerosene; 1,910 kN	Atlas; MA-5A single nozzle LO_2/kerosene; 269 kN	Centaur Twin RL10-A4; deployable nozzle; LO_2/LH_2 184 kN			3.1T 3.7T with 4 Castor-4A strap-on solid boosters	yes	Canaveral	Intelsat 806, SOHO
Delta 4920	Nine Castor 4A; solid; 487 kN × 9	Delta-Thor ELT; single RS-27 engine; LO_2/Kerosene; 40 kN	Delta K; single AJ10-118 K engine; N_2O_4/Aerozine 50; 43 kN		3.4T	1.2T	yes	Canaveral	Insat 1-D
Delta 6925	Nine Castor 4A; Solid; 487 kN × 9	Delta-Thor XLT; single RS 27-A engine; LO_2/Kerosene; 1,052 kN	Delta K; single Aj10-111K engine; N_2O_4/Aerozine 50; 43 kN	PAM-D; solid; 69 kN	3.9T	1.5T	no	Canaveral	Immarsat 2 EUVE
Delta 7925	Nine GEMS; Solid; 502 kN × 9	Delta-Thor XLT-C; single RS-27C; LO_2/kerosene; 1,075 kN	Delta-K; single AJ10-118K engines N_2O_4/Aerozine 50; 43 kN	PAM-D solid; 68 kN	5T	1.8T			Globalstar Iridium Mars Climate Orbiter GPS
Delta IV, EELV-Heavy	Two Delta-RS-68; single RS-68 engine; LO_2/LH_2; 3,378 kN × 2	Delta RS-68; single RS-68 engine; LO_2/LH_2; 3,378 kN	Delta 4; single RL-10B-2 engine; LO_2/LH_2; 112 kN		22.7T	13.2T	yes	Canaveral Vandenburg	Operational 2002–3

United States (cont.)

Launcher	Boosters	Stage 1	Stage 2	Upper Stage(s)	Payload LEO	Payload GTO	Restart	Launch site	Example Payload
Delta IV, EELV-medium-plus 4.2-m fairing diameter	Two GEMS Plus; solid; 640 kN × 2	Delta RS-68; single RS-68 engine; LO_2/LH_2; 3,378 kN.	Delta 3; single RL-10B-2 engine; LO_2/LH_2; 112 kN			5.8T	yes	Canaveral Vandenburg	Operational 2002–3
Delta IV, medium-plus 5.2 m fairing diameter	Two GEMS Plus; solid; 640 kN × 2	Delta RS-68; single RS-68 engine; LO_2/LH_2; 3,378 kN	Delta 3; single RL-10B-2 engine; LO_2/LH_2; 112 kN			4.7T	yes	Canaveral Vandenburg	Operational 2002–3
Delta IV, medium-plus 5.4 m fairing diameter	Four GEMS Plus; solid; 640 kN × 4	Delta RS-68; single RS-68 engine; LO_2/LH_2; 3,378 kN	Delta 3; single RL-10B-2 engine; LO_2/LH_2; 112 kN			6.7T	yes	Canaveral Vandenburg	Operational 2002–3
Pegasus: current model XL	Air launched, L1011 at 12 km altitude	Orion 50S solid; 496 kN	Orion 50 solid; 125 kN	Orion 38 solid; 44 kN	350 kg polar; 475 kg equatorial		no	Anywhere Canaveral Vandenburg Edwards	Alexis, Step, Orbcom satellite constellation
Scout		Algol 3A solid; 520 kN	Castor 2A solid; 2240 kN	3) Antares 3A 62 kN 4) Altair 3A 14 kN	166 kg polar; 220 kg equatorial		no	Vandenburg Wallops San Marco	Ariel V Explorer
Shuttle	twin RSRM; solid; 23 MN	Orbiter plus external tank; three SSME LO_2/LH_2 6,900 kN	Orbiter OMS N_2O_4/MMH	Various solid motors attached to payload in cargo bay	24T	5.9T	yes (OMS)	Canaveral	HST Chandra Space Station Sections
Titan II		Titan II-1; twin	Titan II-2;		3.1T		yes	Canaveral	Gemini

Titan IV B	twin UA1206; solid; 13 MN	LR 87-7 engines; N$_2$O$_4$/ Aerozine 50; 2215 kN	single LR-91-7 engine; N$_2$O$_4$/ Aerozine 50; 454 kN			Vandenburg	NOOA Military
		Titan IVB-1; twin LR-87-11 engines; N$_2$O$_4$/ Aerozine 50; 2,460 kN	Titan IV-2; single LR-91-11 engine; N$_2$O$_4$/ Aerozine 50; 469 kN	Centaur G; twin RL-10A-3A engines; LO$_2$/LH$_2$; 149 kN	17.7T	6.4T	yes	Vandenburg Canaveral	Military
Taurus		Castor 120; solid; 1,628 kN	Pegasus 1; solid; 494 kN	3) Pegasus 2; solid; 120 kN 4) Pegasus 3; solid; 35 kN	1.4T	0.43T	no	Vandenburg	Step Celestis Stex

Appendix 3

Ariane 5*

The Ariane 5 launcher has now taken over as the current model for all Ariane launches. The first commercial launch took place in December, 1999, and placed the ESA spacecraft XMM-Newton faultlessly in orbit. The basic configuration comprises a cryogenic main stage connected axially to an upper stage and payload adaptor, with two solid propellant boosters mounted either side of the main stage. With a lift off mass of 737 t and a height on the launch pad of 54 m, the basic model is capable of placing a spacecraft of 6.64 t into GTO or 9.5 tonnes into LEO, at a cost of approximately 150 million dollars. The launches take place from the Arianespace launch site at Kourou in French Guyana, a few degrees of latitude north of the equator. Two of the three development launches experienced problems, the most spectacular being the destruction of the first flight by aerodynamic forces. At a height of a few tens of kilometres the guidance software attempted to force the vehicle to follow the Ariane 4 trajectory, resulting in unacceptable dynamic loading of the structure causing break-up; the vehicle was destroyed by automatic safety charges. The Ariane 4 software had been passed for use, however there were paths in the code never entered by the Ariane 4 trajectory, and hence untested. Following replacement of the guidance software, the second test flight was partially successful, but fell short of full velocity due to an unforeseen roll control problem. Here the single Vulcain engine powering the main stage imparted some axial angular momentum to the exhaust stream. This was caused by a helical motion of the exhaust gases following the path of the helical cooling channels in the nozzle. The nozzle is cooled by channels that spiral round the nozzle to increase the path length of the hydrogen flowing in them. Under full temperature and pressure, the walls in contact with the hot gas 'dished' slightly creating very shallow helical channels in the inner wall of the nozzle; this imparted the helical motion to the exhaust gases. With such a huge amount of linear momentum developed in the exhaust stream of this high-thrust engine, a tiny fraction of angular momentum was enough to cause axial rotation or roll in the opposite sense. This was beyond the control capabilities of the

---

* The information in this appendix is courtesy of Arianespace.

roll control system and led to sufficient spin to raise the propellant away from the exit ports and up the walls of the chamber; the engine shut down prematurely due to the resultant fuel-out signal. These happenings illustrate the difficulties of developing any new launcher, and the vital importance of test flights to check theoretical predictions that 'everything is OK'. The basic vehicle has since performed properly, following modifications to the Vulcain engine to reduce the roll forces. The XMM-Newton orbital injection was within 19 km of the specified altitude and within 0.001 degrees of the specified inclination. Presently, Ariane takes the biggest share of world commercial launch business.

A3.1 THE BASIC VEHICLE COMPONENTS

The vehicle is illustrated in Plate 20. The single Vulcain engine, is powered by liquid oxygen and liquid hydrogen. It is ignited seven seconds before the two solid boosters allowing full monitoring prior to lift-off. It operates for 589 s and after separation of the upper stage it re-enters and is burnt up in the atmosphere. There is no main structure for this stage; the propellant tank itself provides the mechanical integrity of the stage, the mounting places for the boosters, and the load paths. This is very mass-efficient and helps keep the mass ratio high.

The two solid boosters each contain 238 t of propellant and stand 30 m high. They provide more than 90% of the thrust at lift-off. They burn for 130 s before separating from the still firing main stage. They are recovered for post-flight analysis but not for re-use.

The upper stage Aestus engine is used for orbital injection and is powered by nitrogen tetroxide and monomethyl hydrazine. It burns 9.7 t of propellant, in relatively low thrust operation, 24 kN, nominally into GTO.

The payload is contained in one of two fairings, both with a useable diameter of 4.57 m. The short fairing can accommodate spacecraft up to 11.5 m long while the long fairing can accommodate payloads up to 15.5 m long.

A3.2 EVOLVED ARIANE 5

The vehicle is designed for evolution towards bigger payloads. This involves updates in all three propulsion systems (Table A3.1). The improved cryogenic main stage uses the Vulcain 2 engine designed to use a richer mixture with a higher oxygen percentage, with a higher thrust and a slightly higher exhaust velocity. The main advantage is an improved mass ratio, consequent upon a higher fraction of tank volume that contains the higher density liquid oxygen. The bulkhead between the two propellant tanks is moved to decrease the hydrogen volume and increase the oxygen volume. The improved mass ratio contributes to the higher payload mass capability.

The boosters are modified by replacing the bolted joints between segments with welded joints—reducing casing mass; and an increase in propellant capacity by 2.4 t in the top segment. This increases the launch thrust from 6.85 MN to 7.14 MN.

Two new upper stages are being introduced. The first is a modification of the HM7 B cryogenic engine used on Ariane 4. The change from storable propellants to cryogenic propellants gives a major increase in payload capability, up to 10 t into GTO. The engine is re-startable; it develops 64 kN of thrust and burns 14 t of propellant. The larger engine to be introduced for the upper stage in 2006 is a new design called Vinci; it has the capability to perform multiple re-starts, burns 25 t of propellant and develops a thrust of 155 kN.

The introduction of the new Ariane 5 was not without problems. Cooling of the nozzle for the updated Vulcain 2 engine failed during the demonstration flight, leading to the loss of the mission and the temporary grounding of the fleet. Following reviews, and modifications to both the hardware and the management of the project, the Ariane 5 is again cleared for flight.

Table A3.1. Existing and planned Ariane 5 versions.

	Payload LEO	Payload GTO	Configuration	Height	Mass	Price (m$)
Ariane 5G	9.5 t	6.64 t	Basic core + 2SRB + EPS	54 m	737 t	150
Ariane 5 ES	15.7 t	7.55 t	Extended core Vulcain 2 + 2SRB + EPS-V	53.4 m	767 t	120
Ariane 5 ECA		10.05 t	Extended core Vulcain 2 + 2SRB + ESC-A (HM7-B)	57.7 m	777 t	120
Ariane 5 ECB		12 t	Extended core Vulcain 2	57.7 m	777 t	120

	Boosters	Basic core	Extended core	EPS (EPS-V)	ESC-A	ESC-B
Diameter	3 m	5.4 m	5.4 m			
Length	31.2 m	30.5 m	30.5 m			
Propellant mass	237 t	158 t	170 t	9.7 t (10 t)	14 t	25 t
Total mass	268 t	170 t	182 t	10.9 t (11.2 t)		
Engine		Vulcain	Vulcain 2	Aestus	HM7B	Vinci
Fuel	Solid	LH_2	LH_2	MMH	LH_2	LH_2
Oxidant	Solid	LO_2	LO_2	N_2O_4	LO_2	LO_2
Thrust	5.1 MN	1.12 MN	1.37 MN	29.6 kN	64.2 kN	153 kN
I_{sp} vac		431 s	431 s	321 s	446 s	466 s
Burn time	130 s	600 s	600 s			
Number of engines	2 units	1	1		1	1

Appendix 4

Glossary of symbols

CHAPTER 1

F	Thrust
m	Mass flow rate
v_e	Effective exhaust velocity
V	Vehicle velocity
M_0	Initial vehicle mass
M	Final or current vehicle mass
R	Mass ratio
r	Radial distance of an orbiting vehicle from the centre of the Earth
G	The gravitational constant
$M_\oplus$	Mass of the Earth
h	Angular momentum of orbit per unit mass of the vehicle
ε	Eccentricity of orbit
θ	Angular distance travelled by orbiting vehicle from the azimuth of closest approach
r_0	Radius of closest approach to the centre of the Earth
V_0	Vehicle velocity at closest approach to Earth
r_2	Apogee radius of an elliptical orbit
V_2	Velocity at apogee for an elliptical orbit; velocity of the second stage of a multistage launcher
V_1	Velocity at perigee for an elliptical orbit; velocity of the first stage of a multistage launcher
R_0	Mass ratio for a single-stage launcher
R_1	Mass ratio for the first stage of a multistage launcher
R_2	Mass ratio for the second stage of a multistage launcher
M_S	Structure mass for a stage
M_F	Mass of propellant in a stage
M_P	Mass of payload
L	Payload ratio

304   Appendix 4

γ	Structural efficiency
A, B, C	Fractional mass given to first, second, and third stages
α, β, γ	Fraction of propellant and structural mass assigned to first, second, and third stages
M_{SB}	Mass of solid booster

CHAPTER 2

p	Pressure
A	Cross-sectional area at a location in the nozzle
A_e	Cross-sectional area at exit plane of nozzle
p_e	Pressure at exit plane
F_R	Reaction on the nozzle walls; thrust developed by the nozzle
p_a	Ambient pressure; atmospheric pressure
u_e	True exhaust velocity
c_p	Specific heat of the exhaust gases at constant pressure
T_c	Temperature in the combustion chamber
T_e	Temperature at the exit plane
γ	Ratio of the specific heats for the exhaust gases; adiabatic gas constant
R	Universal gas constant
$\mathfrak{M}$	Mean molecular weight of the exhaust gases
p_e	Pressure at the exit plane of the nozzle
ρ	Density of the exhaust gases
ρ_c	Density in the combustion chamber
A^*	Cross-sectional area of the nozzle at the throat
C_F	Thrust coefficient
p^*	Pressure at the throat of the nozzle
c^*	Characteristic velocity
g	Acceleration of gravity
I_{sp}	Specific impulse

CHAPTER 4

β	Pressure-burning rate index for solid propellant
a	Pressure-burning rate constant for solid propellant

CHAPTER 5

t	Burn time for a rocket engine
s	Distance travelled during a burn

V_i Initial velocity of a vehicle at the start of a burn
s_i Additional distance travelled because of the initial velocity
ψ Thrust-to-weight ratio
θ Pitch angle, measured from the horizontal
V_Z Vertical velocity
V_X Horizontal velocity
γ Flight path angle
α Angle of attack
T Transverse force on rocket due to the atmosphere
L Lift force
D Drag force
C_D Drag coefficient
C_L Lift coefficient
M Mach number
q Dynamic pressure

CHAPTER 6

ξ Power-to-mass ratio for electrical power supply
η Power-to-thrust efficiency for electric thrusters
P_E Electrical power
M_E Mass of electrical power supply
M_P Mass of propellant for electric propulsion
E Electric field
V Electric potential
N Ion density
q Ionic charge
ε_0 Permittivity of free space
j Ion current
M_f Mass of propellant

CHAPTER 7

P Power in the exhaust stream
R/H Radius-to-height ratio for a fission core
k_∞ Multiplication factor for fission, or reproduction constant
η Fuel utilisation factor for fission
ε Fast fission factor
p Resonance escape probability for fission
f Thermal utilisation factor for fission
L_r Neutron diffusion length
B Buckling factor

L_s	Neutron slowing-down length
R	Fission core radius—cylindrical
L	Fission core height—cylindrical
τ	Mean elapsed time for delayed neutrons
v_{esc}	Escape velocity
v_{circ}	Circular velocity

Further reading

One of the problems with additional material is that many basic books on rocketry written during the 1960s and 1970s are now out of print. However, the basic reference (which at the moment *is* still in print) is *Elements of Rocket Propulsion*, by G.P. Sutton (Wiley, 1986), and there are very useful chapters in *Mechanics and Thermodynamics of Propulsion*, by P.G. Hill and C.R. Peterson (Addison-Wesley, 1992). Recent editions of both of these books use SI units. Further information on liquid propellant engines can be found in *The Liquid Propellant Rocket Engine*, by M. Summerfield (Princeton University Press, 1960). Information on launcher dynamics can be found in *Rocket Propulsion and Spaceflight Dynamics*, by Cornelisse, Schoyer and Wakker (Pitman, 1979). *Structures, or Why Things Don't Fall Down*, by J.E. Gordon (Pelican Books, 1979), is an accessible explanation of structures—for example, propellant tanks; and *Principles of Electric Propulsion*, by R.G. Jahn (McGraw-Hill, 1968) provides further reading for Chapter 6 of this book. The history of the Russian and Chinese space programmes is dealt with by Brian Harvey in *The New Russian Space Programme* (Wiley–Praxis, 1996) and *The Chinese Space Programme* (Wiley–Praxis, 1998).

As always nowadays, the Internet is an important source of information: searches, using key-words from within these pages, will reveal many web sites containing copious additional data. *Encyclopaedia Astronautica* is a particularly useful web site for up-to-date information on all aspects of launchers and rocket engines.

Further reading

Index

11D-58, 93

A4, 5, 12, 68
ablative cooling, 107
additives, 102
advanced thermal rockets, 249
aeolopile, 2
Aerobee, 9
aerodynamic forces, 129
aerodynamic lift, 274
aerospike nozzle, 268
Aestus, (Plate 1), 74–76, 95
airless bodies, 133
Apollo 16 launch, (Plate 6)
Apollo lunar transfer vehicle, 67, 90
Apollo programme, 5, 10
Arabs, 2
arc-jet thruster, 155
Ariane 1, 8, 93
Ariane 2, 93
Ariane 3, 93
Ariane 4, 8, 72, 77, 93, 141
Ariane 44, 284
Ariane 5, (Plate 20), 8, 13, 32, 68, 74–75, 78–80, 83, 95, 101, 102, 107, 112, 113, 121, 271, 284, 299–301
Arianespace, 24
Artemis, 168
Astroliner, 273
Atlas, 10, 85, 93
atmosphere, 23, 37, 128, 255

Baikonour, 24
Bepi-Colombo, 194
Braun, Werner von, 2, 3, 4, 5, 9
British army, 2

Cape Canaveral, 24
cavitation, 69, 71
Centaur, 85
cermet, 242
Challenger, 12
Chang Zheng (Long March), 7, 284
characteristic velocity, 52
charge, 97
Chinese space programme, 7
Clark University, 3
CNES (National Centre for Space Studies) (France), 8
Colombia, 12–13
combustion chamber, 62, 106
combustion temperature, 86
composite tanks, 262
Congreve, Sir William, 2
constant pitch (vacuum) trajectory, 137
cooling, 72
  ablative cooling, 107
  dump cooling, 73
  film cooling, 72
Cosmos satellite, 176
Crewed Exploration Vehicle, 13
criticality, 203

DC-X, 273
deceleration grid, 166
Deep Space 1, (Plate 13), 193
  NSTAR ion thruster, (Plate 11, 12), 159–160
Delta, 83, 93
Delta IV, 83–84, 284
Delta-Clipper, 273
Dong Feng (East Wind), 7
drag, 129
dump cooling, 73
Dynosoar, 12

Earth, rotation, 24
Earth-launch trajectories, 135, 139
East Wind (Dong Feng), 7
eccentricity, 281
electric field and potential, 162
electric propulsion, 34, 145, 187
  applications, 189–193
electric thruster, 6, 152
  low-power electric thrusters, 179–180
electrical efficiency, 166
electrical power generation, 180
  nuclear fission power generators, 184–187
  radioactive thermal generators (RTG), 181–184
  solar cells, 180–181
  solar generators, 181
electromagnetic thruster, 157
electrothermal thruster (resisto-jet), 152
elliptical transfer orbit, 22
escape velocity, 22, 281
Esnault-Pelterie, Robert, 197
European Space Agency (ESA), 8
Evolved Expendable Launch Vehicle, 83
exhaust nozzle, 41
exhaust velocity, 38, 43, 145, 149, 252
expander cycle, 70, 78
Express telecommunications satellite, 176

factory joints, 109
Feng Jishen, 2
Field Effect Emission Thruster (FEEP), 179–180
field joints, 109
film cooling, 72

fission *see* nuclear propulsion
flow separation, 264
'four factor formula', 203–205
French space programme, 8
fuel rods, 203

Gagarin, Yuri, 6, 9
Gals telecommunications satellite, 176
geostationary orbit, 190
gimballed mounting, 68
Glushko, Valentin, 6
Goddard, Robert, 4–5, 13, 197
grain, 97, 104
gravity, 18, 117, 120, 278
gravity loss, 191
gravity turn (transition trajectory), 131, 136
gunnery, 14

H-2, 284
Hall effect thruster, 6, 147, 170–176
  D-100, 175
  PPS 1350, (Plate 14), 194–195
  SP-100, 174
  SP-150, 195
  Stationary Plasma Thruster (SPT), 173–176
  Thruster with Anode Layer (TAL), 174–176
  variants, 176
heat sink, 108
heavy launchers, 284
Hero of Alexandria, 1
HM7 B, (Plate 3), 77, 78, 301
horizontal velocity, 123
Houboult, John, 9
hypergolic propellants, 67

ignition, 64, 110
impinging jet injector, 64, 66
inclined motion, 123
Indian space programme, 6
injection, 63
injectors, 64
Institute of Space and Astronautical Science (ISAS) (Japan), 7, 139
intersection joints, 109
International Space Station, 8, 13, 284
ion propulsion, 158

ion thruster, 147
  NSTAR, 159
  theory, 159
Italian Space Agency (ASI), 242

J-2, 10
Jackass Flats, 199
Japanese space programme, 6, 7
Jupiter C, 9
Jupiter Icy Moons Orbiter (JIMO), 187, 189, 237

KIWI reactor, 216–217, 236–239
Korolev, Sergei, 2, 6, 7
Kourou launch site, 299
Kublai Khan, 2

launch dynamics, 115, 133
launch sites, 283
launch trajectories, 23, 121, 125, 126, 129
launcher capability, 284
LEO (lunar transfer orbit), 31
lift, 129
linear aerospike engine, 269
liquid hydrocarbon–liquid oxygen engine, 92
liquid oxygen–liquid hydrogen engine, 90
liquid propellant distribution system, 69
liquid-fuelled engines, 35, 61
Long March (Chang Zheng), 7, 284
low Earth orbit, 190
low-power electric thrusters, 179–180
  Field Effect Emission Thruster (FEEP), 179–180
Luch satellite, 176
lunar transfer orbit (LEO), 31

magnetic nozzle, 177
Mars, 6, 191
mass flow rate, 46
mass ratio, 15, 25
medium launchers, 285
Meteor satellites, 176
Molniya, 6
Mongols, 2
MPS (Moteur à Propergol Solide), 113–114
Mu-3-S-II, 68, 139
multistage rockets, 25

Napoleonic Wars, 2
National Aeronautics and Space Administration (NASA), 9
National Centre for Space Studies (CNES) (France), 8
National Space Development Agency (NASDA) (Japan), 7
Nuclear Engine for Rocket Vehicle Applications (NERVA), (Plate 17, 18), 199, 210, 215, 218–220, 224–225, 236–237, 241
  NERVA 1, 220
neutron flux, 200
Newton's third law, 14, 37, 116
nose-cone, 33
Nova, 9, 197
nozzle, 62, 110
  aerospike nozzle, 268
  magnetic nozzle, 177–178
  nuclear engine, 228–229
  plug nozzle, 266
NRX, 220, 237, 239–240
NSTAR ion engine, (Plate 11, 12), 159–160, 193
nuclear fission power generators, 184–187
  for JIMO, 187, 189
  SNAP 10-A, 185–186
  Topaz, 186
nuclear propulsion, 197–247
  see also nuclear thermal rocket engine; NERVA
  barn, 208
  buckling factor, 206–207
  control of neutron flux, 209–210
  criticality, 203–205
  delayed neutrons, 211–212
  diffusion length, 206
  energy, 198
  exhaust velocity, 216–218
  fast fission factor, 203–204
  fission basics, 199–201
  'four factor formula', 203–205
  fuel elements, 214–216
  fuel utilisation factor, 203–204
  heterogeneous reactor, 202
  homogeneous reactor, 202
  hydrogen storage, 234–235
  missions, 246–247
  moderator, 202

312 Index

multiplication factor, 203
neutron leakage, 205–209
operating temperature, 218–221
plutonium, 202
power, 198
principles of, 213–214
prompt neutrons, 211
reactor dimensions, 205–209
reflection, 210–211
resonance excape probability, 204
safety issues, 243–246
slowing-down length, 206
sustainable chain reaction, 202
thermal stability, 212–213
thermal utilisation factor, 204
thrust, 198
uranium, 200–202
nuclear thermal rocket engine, 221
   applications of nuclear engines, 229–230
   control drums, 226–227
   development status, 235–241
   hot bleed cycle, 225–226
   interplanetary journeys, 232–234
   interplanetary manoeuvres, 231–232
   nozzle, 228–229
   operational issues, 230–231
   propellant flow and cooling, 224–226
   radiaiton management, 221–224
   start-up and shut-down, 227–228
   thrust generation, 228–229
   topping cycle, 225

'o'-ring, 109
Oberth, Herman, 2–4
orbital injection, 137
orbital motion, 277
orbits, 17
Osumi, 7

Pakistani space programme, 6
parallel injector, 65
payload mass, 25
Peenemunde, 5, 12
PEEWEE reactor, 237, 239
Pegasus, 143
Phoebus, 237, 239
pitch angle, 124

plasma thruster, 168
plug nozzle, 266
plutonium, 202
pogo, 71
propellant flow, 74
propellant physical properties, 88
Proton, 93, 95, 284
pulse detonation engine, 271
pulsed magnetoplasmadynamic thruster, 170

R-1, 7
R-2, 7
R-7, 31
radioactive thermal generators (RTGs), 181–184, 199
   safety issues, 245–246
radiofrequency thrusters, 177–178
   VASIMIR, 177–178
range, 117, 122
RD 100, 6
RD 170, 93
RD 200, 6
RD 253, 95
RD 300, 6
Redstone, 9
reference area, 130
resisto-jet (electrothermal thruster), 152
RL 10, (Plate 7), 85–86
rocket equation, 2, 14, 115, 259
Rohini, 8
ROKOT, 285
rotary engine, 271
rotation of the Earth, 24
Roton, 273
RS 27, 93
RS 68, 83–84
Russian space programme, 6

Sanger, Eugene, 12
Saturn V, (Plate 6), 6, 9, 12, 16, 30, 31, 73, 93
shipwrecks, 2
shroud, 33
single stage to orbit (SSTO), 255, 261, 273
small launchers, 285
SMART-1, (Plate 15), 194–195
solar cells, 180–181
   Solar Electric Propulsion (SEP), 181

solar generators, 181
solid propellant composition, 100
solid-fuelled motors, 35, 87
Soyuz, 4
space charge limit, 160
space plane, 12
Space Shuttle, 12–13, 22, 32, 68, 101, 284
Space Shuttle main engine (SSME), (Plate 4, 8, 9), 33, 66, 74, 81–84, 91
Space Shuttle SRB, 111
specific impulse, 57
Stationary Plasma Thruster (SPT), 173–176
Sputnik 1, 9
staged combustion system, 81
station keeping, 176, 189
strap-on boosters, 31
structural mass, 25, 150

Thruster with Anode Layer (TAL), 174–176
thermal engines, 35
thermal protection, 107
thermodynamic thrust equation, 51
thermodynamics, 42
thrust, 37
   coefficient, 52, 252
   equation, 41, 51
   profile, 104
   stability, 103
   vector control, 67, 111
Tipoo Sultan, 2
Titan, 12, 85
Titan III, 284
Titan IV, (Plate 10), 284
topping cycle, 70
trajectories, 23, 131, 135, 136, 139
tri-propellant engines, 269
Tsander, Friedrich, 12
Tsiolkovsky, Konstantin E., 2–4, 14, 25, 115, 197
turbines, 71
two-phase flow, 102

United States space programme, 8
Uranium 200–202, 214–215
see also nuclear propulsion
   enriched, 202
   fission, 200–201
   melting/sublimation points, 214
   sustainable chain reaction, 202
US Army, 9

V1, 270
V2, 6
vacuum (constant pitch) trajectory, 137
Van Allen radiation belts, 9
Vanguard, 9
VASIMIR, 177–178
velocity, 15, 120, 149
   characteristic, 52
   exhaust, 38
   increment, 25
Venture Star, 273
Venus, 6
Verein für Raumschiffahrt, 3
Verne, Jules, 3
vertical motion, 120
Vinci, 78–79
Viking, (Plate 2), 8, 75, 93, 95
Viking 4B, 59, 60, 77
Viking 5C, 59, 77
Viking 6, 75
Vostok, 4
Vulcain, (Plate 5), 78, 92, 299
Vulcain 2, 80–81, 83, 300–301

Wan Hu, 2
War of Independence, 2

XLR 105-5, 93
XMM-Newton, 300

Ysien Hsue-Shen, 7

Zenit, 284
Zhou Enlai, 7